Principles of Indoor Positioning and Indoor Navigation

For a listing of recent titles in the
Artech House GNSS Technology and Applications Library,
turn to the back of this book.

Principles of Indoor Positioning and Indoor Navigation

Li-Ta Hsu
Guohao Zhang
Weisong Wen

ARTECH
HOUSE
BOSTON | LONDON
artechhouse.com

Library of Congress Cataloging-in-Publication Data
A catalog record for this book is available from the U.S. Library of Congress.

British Library Cataloguing in Publication Data
A catalog record for this book is available from the British Library.

ISBN-13: 978-1-63081-977-4

Cover design by Andy Meaden

685 Canton St.
Norwood, MA

10 9 8 7 6 5 4 3 2 1

Contents

Preface

I am very lucky to have the opportunity of living in different cities during my career. This is exactly the same time that we, the general public, navigate using smartphones. I apparently am not the only one who has had disorientating experiences when navigating through the cities' canyons and indoor venues using a smartphone. With a sense of contributing to society, urban navigation became the research topic of the research group that I founded in 2016, the Intelligent Positioning and Navigation Laboratory (IPNL). We, including my two brilliant coauthors Guohao Zhang and Weisong Wen, aim to research and develop the state-of-the-art positioning and navigation methods, algorithms, systems, and technologies for people with smart devices (phones, watches, and glasses) and eventually for unmanned autonomous systems (drones and driverless cars). With various opportunities (academic challenges and industrial collaborations) presented to us, we examined most of the sensors and learned different filtering, estimation, and optimization theories used in positioning and navigation in the past 10 years. We gradually gained an understanding of the challenges and limitations of different approaches.

I was invited by Lisa Beaty and John Raquet from the Institute of Navigation (ION) to teach the very first Indoor Positioning and Navigation tutorial in ION GNSS+ 2020. Even though we did not publish or present many research papers on indoor positioning, the research foundation that we showed in the ION community gained trust. We did extensive studies on various indoor positionings and sensor fusion, which led us to publish the article,

"Tightly Coupled Integrated Navigation System Via Factor Graph for UAV Indoor Localization" in *Aerospace Science and Technology* in 2021. Luckily, this article attracted the attention of Casey Gerard from Artech at the end of 2021, when we believed that there was a strong market need for people who wanted to learn indoor positioning and navigation.

Reference books are always important for students, researchers, and engineers. There are many great books regarding GPS, GNSS positioning, navigation, and sensor fusions published by Artech House. As the readers of these important reference books of our navigation community, we greatly benefit from them, which lays out a solid foundation for our research and career. So we believed that we had to contribute our knowledge and share it with the navigation community. It took us 4 years to complete the book. We had to make sure that it covered most of the indoor positioning and navigation theories from the perspective of math and physics. This is such an interdisciplinary field, and people from aerospace, geodesy, electrical, electronic engineering, mechatronics, robotics, computer vision, computer science, and data sciences are working on this grand topic. New principles will be discovered and new theories will be proposed. As advised by Gavin Jamieson, our Artech House commissioning editor, at some point, we must let it go.

Our design for this book is that we prepared all the keys and fundamental mathematical equations and derivations that will not change with time. To facilitate the learning experience, we will provide a living GitHub repository that not only will include the open-source code, simulation, and data, but also will act as a platform to form a community for the readers to interact and make requests.

If you are a self-learning student, we hope you will enjoy reading this book and making use of the open-source code for your research and development. If you are a university lecturer, our chapter arrangement could provide you with a hint on how to structure your weekly courses (and could be separated into 12-week lectures and laboratory work that are facilitated by our open-source data and codes).

If you have any questions regarding the book, please do not hesitate to contact us. We want to hear your feedback and learn how we can help you to learn and grow with this fascinating topic that will utterly be important for our future society.

Li-Ta Hsu
November 2025

Acknowledgments

The writing of this book has benefited immeasurably from the insight and incredible energy of the Intelligent Positioning and Navigation Laboratory.

First and foremost, I am deeply indebted to Dr. Shiyu Bai. His tireless assembly of references, datasets, and background material formed the backbone of our literature map and saved countless hours of preliminary digging. Beyond logistics, his critical eye caught inconsistencies in early drafts, and his thoughtful suggestions often pushed chapters toward greater conceptual clarity.

I owe particular thanks to Dr. Feng (Darren) Huang and Mr. Xikun Liu, whose contributions went beyond day-to-day support. They took the lead in designing many of the book's key figures and were deeply involved in deriving and rigorously validating the theoretical results that underpin each chapter. I am also grateful to Dr. Qing Liang for his consistently sharp input on algorithm development. His attention to references and technical interpretation ensured that each methodological choice is both transparent and well-supported.

Finally, I wish to acknowledge the broader IPNL team, Jiachen Zhang, Xi Zheng, Lyu Zhen, Yingying Wang, Xin Wang, and Runzhi Hu, for their collective diligence in reviewing draft manuscripts, and flagging obscure edge cases that only emerge when fresh eyes meet complex systems. Their steady feedback loop sharpened the book's technical rigor and improved its readability for newcomers to the field. To all of you in IPNL, thank you for turning solitary authorship into a genuinely collaborative endeavor.

Finally, this book was made possible by the *Financial Support for Book Writing* scheme of the Research Committee, The Hong Kong Polytechnic University. The grant underpins the creation and long-term maintenance of the open-source codebase and data sets released alongside this book (see https://github.com/IPNL-POLYU/IPIN-Examples).

Last but not least, Casey, Isabel, Leona, Julia, and Gavin at Artech, thank you so much for your tireless work and great patience with us. We are blessed to have you guide us through the challenges that we faced. We hope this book can also make you proud.

1

Introduction to Indoor Positioning and Navigation Systems

The Global Navigation Satellite System (GNSS) has directed modern navigation, profoundly influencing our daily lives. Its ubiquity has led some to lose their sense of direction and map-reading skills. However, its effectiveness is challenged in indoor environments where GNSS signals are weak or nonexistent. This has led to a noticeable increase in disorientation within expansive and multileveled indoor spaces. Cities worldwide are dotted with such complex structures, ranging from the Pentagon in the United States and The Shard in London to the enormous Beijing Daxing International Airport and the Tokyo International Forum. Chances are, you are familiar with similar mazes in your vicinity.

Interestingly, according to the National Human Activity Pattern Survey (2001), we spend approximately 90% of our time indoors. This statistic underscores a significant challenge: providing seamless positioning and navigation services from outdoor settings such as train stations to indoor venues such as shopping malls, parking garages, hospitals, and office buildings. This challenge has become a focal point in the navigation community.

The transformative potential of Indoor Positioning and Indoor Navigation (IPIN) is immense, as underscored by the sector's impressive growth trajectory. Market reports in 2025 project compound annual growth rates (CAGR) ranging from 22.52% [1] to as high as 41.9% [2] in the coming years. Initial estimates valued the IPIN market between $11.31 billion and $20.38 billion in 2023 [2]. Fueled by increasing demand and technological advancements, the market is expected to reach $76.72 billion [3] by 2028 and further expand to between $109.2 billion [4] and $174.02 billion [2] by 2030. It is important to note that these market reports primarily focus on radio frequency (RF)-based indoor positioning technologies, not fully encompassing the IPIN technologies enabling mobile robots in indoor environments. The introduction of indoor mobile robots is revolutionizing warehouse logistics, driven by the demand for automation, e-commerce growth, and labor shortages. The global warehouse robotics market was valued at $4.31 billion in 2022 [5]. This market is expected to grow, with projections estimating the autonomous mobile robot market for logistics and warehousing to reach $18.56 billion by 2032 [6]. This reflects a CAGR of 19.6% for warehouse robotics from 2023 to 2030 [5], and 14.9% for the overall mobile robots market from 2023 to 2029 [7].

While the ideal solution might be a universal indoor GNSS capable of covering all types of indoor environments, such technology remains ambiguous. Current inertial navigation systems, for instance, still face the challenge of drifting over time. It will take a long time for quantum inertial navigation systems, which use ultracold atoms and laser pulses to measure acceleration and rotation with extreme precision, enabling highly accurate positioning without relying on external signals, to take place in IPIN applications. For a detailed look at the performance of existing positioning technologies, refer to Figure 1.1.

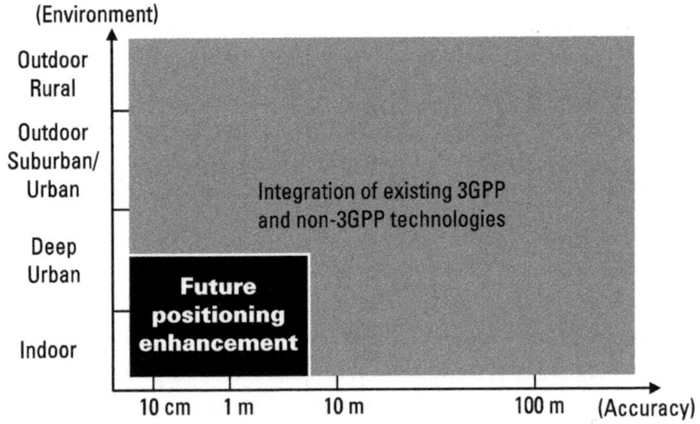

Figure 1.1 A study on the currently available positioning performance in different environments according to the 3GPP TR 22.872 [8].

Figure 1.1, extracted from a technical report by the Third Generation Partnership Project (3GPP) [8], an ensemble of standards organizations dedicated to developing protocols for mobile telecommunications, highlights the ongoing challenges in indoor and deep urban environments. These challenges have propelled the development of new radio communication protocols. While the impact of these protocols is significant, it is clear that no single solution can address the entire spectrum of positioning challenges. Consequently, the selection of different indoor positioning systems to fit different applications based on the user scenarios became utterly important.

The primary aim of this book is to equip the reader with a solid understanding of the fundamental principles, theories, and algorithms underlying various indoor positioning and navigation systems. Figure 1.2 illustrates the process of positioning estimation, a key focus of this book. We will dive into the modeling and estimation aspects of these systems, offering a practical approach to learning. To enhance the learning experience, the equations introduced in this book are complemented by open-source codes and datasets that we provide on GitHub [9]. Note that the primary focus of this book is not on introducing the sensors, technologies, or performance metrics of indoor positioning systems. For readers interested in these aspects, we recommend a comprehensive resource [10]. Our goal is to provide a foundational understanding of the algorithms and models that drive these systems, empowering readers to navigate this evolving field effectively.

1.1 How to Select an IPIN System for an Application

IPIN applications can be categorized based on their underlying technologies. Table 1.1 demonstrates the typical accuracy, primary applications, market values, and CAGR of these technologies. As can be seen, location-based services (LBS) empowered by Wi-Fi, Bluetooth Low Energy (BLE), ultrawideband (UWB), and indoor mobile robot are taking the most market share in 2025. Because of the accuracy and cost of implementations, the use cases (which generate the market values) are different too.

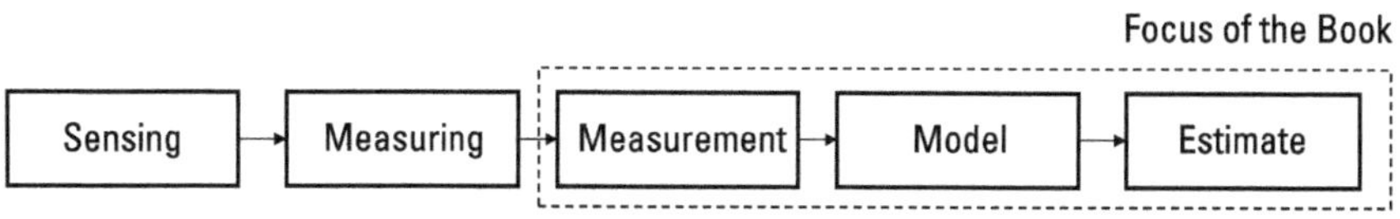

Figure 1.2 A general flow of a positioning system.

Table 1.1
Overview of IPIN Technologies, Applications, and Market Values

Technology	Typical Accuracy (meters)	Primary Applications (Indoor)	2024 Market Size	CAGR (2024–2030)
Wi-Fi (excluding round-trip time (RTT)) [2, 11]	~3–5m (with time difference of arrival (TDOA))	Asset tracking in hospitals and warehouses; smartphone wayfinding in large venues	≈30%–35% of IPIN market (~$8 to $9 billion)	~15% (moderate growth)
Bluetooth Low Energy (BLE) [2, 11, 12]	~3–5m (RSSI); <1m with angle of arrival (AOA)	Indoor navigation (malls, airports, museums); proximity marketing with beacons	≈35% of market (~$9 billion in 2023)	High (e.g., ~30%+)
UWB [2, 13, 14]	0.1–0.3m (centimeter-level)	High-precision asset tracking (warehouses), industrial tools and medical equipment tracking	~20%–22% of market (~$5 billion)	~18.9% (fast-growing)
5G cellular [11, 15, 16]	~1–3m (with new radio 5G (NR))	Emerging for factory automation and smart buildings; integrates with IoT for positioning where 5G is deployed	Minimal in 2024 (pilot phase)	N/A (anticipated ~40%+ as adoption grows)
Infrared (IR) [17]	Room-level (≈<1m)	Patient tracking and personnel locating in hospitals (line-of-sight ID badges); secure access control	Niche segment (<5%)	Low (steady in niche uses)

RSSI: Received signal strength indication.

Table 1.1
(Cont.)

Technology	Typical Accuracy (meters)	Primary Applications (Indoor)	2024 Market Size	CAGR (2024–2030)
Ultrasonic [18]	Centimeter-level (~0.1–0.3m)	Healthcare asset tracking (e.g., tags with ultrasound); some industrial safety applications (no RF interference)	Niche segment (<5%)	Low (steady in niche uses)
Visible light (VLC) [19]	≈0.3m (30 cm)	Retail indoor navigation (LED lighting-based wayfinding); occupancy analytics in smart offices	~5% of market (small but growing)	High (rapid growth from small base)
Magnetic field [20]	~1m (with smartphone sensors)	Smartphone navigation in large venues (no extra hardware), for example, malls, airports (often combined with other sensors for stability)	Niche (few providers; part of "others")	Moderate (specialized use)
Indoor mobile robots (AMRs/AGVs) [21, 22]	~0.02–0.05m (2–5 cm)	Autonomous navigation in warehouses and logistics centers (inventory robots, forklifts); hospital delivery robots (using lidar-SLAM, vision)	~$9 billion (global indoor robots 2023)	~15.1% (2024–2031)

*The market values are roughly estimated based on our search of available market reports, and the year of estimation is 2025.

†LED: Light emitting diode. AMR: Autonomous mobile robot. AGV: Automated guided vehicle.

The market demand for IPIN technologies is a key driver advancing their innovations in accuracy, scalability, and integration. With industries such as retail, healthcare, logistics, and smart buildings increasingly relying on indoor location data, companies are investing in more precise, reliable, and cost-effective solutions. Retailers seek better customer engagement through location-based services, while hospitals require real-time tracking of medical equipment and patients. The rise of automated warehouses and robotics is further pushing the need for high-precision positioning, with UWB and light detection and ranging (lidar)-based solutions becoming critical for navigation and tracking. Additionally, the emergence of fifth generation (5G) and artificial intelligence (AI)-driven sensor fusion is enabling more robust hybrid positioning systems, integrating Wi-Fi, BLE, and machine learning to refine accuracy. Smart city initiatives are also accelerating IPIN adoption, with governments investing in seamless indoor-outdoor navigation, particularly for transport hubs and public buildings. The global IPIN market is projected to exceed $174 billion by 2030, growing at over 30% CAGR, demonstrating strong momentum. So you are on the right track to read this book.

The reason that you are picking up this book is very likely that you would like to develop a system making use of IPIN technologies. The first question that pops in our mind is how to select the "right" technology for my system and application. This is a complex process. There are many considerations required, including: (1) what the service area is, (2) what the required accuracy and availability are, (3) what the acceptable user terminal cost is, (4) what the acceptable infrastructure cost is, and (5) what the engineering cost is in terms of the human resources needed. The list can go on and on if we look at the problem in more detail. These five points are our suggested primary questions. If you are interested in more details, Dr. Nel Samama's book provides a systematic approach, defining parameters such as hardware, real implementation, type and performance, and physical aspects to aid in technology selection [10]. It is a must-read in our opinion.

After the five questions are answered for a targeting application. Then the selection of "sensors on terminal," "infrastructure," and corresponding "positioning and fusion (combing several sensors' data) algorithms" is required. Their relationship is depicted as Figure 1.3.

In this book, we focus on the physical aspects of IPIN technologies, including types of physical measurements, signal processing algorithms, position estimation, and the physical quantities associated with these measurements. It is most beneficial for readers who have already identified the technologies relevant to their applications. To assist readers in understanding

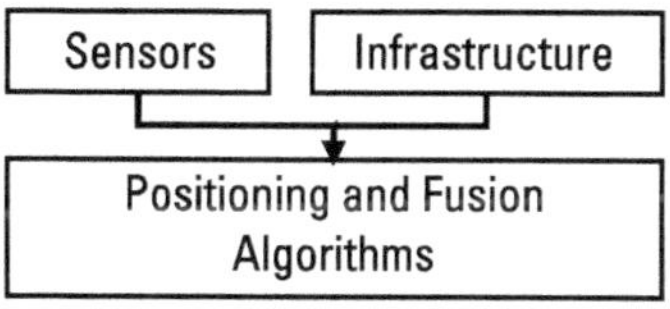

Figure 1.3 A general flow of a positioning system.

how to select appropriate IPIN technologies for their specific applications, we provide several illustrative examples. These examples demonstrate the decision-making process by considering key factors such as sensors, infrastructure, and algorithms, which are fundamental components of any indoor positioning system. Sensors refer to the devices used to detect and collect positioning-related signals, such as smartphones equipped with BLE. Infrastructure includes the deployed hardware that supports positioning, such as BLE beacons. By presenting real-world application scenarios, we aim to clarify how these elements influence the choice of technology and ensure that the selected system meets the intended requirements in terms of cost, accuracy, and scalability. The following sections serve this purpose.

Please note that cost and service area are hard to clearly define, as they vary depending on the application, deployment scale, and technological constraints. In general, a small area refers to localized environments such as individual shops or small offices, typically covering less than 500 square meters. A medium area encompasses spaces such as warehouses, shopping malls, or multifloor buildings, ranging from a few hundred to several thousand square meters. A large area extends to city-wide environments, including entire urban districts or transit hubs, often covering tens to hundreds of thousands of square meters. Similarly, cost classification follows a relative scale: low-cost solutions leverage existing infrastructure (e.g., BLE beacons or Wi-Fi), requiring minimal additional investment; medium-cost solutions involve dedicated hardware (e.g., UWB beacons, vision markers) but remain scalable; and high-cost solutions demand specialized infrastructure (e.g., lidar, extensive cloud computing) with significant upfront and maintenance expenses. For a more detailed discussion, readers may refer to another book [10].

1.1.1 Example 1: Proximity Marketing in a Shopping Mall

Proximity marketing in retail requires an efficient, low-cost, and scalable indoor positioning system. The selection of sensors, infrastructure, and algorithms aligns with five key requirements (as shown in Figure 1.4 and below):

1. *Service area:* The system should operate in the vicinity of a specific shop, covering small to medium areas.
2. *Performance:* It must provide ~3-m accuracy with 99% reliability, ensuring customers receive location-specific promotions without excessive errors.
3. *User terminal cost:* The solution should work on existing smartphones, eliminating the need for additional devices.
4. *Infrastructure cost:* The deployment must be low-cost, with minimal installation and maintenance expenses.
5. *Engineering cost:* Development should be a one-time effort, avoiding complex integrations or expensive recalibration.

Sensors

One example of sensors is smartphone BLE. Smartphones are selected as sensors because they are widely available, and BLE support is standard across modern devices. BLE is a low-power technology, ensuring that the system does not significantly impact battery life, while also being compatible with both Android and iOS, ensuring high adoption rates.

Infrastructure

One example of infrastructure is BLE beacons. BLE beacons are chosen due to their low cost, ease of installation, and scalability. Compared to Wi-Fi or UWB, BLE beacons are affordable and battery-powered and require minimal maintenance. They effectively cover small zones with a 3-m precision, making them ideal for shop-level engagement without the high cost of advanced positioning technologies.

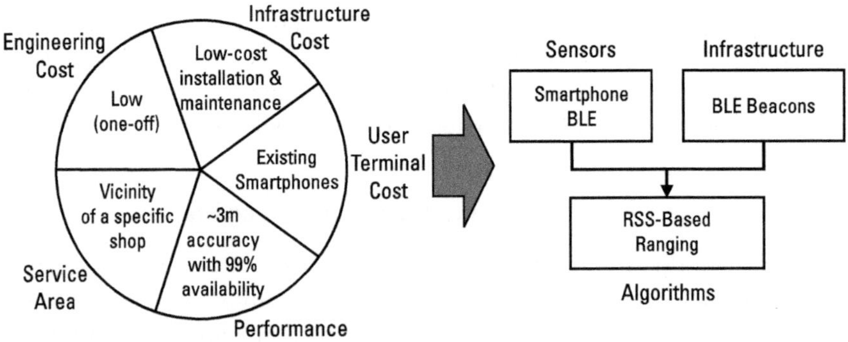

Figure 1.4 IPIN technology selection for proximity marketing in a shopping mall.

Algorithms

One example of an algorithm is received signal strength (RSS)-based ranging. RSS-based ranging is selected for its simplicity and low processing demand. It estimates a user's proximity to the beacon based on signal attenuation, enabling zone-based marketing without requiring complex calculations. While advanced methods like AOA exist, RSS is more practical, requiring only standard BLE beacons without directional antennas.

Finally, this selection is therefore a cost-effective, scalable, and user-friendly solution for proximity marketing in retail spaces. Let's try a second one.

1.1.2 Example 2: Inventory Management

Inventory management in warehouses or large indoor facilities requires high-accuracy tracking to monitor goods and assets efficiently. The five key requirements are shown in Figure 1.5 and as below:

1. *Service area:* The system must cover all areas inside a building, including storage rooms and aisles, ensuring full facility-wide tracking.
2. *Performance:* The system must achieve ~1-m accuracy in 2-D and floor-level identification (99%) to precisely locate assets in complex environments.
3. *User terminal cost:* The solution should be medium-cost, with custom tracking devices that balance affordability and precision.
4. *Infrastructure cost:* A medium-cost setup with densely deployed UWB beacons ensures accuracy while keeping maintenance manageable.
5. *Engineering cost:* Development requires full-time engineers, making the engineering cost moderate but justified for precise tracking.

Sensors

One example of sensors is UWB/inertial measurement unit (IMU) integrated trackers. UWB sensors are selected because they offer centimeter-level precision in time-of-flight positioning. Compared to BLE or Wi-Fi, UWB provides higher accuracy and better robustness against interference from metal shelves and equipment. The IMU is added to improve tracking in areas with limited UWB coverage, such as narrow aisles or dynamic environments where continuous tracking is required.

Infrastructure

One example of infrastructure is densely distributed UWB beacons. A dense network of presurveyed UWB beacons is deployed to ensure real-time, high-accuracy tracking across the facility. Unlike radio frequency identification (RFID), which requires close-range scanning, UWB beacons allow seamless real-time monitoring of moving inventory without manual scanning.

Algorithms

One example of algorithms was time difference of arrival (TDOA) + pedestrian dead-reckoning (PDR)/vehicle dead-reckoning (VDR) + extended Kalman filter (EKF). The TDOA method is used because it enables high-precision tracking without requiring the tracker to actively transmit signals, reducing energy consumption. PDR/VDR helps to track movement in beacon-limited areas, while the extended EKF fuses multisensor data for smoother, more reliable tracking.

1.1.3 Example 3: Analytics for Foot Traffic on a City Scale

Foot traffic analytics across a city requires scalable, cost-efficient, and cloud-based positioning to gather insights from a large population. The five key requirements are shown in Figure 1.6 and as below:

1. *Service area:* The system must cover all indoor areas inside a city, including shopping centers, transit hubs, offices, and public buildings.
2. *Performance:* The system aims for ~10-m accuracy in 2-D (80%) and floor-level positioning (80%), balancing feasibility and large-scale usability.
3. *User terminal cost:* The solution must work with existing smartphones, ensuring widespread adoption without requiring additional devices.
4. *Infrastructure cost:* A low-cost approach using crowdsourced Wi-Fi radio maps avoids the need for expensive beacon deployments.

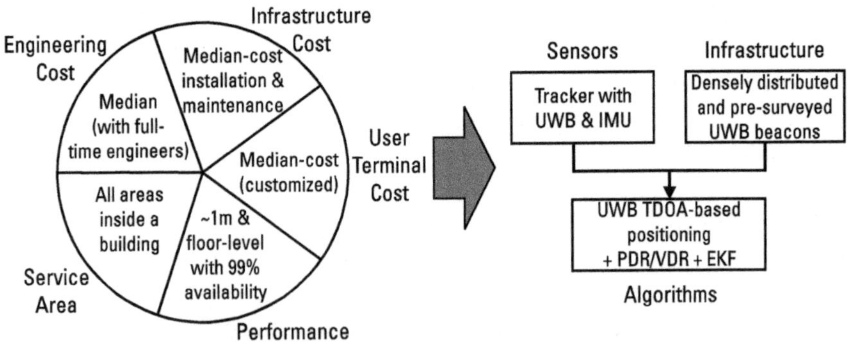

Figure 1.5 IPIN technology selection for inventory management.

5. *Engineering cost:* The solution requires a dedicated engineering team and cloud-based industrial servers, making development costs high but justified for city-scale analytics.

Sensors

Examples of sensors include smartphone Wi-Fi, IMU, and a barometer. Smartphones are chosen as sensors because they are ubiquitous and equipped with Wi-Fi, IMUs, and barometers. Wi-Fi scanning provides location fingerprints, the IMU assists in movement tracking, and the barometer detects floor-level changes in multistory buildings. This combination enables passive tracking without user intervention.

Infrastructure

An example of infrastructure is a crowdsourced Wi-Fi RF Feature Map. Instead of deploying dedicated infrastructure, the system leverages Wi-Fi-based RF mapping collected via crowdsourcing. This approach continuously updates the location database with minimal infrastructure costs, ensuring that the system remains adaptive to environmental changes (e.g., newly constructed buildings or modified layouts).

Algorithms

Examples of algorithms are DL-based RF feature matching + DL-pedestrian dead reckoning (PDR) + barometer + particle filter. A DL-based RF feature matching technique is used to analyze Wi-Fi signal patterns across the city. DL-PDR further refines position estimates by analyzing smartphone motion patterns. The barometer improves floor-level estimation, and particle filters enhance robustness by fusing multiple data sources to smooth positioning outputs.

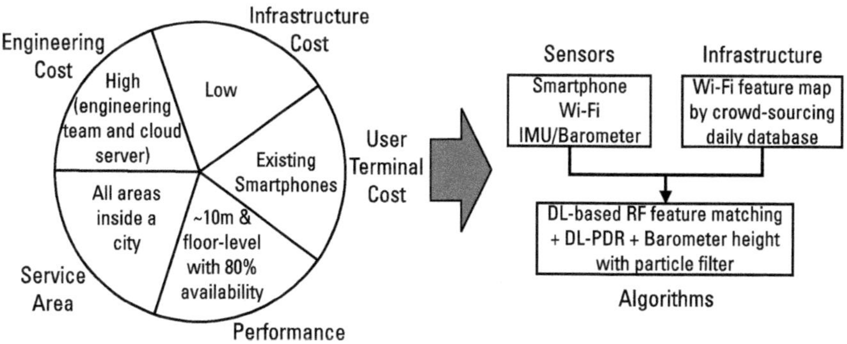

Figure 1.6 IPIN technology selection for analytics for foot traffic in city scale.

1.1.4 Example 4: Automatic Parking System for Intelligent Vehicles

An automatic parking system for intelligent vehicles requires high-accuracy positioning, real-time updates, and reliable infrastructure to guide vehicles autonomously in parking lots or garages. The five key requirements are shown in Figure 1.7 and as below:

1. *Service area:* The system must cover most indoor parking areas, including underground garages, multilevel parking lots, and smart parking spaces in urban environments.
2. *Performance:* It must achieve <0.3-m accuracy in 2-D positioning and precise lane-level tracking to ensure vehicles park safely and efficiently.
3. *User terminal cost:* The solution must integrate with existing intelligent vehicle sensors, avoiding the need for additional user-side hardware.
4. *Infrastructure cost:* A medium-cost deployment using a combination of vision-based markers and lidar maps ensures accuracy without excessive cost.
5. *Engineering cost:* The system requires a dedicated engineering team and AI-based cloud computing, making the initial development cost high but justified for scalability and automation.

Sensors

Examples of sensors include vehicle, lidar, cameras, and IMU. Autonomous vehicles come equipped with lidar, stereo cameras, and IMUs. Lidar and vision-based sensors detect lane markings, parking spots, and obstacles. IMUs improve dead reckoning, enabling smooth transitions when switching between sensor modalities.

Infrastructure

Examples of infrastructure are vision markers and pre-mapped lidar database. Vision-based augmented reality (AR) markers or Quick Response (QR) codes assist in visual navigation when aligning with parking spaces. Pre-mapped lidar databases allow vehicles to compare real-time sensor data with stored environmental layouts for more accurate localization.

Algorithms

Examples of algorithms are lidar/visual simultaneous localization and mapping (SLAM) and Map Matching + Factor Graph Optimization (FGO) sensor fusion. SLAM continuously updates vehicle position using environmental features. Map matching provides the position of the vehicle regarding the presurveyed map. FGO sensor fusion combine multiple inputs, optimizing the positioning to provide robust and accurate real-time positioning for autonomous parking (Figure 1.7).

This expenditure is very high, deemed acceptable because this application represents an innovative solution and necessitates a long-term investment to establish its viability. There is optimism that, as this application gains wider acceptance among the public, the costs will decrease.

1.2 Sensing Technologies Based on RF

IPIN systems rely on a variety of measurements and their corresponding models to accurately estimate users' positions. These sensors, broadly classified into RF-based and physical sensing technologies, serve distinct purposes. RF-based sensing is typically used for point positioning, while physical sensing technologies are more suited for dead reckoning. We will quickly introduce these technologies in the following sections.

1.2.1 Measurements

In IPIN, the most prevalent RF technologies include BLE and BLE Channel Sounding (CS), various iterations of Wi-Fi (802.11a/b/g/n, known as Wi-Fi 4/5/6), Wi-Fi RTT (802.11mc/az), 5G, and UWB (802.15.4a/z). Each of these technologies generates specific types of measurements, pivotal for IPIN applications as shown in Table 1.2. These measurements included:

- *Received signal strength:* This provides a measure of the power level received by the receiver from the transmitting device.

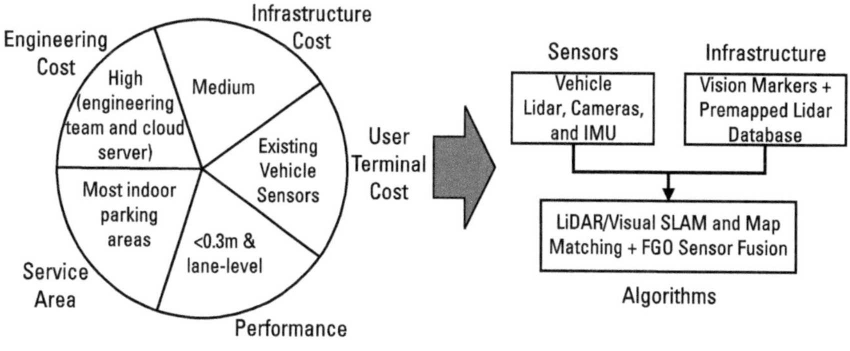

Figure 1.7 IPIN technology selection for analytics for automatic parking system.

Table 1.2
RF Technologies and Their Measurement Capabilities for Indoor Positioning

Technology	RSS	CSI/CIR	RSS-Based Ranging	TOA (One-Way) Ranging	Two-Way TOA Ranging	TDOA Ranging	AOA/AOD
BLE [23]	Yes	No (standard mode)	Yes (via path loss, but coarse; ~3–5m error)	No	No	Limited (requires synchronized multireceiver setups)	Yes (with BLE 5.1 AOA support; submeter possible)
BLE Channel Sounding (CS) [24]	Yes (ancillary)	Yes (phase-based CSI/CIR extraction)	Yes (using phase-based ranging; ~10–30-cm accuracy)	Yes (via phase-based time-of-flight (TOF) measurement)	Yes (via two-way tone exchange)	No	No
Wi-Fi (802.11a/b/g/n; WiFi 4/5/6) [25]	Yes (RSSI available)	Yes (CSI can be extracted with modified NICs)	Yes (by pathloss models; typical error 5–10m)	No (one-way TOA not natively supported)	No (not defined in standard)	Possible (with synchronized sniffers; research use)	Yes (with MIMO arrays and dedicated algorithms)

Table 1.2
(Cont.)

Technology	RSS	CSI/CIR	RSS-Based Ranging	TOA (One-Way) Ranging	Two-Way TOA Ranging	TDOA Ranging	AOA/AOD
Wi-Fi RTT (802.11mc/az) [26, 27]	Yes	Implicitly used for timing (not exposed as CSI)	Yes (often combined with RSS)	Not applicable (uses two-way FTM)	Yes (using Fine timing measurement [FTM]; typical accuracy ~1–2 m indoors)	Yes (multiAP FTM can derive TDOA)	Yes (planned for 802.11az AOA/AOD support)
5G NR [28, 29]	Yes	Yes (CSI used for MIMO/beamforming)	Not typically used for ranging	Yes (using downlink PRS-based TOA)	Yes (can be implemented via RTT exchanges)	Yes	Yes (using massive MIMO arrays for AOA/AOD)
UWB (802.15.4a/z) [30–32]	Yes (available, but not primary for ranging)	Yes (CIR is derived via pulse correlation)	Not typically used (range derived from timing is superior)	Yes (if clocks are synchronized)	Yes (standard two-way ranging; typical accuracy ~10 cm in LOS)	Yes (if anchors are synchronized, enabling TDOA)	Yes (via AOA with multiantenna setups)

- *Channel state information/channel impulse response (CSI/CIR):* This provides detailed information about the channel properties and the signal's interaction with the environment.
- *RSS ranging:* This estimates the distance based on the signal strength.
- *Time of arrival (TOA) ranging:* This determines the distance to a transmitter by measuring the time that it takes for a signal to travel from the transmitter to the receiver.
- *Two-way TOA ranging:* This is an enhancement of TOA, involving a two-step communication process for improved accuracy.
- *Time difference of arrival (TDOA) ranging:* This calculates the position of a transmitter by measuring the difference in time at which a signal arrives at multiple receivers.
- *Angle of arrival/angle of departure (AOA/AOD):* This measures the angle at which the signal arrives at or departs from the receiver, aiding in position estimation.

Table 1.2 highlights the diverse measurements these technologies provide for indoor positioning. RSSI is available across all technologies but is primarily useful for coarse location tracking due to its susceptibility to environmental factors. CSI/CIR is exposed in Wi-Fi and UWB, offering rich multipath data that can be used for fine-grained positioning. TOA and TDOA are leveraged in Wi-Fi RTT, 5G, and UWB, enabling precise ranging by measuring signal propagation time. Two-way TOA is supported in BLE channel sounding (CS) and UWB, allowing accurate distance estimation without strict clock synchronization. AOA and AOD are particularly powerful in BLE (AOA in Bluetooth 5.1+), Wi-Fi (via multiple input multiple output (MIMO)), and 5G (via massive MIMO arrays), enabling direction-based positioning. As known, the resolution of these ranging measurements is closely linked to the bandwidth used by the RF technologies. Generally, larger bandwidth leads to better resolution due to improved time-domain resolutions, enhanced multipath resolution, and superior frequency-domain characteristics.

These diverse measurements are crucial for different positioning techniques, including trilateration, fingerprinting, and hybrid sensor fusion, depending on accuracy, power, and deployment constraints. The next section explores how these measurements are utilized to improve indoor positioning accuracy across various applications.

1.2.2 RF-Based Positioning Method: Geometry Model

Chapter 4 introduces point positioning using geometry model-based methods with measurements from RF technologies. One common method is RSS ranging, known for its simplicity and cost-effectiveness. However, it typically offers less accuracy compared to TOA ranging measurements, which provides more precise distance estimation by directly measuring signal travel time. TOA ranging, although more accurate, is sensitive to obstacles and nonline-of-sight conditions and demands sophisticated implementation, particularly in terms of clock synchronization between beacons.

To avoid synchronization challenges, two-way TOA ranging simplifies implementation by eliminating the need for synchronized clocks. This makes it more feasible for widespread use, especially in the consumer electronics world. While TOA excels in ideal conditions, the ease of use and reduced complexity of two-way TOA make it a popular choice, despite potential accuracy reductions due to processing delays.

TDOA ranging, in contrast, offers enhanced accuracy over larger areas by triangulating the time difference of signal arrival at multiple receivers. This method requires at least three receivers and precise timing for two-dimensional (2-D) positioning, which increases its complexity and infrastructure demands. TDOA can be implemented in both uplink and downlink manners. Uplink TDOA, where the user equipment (UE) transmits signals to multiple base stations, reduces complexity for the UE but necessitates a dense, coordinated network infrastructure. This approach grants users more control, making it versatile for various applications, especially those that focus on privacy. Conversely, downlink TDOA increases complexity for the UE, as it must process signals from multiple transmitters, with the network controlling the positioning process. This method is more effective in dense urban settings with numerous beacons.

Finally, AOA determines the direction from which a signal is received by measuring the phase difference of the incoming signal across multiple antennas. AOA calculates the signal's arrival angle, which, when combined with data from multiple receivers or observed over time, can triangulate the transmitter's position. AOA typically requires an array of antennas for accurate angle measurement and is most effective in environments with clear signal paths. However, its accuracy can be compromised by signal reflections in complex indoor spaces, necessitating advanced antenna design and signal processing. As such, AOA requires further improvements before its practical application in indoor positioning systems can be realized.

1.2.3 Positioning Method: Feature Matching

Chapter 5 introduces the feature matching positioning method, particularly focusing on RSS, a common feature utilized by RF technologies. Although RSS inherently decreases with distance, making it a straightforward metric, its accuracy can be compromised by environmental factors and physical obstructions. Yet this variability can also be advantageous, as different locations often exhibit unique RSS profiles due to reflections, diffractions, or blockages. This uniqueness makes RSS a valuable feature in fingerprinting methods. As most devices, including Wi-Fi, BLE, and UWB, inherently measure RSS, the need for specialized user terminals is eliminated. This approach leverages existing Wi-Fi networks, thus reducing infrastructure costs. The fingerprinting method using RSS involves compiling signal strengths to create a signal map. While this method is less accurate than more complex techniques and may require periodic recalibration, its affordability, compatibility, and simplicity render it a practical choice for many indoor positioning applications.

RSSI provides a measure of signal strength but lacks detailed information about the signal's propagation characteristics, making it susceptible to environmental interference. In contrast, CIR captures the time-domain response of the channel, detailing how signals interact with obstacles and surfaces, providing a more comprehensive view of multipath effects. As a result, CSI and CIR offer improved accuracy and robustness over RSS for fingerprinting methods. CSI provides detailed data on the communication channel, including amplitude and phase for each subcarrier in orthogonal frequency division multiplexing (OFDM) systems such as Wi-Fi 4/5/6 and 5G. This richness enables a more detailed understanding of environmental characteristics, resulting in more distinct and accurate location fingerprints. CIR, meanwhile, offers a precise time-domain representation of multipath propagation, detailing the arrival times and paths of signal components. This clarity in signal interaction with the physical environment enhances location estimation. Both CSI and CIR are less prone to issues such as small-scale fading and offer stability in dynamic environments, thanks to their capability to dissect and utilize multipath characteristics. The phase information in CSI, in particular, allows for more precise calculations of the signal's origin, significantly improving accuracy over RSS, which only measures signal strength. However, it is essential to weigh the trade-offs. CSI and CIR require more advanced hardware and software, higher computational power, and frequent calibration with sophisticated algorithms to maintain accuracy, particularly in dynamic environments. Their computational intensity comes from the need for high-resolution signal processing to extract fine-grained amplitude and phase information across multiple subcarriers or time-domain samples. This involves complex matrix operations, phase unwrapping, and noise filtering, demanding significant processing power, especially for real-time positioning. Additionally, CSI and CIR data require substantial memory and high data transfer rates, as raw channel information must be continuously updated and analyzed to adapt to environmental changes, further adding to the computational burden. Despite their complexity and resource intensity, the enhanced accuracy and reliability of CSI and CIR make them the preferred choices for applications where precision is up most important, balancing complexity with substantial improvements in indoor positioning accuracy and potentially introducing indoor sensing applications.

1.3 Physical Sensors

Physical sensors, including IMUs, barometers, magnetometers, cameras, and lidar, provide crucial data for indoor positioning by detecting motion, orientation, altitude, and environmental features. Unlike RF-based technologies, which rely on external signals for positioning, physical sensors operate independently, making them useful for infrastructure-free or sensor-fusion-based navigation in complex indoor environments.

Table 1.3 provides measurement capabilities of the physical sensors. The IMUs measure acceleration and angular velocity, enabling motion tracking and dead reckoning. While effective for short-term positioning, IMUs suffer from cumulative drift over time, requiring external corrections.

Cameras and lidar capture environmental features for mapping and localization. Cameras rely on visual feature detection and SLAM, making them ideal for AR and robotics. However, they require good lighting and rich textures, struggling in featureless environments. lidar actively scans the surroundings using laser pulses, generating precise three-dimensional (3-D) maps for positioning. Lidar-based localization is highly accurate and resistant to lighting changes, but comes with higher power consumption, data processing demands, and cost.

The key difference between RF-based and physical sensor-based positioning lies in their reliance on external infrastructure. RF methods require network anchors for positioning using TOF, AOA, or signal strength measurements. In contrast, physical sensors work independently, using motion tracking and environmental cues. However, RF technologies typically provide absolute position estimates, while physical sensors offer relative motion tracking, requiring external correction to maintain accuracy over time.

In practical applications, hybrid positioning, combining RF-based localization with physical sensor data, delivers the most robust and seamless indoor positioning, leveraging the strengths of both absolute RF-based positioning

Table 1.3
Physical Sensor Technologies and Their Measurement Capabilities for Indoor Positioning

Sensor	Measurement Type	Accuracy	Computational Cost	Power Consumption	Deployment Complexity
IMU [33, 34]	Relative motion tracking (dead reckoning, orientation)	High short-term; accumulative drift over time (~5% distance error/min)	Low (real-time sensor fusion, Kalman filtering)	Low (submilliwatt microelectromechanical system (MEMS) sensors)	Low (found in smartphones, requires calibration)
Barometer (pressure sensor) [35, 36]	Altitude estimation (floor detection)	±0.06–0.08 hPa (~0.5–0.7m); requires calibration	Low (direct conversion of pressure to altitude)	Low (~microamps for barometric pressure sensors)	Moderate (sensitive to weather changes, needs reference calibration)
Magnetometer (compass) [37]	Heading determination, geomagnetic fingerprinting	±1–5° heading; affected by indoor magnetic distortions	Low (heading from magnetic field vector)	Low (milliamps, negligible for mobile use)	Moderate (affected by ferromagnetic interference, requires calibration)
Camera (visual sensor) [38, 39]	Visual SLAM, feature tracking, marker-based positioning	Decimeter-level in good condition; drift over time without external reference	High (real-time SLAM, visual feature tracking)	High (image processing and sensor operation)	Moderate (depends on lighting, features, requires calibration)
Lidar [40–42]	3-D mapping, lidar-SLAM, structural localization	±2–3 cm in range; very accurate with good mapping	High (point cloud processing, SLAM, geometric feature extraction)	High (~5–8W for 360° scanning lidar)	High (costly hardware, setup requires calibration and mapping)

and continuous motion tracking from physical sensors. In this book, we focus on the physical sensors that can be used, mainly for dead reckoning and mapping including IMUs, cameras, and lidar.

1.3.1 Proprioceptive Sensors

First, IMUs are a cornerstone in this domain. Operating independently of external signals, which may be unreliable or unavailable in some cases, IMUs rely on internal sensors, accelerometers, gyroscopes, and magnetometers, to detect linear acceleration, rotational changes, and orientation relative to the Earth's magnetic field. This independence is especially valuable in indoor environments, where external signals are obstructed or nonexistent. IMUs function continuously, providing real-time updates on changes in position and orientation. This is particularly crucial in dynamic indoor environments where obstacles and pathways frequently change. The high sampling rate of IMUs enables the detection of rapid movements and subtle directional changes, enhancing the responsiveness and accuracy of navigation systems.

In addition to IMUs, barometers, magnetometers, and wheel encoders for vehicle odometry further enhance indoor positioning capabilities. Barometers measure air pressure variations, making them useful for floor-level detection in multistory buildings, an aspect that IMUs alone cannot determine. By detecting small altitude changes, barometers help differentiate between floors in high-rise environments, improving vertical positioning accuracy. Magnetometers provide absolute heading by detecting the Earth's magnetic field, complementing IMUs, which suffer from orientation drift over time. However, magnetometers require careful calibration as they are susceptible to indoor magnetic interference from steel structures and electronic devices. Wheel encoders, commonly used in vehicles and robotics, measure wheel rotations to estimate traveled distance, offering odometry-based positioning. They are particularly effective in structured indoor environments such as warehouses and underground parking garages, where they provide drift-free distance estimation over short durations. However, odometry alone cannot determine orientation and is prone to slippage errors, making it most effective when fused with IMUs and magnetometers for a more stable navigation solution.

Furthermore, the compact and robust nature of modern IMUs, barometers, magnetometers, and wheel encoders facilitate their integration into a wide range of devices, including smartphones, wearables, autonomous robots, and vehicles. This adaptability has broadened their application across indoor navigation, industrial automation, and emergency response systems, reinforcing their role in indoor positioning solutions. The applications and details of using the proprioceptive sensors in indoor positioning systems are covered in Chapter 6.

1.3.2 Cameras

Cameras have emerged as a pivotal technology in indoor navigation, primarily due to their capacity to capture visual information. They provide rich, detailed imagery, enabling a deep understanding of and interaction with the environment. This is how we human beings used to navigate ourselves. Utilizing computer vision techniques and machine learning algorithms, cameras can identify features, objects, and patterns within an indoor setting. This capability is crucial for tasks such as creating visual maps, performing localization (determining the user's precise location within the map), and navigating by recognizing landmarks or following visual cues.

In varying lighting conditions, traditional cameras might encounter challenges, but advancements such as infrared or low-light cameras have broadened their applicability. These enhanced cameras ensure consistent visual data capture, even in environments where lighting conditions are unpredictable or cannot be controlled. This adaptability is essential for maintaining reliable navigation assistance in diverse settings.

Additionally, cameras play a significant role in more advanced applications, such as AR. In AR, digital information is superimposed onto the camera feed, providing users with navigational cues and contextual information. This integration not only enriches the user experience but also offers intuitive navigation aids that seamlessly blend with the physical environment.

Furthermore, the widespread availability of cameras, particularly in smartphones and other portable devices, positions them as an accessible and cost-effective option for indoor navigation. Their integration into commonly used devices ensures that a broad user base has access to sophisticated navigation tools without the need for additional hardware. This ubiquity and ease of access make cameras a practical choice for a variety of indoor navigation applications.

1.3.3 Lidar

Lidar technology offers precision and the ability to create accurate 3-D representations of environments. This technology operates by emitting pulsed laser light and then measuring the time taken for these light reflections to return. Through this process, lidar sensors are able to construct detailed 3-D maps of their surroundings, providing essential depth data that aids in understanding spatial layouts, identifying obstacles, and facilitating autonomous navigation tasks.

A key advantage of lidar is its operational robustness under various lighting conditions. In contrast to cameras, lidar does not depend on ambient light for functionality, as it generates its own laser light source. This attribute ensures that lidar remains reliable and consistent, regardless of the existing lighting conditions within an indoor setting. This feature is particularly advantageous in situations where other navigation aids, such as RF-based technologies, are either unavailable or unreliable.

Lidar's capability to precisely measure distances and construct spatial maps renders it an ideal solution for applications that demand accurate navigation and effective obstacle avoidance. Its high resolution and accuracy are particularly valuable in the field of robotics, where a detailed understanding of the environment is crucial for safe and efficient operations. Robots equipped with lidar sensors can effectively navigate, plan routes, and interact with their surroundings, thanks to the detailed spatial data provided by these sensors. It is suitable for a variety of applications, particularly in complex and dynamically changing environments.

1.3.4 Integration

Both cameras and lidar are pivotal technologies for indoor navigation, each with unique strengths and contributions to the navigation process. Cameras provide visual context, color information, and feature recognition, making them highly effective for environmental understanding, landmark identification, and user-friendly navigation interfaces. In contrast, lidar excels in precisely mapping structural layouts and remains reliable regardless of lighting conditions, making it crucial for precision navigation and obstacle detection.

In many advanced indoor navigation systems, a combination of cameras, lidar, and IMUs is employed, a technique known as sensor fusion. Cameras and lidar complement each other by integrating detailed visual data with precise depth information, resulting in greater accuracy, robustness, and functionality than either technology alone. IMUs play a critical role in this integration, particularly in visual SLAM and lidar-based mapping. In visual SLAM, IMUs aid in depth reconstruction by stabilizing camera motion and improving depth estimation, especially in feature-poor environments. Additionally, lidar-based systems leverage IMUs to compensate for motion distortion in point clouds, ensuring accurate 3-D representations, particularly when the sensor is in motion.

The integration of IMUs, lidar, and camera technologies is revolutionizing indoor navigation across various industries. Autonomous robots in logistics use them for efficient warehousing, while service robots in hospitals and hotels rely on them for item delivery and visitor guidance. AR and virtual reality devices benefit from enhanced environmental mapping, creating more immersive experiences. In smart buildings, these technologies optimize space utilization and enhance security. Retailers implement them for in-store navigation and personalized customer experiences, while emergency responders utilize them for navigating complex environments during crises. In healthcare, they assist in patient monitoring and aiding the visually impaired, and, in intelligent vehicle systems, they play a crucial role in autoparking in underground garages. Details on the applications and workings of visual and lidar SLAM technologies are provided in Chapter 7.

1.4 Overview of This Book

1.4.1 Chapter Arrangements

The primary aim of this book is to introduce the theoretical foundations and algorithmic strategies used in the positioning engines of IPIN systems. This involves the analysis and application of measurements received and the derivation of models for position estimation.

Chapter 2 lays the foundational groundwork, introducing essential concepts in estimation. It covers coordinate systems and their transformations, state-space representation, and linear/nonlinear least squares methods along with popular solvers. This chapter is crucial for understanding the basic principles that underpin the subsequent chapters.

Chapter 3 introduces the estimation algorithms and filtering techniques commonly used in indoor positioning. It covers weighted least squares (WLS), Kalman filters, FGO, and particle filters, comparing their strengths and tradeoffs. These methods form the core mathematical tools for refining location estimates and mitigating uncertainties.

The book underscores the complementary nature of RF-based technologies and physical sensors in indoor positioning systems. RF technologies typically provide point positioning, while physical sensors offer relative positioning in a temporal sense. This difference is illustrated in Figure 1.8, highlighting the synergy between these two approaches. Chapters 4 to 7 explore various indoor positioning systems in detail. Each chapter contributes uniquely to the understanding of these systems, as illustrated in Figure 1.9. It is important to

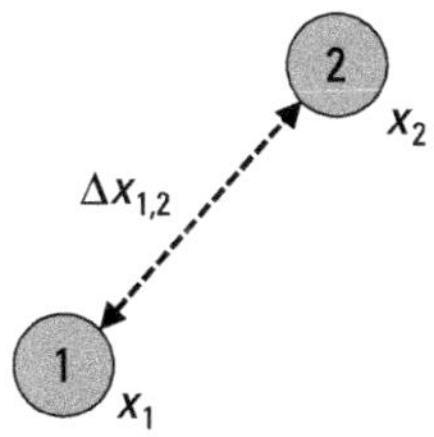

Figure 1.8 Sensor fusion for point positioning and relative positioning (dead reckoning). The subscript denotes the epoch of time.

note that the measurement engines (that generates the measurements for the positioning algorithm), although crucial in indoor positioning systems, are not the primary focus of this book.

Chapter 4 explores positioning methods based on RF signals, including Wi-Fi, BLE, UWB, 5G, and GNSS in constrained environments. It explains trilateration, and multilateration and time-based techniques while discussing accuracy, reliability, and challenges for each positioning method.

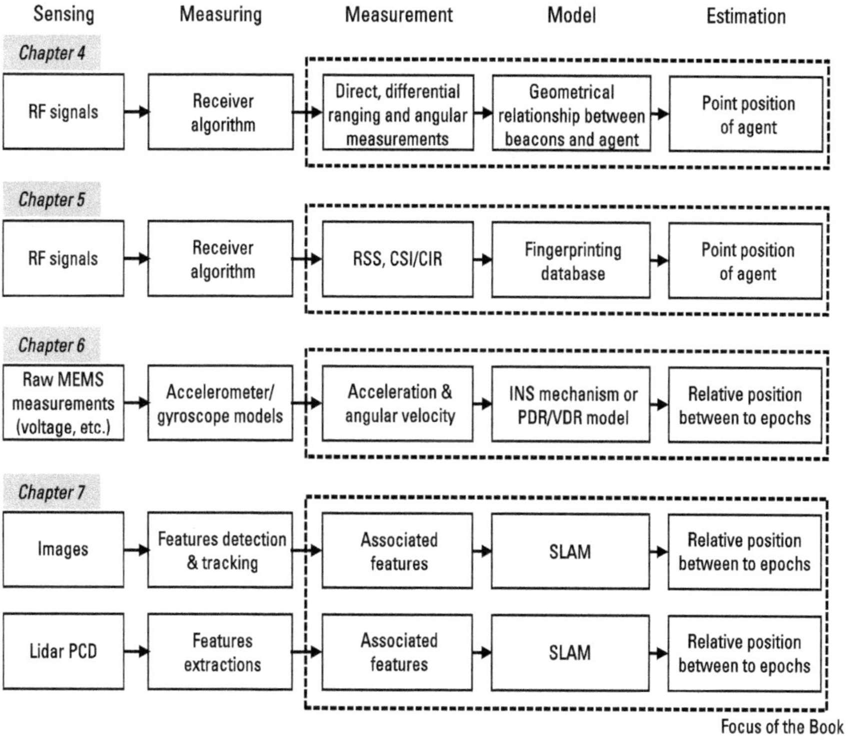

Figure 1.9 Frameworks of positioning engines introduced in different chapters.

Chapter 5 focuses on fingerprint-based indoor positioning, a technique that relies on signal pattern recognition rather than geometric models. It explains how Wi-Fi, magnetic field, and hybrid fingerprints are processed, and matched to pre-built databases. The chapter also addresses machine learning approaches for improving fingerprint-based localization accuracy.

Chapter 6 discusses proprioceptive sensors such as IMUs (accelerometers and gyroscopes), barometers, magnetometers, and odometry sensors for self-contained navigation. It explains PDR, motion tracking, and sensor drift correction, highlighting their role in infrastructure-free positioning.

Chapter 7 introduces SLAM techniques, which allow systems to localize while simultaneously building a map of the environment. It covers visual SLAM, lidar-based SLAM, and approaches for their fusion, detailing how these methods enable navigation in unknown environments without relying on external infrastructure.

Chapter 8 explores the fusion of multiple sensors to enhance positioning accuracy and robustness. It discusses integrating IMUs, lidar, cameras, and RF signals using sensor fusion techniques such as EKF, particle filter, and FGO. The chapter emphasizes how combining data from diverse sources improves indoor positioning.

Chapter 9 presents state-of-the-art advancements in indoor positioning and navigation. It discusses emerging techniques such as AI-driven localization, deep learning-based SLAM, cooperative positioning, and edge computing for real-time navigation. The advancements that the Internet of Things (IoT) brings to the domain of indoor positioning systems will be discussed as well.

1.4.2 Open-Sourced Codes and Data

Beyond providing theoretical insights, this book emphasizes learning by doing, enabling readers to bridge the gap between mathematical models, algorithms, and numerical simulation/real-world implementation. To achieve this, open-source code and datasets will be made available through a dedicated GitHub repository, allowing readers to experiment with, modify, and apply key indoor positioning and navigation techniques while engaging with a broader community, via https://github.com/IPNL-POLYU/IPIN-Examples. The examples are organized aligning each chapter of this book, in which the detailed structure can be found in the "README" of this repository.

Understanding indoor positioning requires more than just equations, it demands practical experience in implementing and testing algorithms on real data. Concepts such as Kalman filtering, SLAM, fingerprint matching, and sensor fusion become significantly more intuitive when learners can visualize how raw sensor inputs are transformed into meaningful position estimates. By directly interacting with the code, readers can observe errors, tune parameters, and develop a deeper understanding of how theoretical models behave in dynamic indoor environments.

The GitHub repository will host implementations in Python, MATLAB, or C++, covering algorithms discussed in each chapter. Additionally, real-world/simulated datasets from IMUs, lidar, cameras, Wi-Fi, UWB, and BLE will be provided, allowing readers to test positioning techniques without needing specialized hardware. Hands-on exercises and coding challenges will be included to reinforce key concepts and encourage experimentation.

Crucially, open-sourcing the data and code creates an opportunity for readers to interact, ask questions, and collaborate with both peers and authors. GitHub's issue tracking and discussion forums will enable learners to report challenges, share insights, and seek clarification, fostering a collaborative learning environment. This engagement not only helps individual learners but also contributes to the broader IPIN community, facilitating knowledge-sharing, algorithm improvements, and novel research directions. By following a code-first, community-driven approach, readers can validate mathematical derivations, debug algorithmic choices, and develop intuition for complex topics while benefiting from collective expertise. Through this interactive learning experience, the book not only imparts theoretical knowledge and practical skills but also strengthens the global IPIN research and development community.

References

[1] 360iResearch, "Indoor Positioning & Indoor Navigation Market by Technology (Bluetooth Low Energy (BLE), Infrared Sensors, Magnetic Positioning), Solution Type (Architecture, Positioning Analytics, Proximity Solutions), Application, End User, Industry Vertical, Deployment Mode, Device, Component, Connectivity Type—Global Forecast 2025–2030," 2025, https://www.360iresearch.com/library/intelligence/indoor-positioning-indoor-navigation.

[2] Grand View Research, "Indoor Positioning and Navigation Market Size, Share, & Trends Analysis Report by Component (Software, Hardware, Service), by Technology, by Application, by End-Use, by Region, and Segment Forecasts, 2024–2030," 2024, https://www.grandviewresearch.com/industry-analysis/indoor-positioning-navigation-market-report#.

[3] GII Global Information, "Indoor Positioning and Indoor Navigation (IPIN) Global Market Report 2024," 2024, https://www.giiresearch.com/report/tbrc1464116-indoor-positioning-indoor-navigation-ipin-global.html.

[4] The Business Research Company, "Indoor Positioning and Indoor Navigation (IPIN) Global Market Report 2025—By Component (Software, Hardware, Services), by

Technology (Ultra-Wideband Technology, Bluetooth Low Energy, Wi-Fi, Other Technologies), by Application (Asset and Personnel Tracking, Location-Based Analytics, Navigation and Maps, Other Applications), by End Use Industry (Healthcare, Retail, Manufacturing, Travel and Hospitality, Office Spaces, Public Spaces, Logistics and Warehouses, Other End Use Industries)—Market Size, Trends, and Global Forecast 2025–2034," 2025, https://www.thebusinessresearchcompany.com/report/indoor-positioning-and-indoor-navigation-ipin-global-market-report.

[5] Grand View Research, "Warehouse Robotics Market Size, Share & Trends Analysis Report by Product, by Function (Pick & Place, Transportation, Packaging), by Payload Capacity, by Component, by Software, by Application, by Region, and Segment Forecasts, 2023–2030," Warehouse Robotics Market Size, Share & Trends Report, 2022, https://www.grandviewresearch.com/industry-analysis/warehouse-robotics-market.

[6] SNS Insider, "Autonomous Mobile Robots for Logistics and Warehousing Market Set to Achieve USD 18.56 Billion by 2032 Driven by Growing Automation Demand," 2024, https://www.globenewswire.com/news-release/2024/09/26/2953889/0/en/Autonomous-Mobile-Robots-for-Logistics-and-Warehousing-Market-Set-to-Achieve-USD-18-56-Billion-by-2032-Driven-by-Growing-Automation-Demand.html.

[7] MarketsandMarkets, "Mobile Robots Market Size, Share & Industry Growth Analysis Report by Type (Professional and Personal & Domestic Robots), Application (Agricultural, Cleaning, Educational, Healthcare, Manufacturing, Warehousing & Logistics) and Region (North America, Europe, APAC, ROW)—Global Forecast to 2028," 2023, https://www.marketsandmarkets.com/Market-Reports/mobile-robots-market-43703276.html.

[8] 3GPP, "Technical Specification Group Services and System Aspects; Study on positioning use cases; Stage 1," Third Generation Partnership Project, 2018.

[9] Hsu, L. -T., G. Zhang, and W. Wen, "Indoor-Positioning-and-Indoor-Navigation—Examples-Codes," 2024, https://github.com/IPNL-POLYU/Indoor-Positioning-and-Indoor-Navigation—Examples-Codes.

[10] Samama, N., *Indoor Positioning: Technologies and Performance*, New York: John Wiley & Sons, 2019.

[11] MarketsandMarkets Research, "The Rise of Indoor Location Market: $31.4 billion Industry Dominated by Tech Giants—Zebra Technologies (US), Cisco (US), Google (US), Microsoft (US), HPE (US), Apple & Esri," 2024, https://www.globenewswire.com/news-release/2024/08/26/2935606/0/en/The-Rise-of-Indoor-Location-Market-31-4-billion-Industry-Dominated-by-Tech-Giants-Zebra-Technologies-US-Cisco-US-Google-US-Microsoft-US-HPE-US-Apple-Esri.html.

[12] Favendo, "Indoor Location Tracking—Five Technologies in Brief," 2020, https://www.favendo.com/asset-tracking/indoor-location-5-technologies.

[13] MarketResearch.com, "Ultra-Wideband (UWB)," Global Industry Analysts, 2025, https://www.marketresearch.com/Global-Industry-Analysts-v1039/Ultra-Wideband-UWB-39564301/.

[14] FINDIT, *Ultra Wideband Market Size, Share & Trends Analysis Report*, 2025, https://findit.com/vbrptwrbhnvsygp/RightNow/ultra-wideband-market-size-share-trends-analysis.

[15] Huawei Technologies Co., "Huawei Launches the Industry's First Commercial 5G Indoor Positioning Solution to Advance toB," 2022, https://www.huawei.com/en/news/2022/4/5g-positioning-2022.

[16] Van de Velde, S., "Is 5G Positioning Ready for Industrial Asset Tracking?" 2023, https://www.pozyx.io/newsroom/is-5g-ready-or-not-for-industrial-asset-tracking.

[17] Link Labs, *The Ins and Out of An Indoor Positioning System*, 2018, https://www.link-labs.com/blog/indoor-positioning-system.

[18] Rozaev, N., "How Ultrasonic Local Positioning System Works," 2023, https://navigine.com/blog/how-ultrasonic-indoor-positioning-works.

[19] Interact, "Enhance the Intelligence of Illuminated Spaces with an "Indoor GPS,'" https://www.interact-lighting.com/en-us/capabilities/indoor-navigation.

[20] Haverinen, J., "The Rise of Indoor Positioning: A 2016 Global Research Report on the Indoor Positioning Market," *IndoorAtlas*, VansonBourne2016.

[21] Pastor, A., "AGV Navigation: Methods, Comparison, Pros and Cons—Illustrated Guide," https://www.agvnetwork.com/types-of-navigation-systems-automated-guided-vehicles.

[22] ResearchAndMarkets, "Global Autonomous Mobile Robots Market 2023–2030: Emergence of Artificial Intelligence Technology in Autonomous Mobile Robots Drives Growth," 2023, https://www.globenewswire.com/news-release/2023/04/24/2653189/28124/en/Global-Autonomous-Mobile-Robots-Market-2023-2030-Emergence-of-Artificial-Intelligence-Technology-in-Autonomous-Mobile-Robots-Drives-Growth.html.

[23] Bluetooth Special Interest Group, "Bluetooth Core Specification v5.1," 2019, https://www.bluetooth.com/specifications/bluetooth-core-specification/.

[24] Hillyard, J., C. Bachmann, and L. Huang, "Bluetooth Channel Sounding—A Step Towards 10-cm Ranging Accuracy for Secure Access, Digital Key, and Proximity Service," 2023, https://www.bluetooth.com/blog/bluetooth-channel-sounding-for-accurate-ranging/

[25] Halperin, D., et al., "Tool Release: Gathering 802.11n CSI Data with the Intel 5300 Wireless NIC," *ACM SIGCOMM Computer Communication Review*, Vol. 41, No. 1, 2011, pp. 53–59.

[26] Mohsen, M., H. Rizk, and M. Youssef, "Privacy-Preserving by Design: Indoor Positioning System Using Wi-Fi Passive TDOA," *2023 24th IEEE International Conference on Mobile Data Management (MDM)*, 2023, pp. 221–230.

[27] Horn, B. K. P., "Doubling the Accuracy of Indoor Positioning: Frequency Diversity," *Sensors (Basel)*, Vol. 20, No. 5, March 9, 2020.

[28] Dwivedi, S., et al., *5G Positioning: What You Need to Know*, Ericsson, 2020, https://www.ericsson.com/en/blog/2020/12/5g-positioning--what-you-need-to-know.

[29] Mukkavilli, K. and Zhang, X., "5G: Bringing Precise Positioning to the Connected Intelligent Edge," Qualcomm, 2021.

[30] IEEE Standards Association, "IEEE 802.15.4a: Ultra-Wideband (UWB) PHY for High-Precision Ranging," 2007, https://standards.ieee.org/standard/802_15_4a-2007.html.

[31] Muto, Y., "UWB Terminology," 2023 https://www.firaconsortium.org/resource-hub/blog/uwb-terminology.

[32] Qorvo, "DWM1000 Ultra-Wideband Module Product Brief," 2018, https://www.qorvo.com/products/p/DWM1000.

[33] Collin, J., et al., "Indoor Positioning System Using Accelerometry and High Accuracy Heading Sensors," *Proceeding GPS2003*, Alexandria, VA, January 1, 2003.

[34] Sahli, H., et al., "Small Pipeline Trajectory Estimation Using MEMS Based IMU," *Proceedings of the 27th International Technical Meeting of The Satellite Division of the Institute of Navigation (ION GNSS+ 2014)*, 2014.

[35] Sewio Documentation, "Z-Axis Based on Barometer," https://www.sewio.net/sewio-rtls-2-5-first-rtls-in-the-world-to-enable-full-visibility-into-narrow-aisles-with-unmatched-cost-effectiveness/

[36] Bosch Sensortec, "BMP280: Barometric Pressure Sensor Datasheet," 2015, https://www.bosch-sensortec.com/.

[37] Google Android, "Android Open Source Project: Sensor Types," 2025, https://source.android.com/docs/core/interaction/sensors/sensor-types.

[38] Li, M., et al., "A Precise Indoor Visual Positioning Approach Using a Built Image Feature Database and Single User Image from Smartphone Cameras," *Remote Sensing*, Vol. 12, No. 5, 2020, p. 869.

[39] Luo, W., et al., "Indoor Smartphone SLAM with Learned Echoic Location Features," *Proceedings of the 20th ACM Conference on Embedded Networked Sensor Systems*, 2022, pp. 489–503.

[40] Velodyne Lidar, "Surround Sensor Product Guide," 2019.

[41] Wu, Y., et al., "An FPGA Based Energy Efficient DS-SLAM Accelerator for Mobile Robots in Dynamic Environment," *Applied Sciences*, Vol. 11, No. 4, p. 1828, 2021.

[42] Guarnera, C., "Lidar Indoor Mapping: Revolutionizing Indoor Navigation," 2024, https://www.bluefalconaerial.com/lidar-indoor-mapping.

2

Fundamentals of Indoor Positioning and Navigation Systems

This chapter introduces the fundamental knowledge required to address the estimation problems encountered in IPIN applications. Section 2.1 covers the coordinate systems commonly adopted and the transformations among these systems, which are essential for accurate position representation and system integration. Section 2.2 focuses on the definition and representation of attitude, describing how an object's orientation is characterized using different mathematical forms, including Euler angles, rotation matrices, and quaternions. These foundational topics set the stage for the estimation methods and algorithms discussed in later chapters.

2.1 Coordinate Systems and Transformations

The fundamental question to be solved for positioning is: "Where am I?" To answer this question, one needs a system to indicate our position with respect to a specific referencing position by coordinate values (i.e., a coordinate system). Besides, different applications or sensors have their own preferring coordinate system, which are usually not the same. Therefore, it is necessary to establish transformations between different coordinate systems. This section covers the essential knowledge of coordinate systems for IPIN, whereas a comprehensive introduction can be found in [1]. An overview of the coordinate systems for IPIN is shown in Figure 2.1. The coordinate systems are categorized into local coordinate system and global coordinate system, depending on whether the system is applicable in a certain area around a reference or universally applicable with respect to the Earth.

2.1.1 Local Body Coordinate System

A popular coordinate system for IPIN applications is the local body coordinate system [2], which describes the movement of an object from its own point of view or describes the position of an object from another object's point of view. The local body coordinate system is a Cartesian coordinate system fixed with respect to a reference body, either being static or under movements. An example is shown as Figure 2.2. The origin B can be defined as the center of mass of the body. For IPIN applications, users are often concerned with objects located above themselves or above a reference point, such as ceiling-mounted beacons. Consequently, the Z-axis typically points upward from the origin. Following the right-hand rule and the commonly used convention, the X-axis is defined as pointing to the right of the body, while the Y-axis points forward. The position of the target object A can be represented by its coordinate values along these three axes with respect to the origin, as follows:

$$\mathbf{x}_{\mathrm{BODY}} = \begin{bmatrix} x_{\mathrm{BODY}} & y_{\mathrm{BODY}} & z_{\mathrm{BODY}} \end{bmatrix}^{\mathrm{T}} \tag{2.1}$$

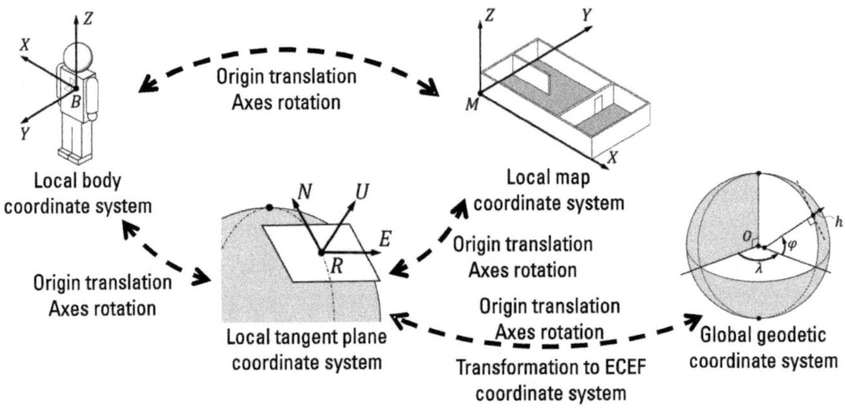

Figure 2.1 Overview of the coordinate systems and transformations.

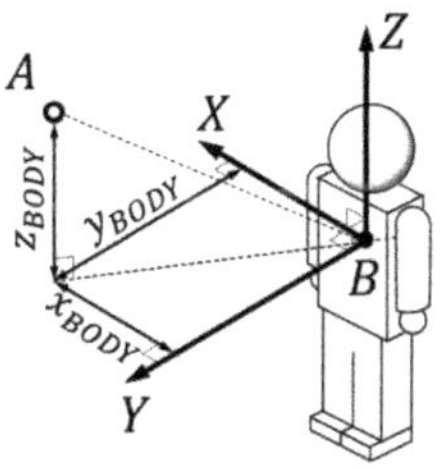

Figure 2.2 Local body coordinate system.

Similarly, the movement of an object can be represented by its projection component on each axis. Noted that definition of local body coordinate system here is different from that in aviation applications, where Z-axis is pointing down to better describe objects below an aircraft.

2.1.2 Local Map Coordinate System

IPIN applications usually prefer a local coordinate system that can represent the position information with respect to the local environment feature. Such a coordinate system is usually constructed based on a "map" of the related area (e.g., the building plan), as Figure 2.3 shows. Similar to the local body coordinate system, a reference point M is selected as the origin to build a Cartesian coordinate system that the axes are along with the map features, such as the walls. Then the position of an object can be represented on this map by the coordinate values, as follows:

$$\mathbf{x}_{\mathrm{MAP}} = \begin{bmatrix} x_{\mathrm{MAP}} & y_{\mathrm{MAP}} & z_{\mathrm{MAP}} \end{bmatrix}^{\mathrm{T}} \tag{2.2}$$

For many cases, the XY-planes of the local map coordinate system and the local body coordinate system are parallel, with a yaw rotation about Z-axis.

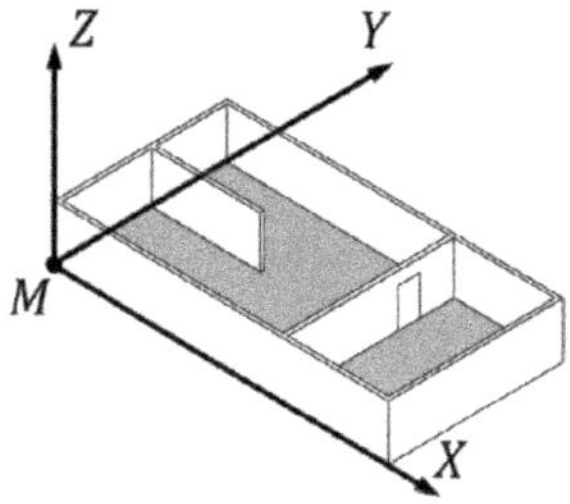

Figure 2.3 Local map coordinate system.

Then, for an object A, the coordinate transformation from the local map coordinate system to the local body coordinate system is achieved as follows:

$$\mathbf{x}_{\mathrm{BODY},A} = \begin{bmatrix} \cos\psi & \sin\psi & 0 \\ -\sin\psi & \cos\psi & 0 \\ 0 & 0 & 1 \end{bmatrix} \left(\mathbf{x}_{\mathrm{MAP},A} - \mathbf{x}_{\mathrm{MAP},B} \right) \tag{2.3}$$

where ψ is the yaw angle from the X and Y axes in the local map coordinate system to align with that in the local body coordinate system. $\mathbf{x}_{\mathrm{MAP},A}$ and $\mathbf{x}_{\mathrm{MAP},B}$ are the position of object A and the local body coordinate origin B in the local map coordinate system, respectively. When the XY-planes of the local map coordinate system and the local body coordinate system are not parallel, an additional rotation is required to complete the coordinate transformation. The detailed representation and computation of these rotations (attitude angles) will be introduced in Section 2.2.

2.1.3 Local Tangent Plane Coordinate System: East, North, Up and North, East, Down Coordinate System

Some applications prefer to describe the position of an object with respect to the geographical features on the Earth's ground like the directions of East and North. For convenience, if such an application is within a certain area (e.g., within hundreds of meters), the ground is approximated as a tangent plane on a selected reference location on the Earth, neglecting the ground curvature. Then a local Cartesian coordinate system is constructed on this reference location, in which the position of the target object is expressed by coordinate values with respect to this reference point. A widely used local coordinate system is the East, North, Up (ENU) coordinate system, as Figure 2.4 shows. On the selected reference point R, a plane tangent to the Earth's ellipsoidal surface can be obtained. Then two orthogonal axes of a Cartesian coordinate system are resolved on this tangent plane, including the N-axis pointing to the North Pole and the E-axis pointing to the East direction with respect to the reference point. Following the right-hand rule, the third axis is the U-axis perpendicular to the tangent plane, pointing outwards from the ellipsoid. Then the position of object A can be represented by the coordinate values x_{ENU}, y_{ENU}, and z_{ENU} on the E, N, and U axes respect to the reference point, using

$$\mathbf{x}_{\mathrm{ENU}} = \begin{bmatrix} x_{\mathrm{ENU}} & y_{\mathrm{ENU}} & z_{\mathrm{ENU}} \end{bmatrix}^{\mathrm{T}} \tag{2.4}$$

By approximating Earth's surface as the flat plane parallel to the tangent plane, the E-axis and N-axis are parallel to the ground. Then the coordinate

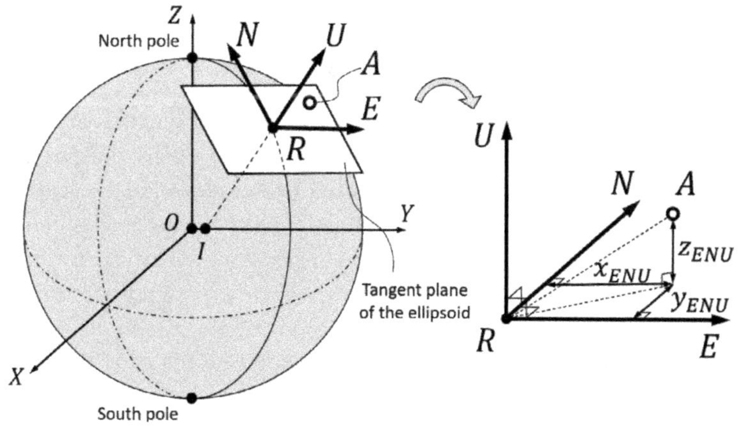

Figure 2.4 ENU coordinate system.

values can intuitively indicate how far an object on the ground is located with respect to the reference position in terms of East and North directions.

An alternative local tangent plane coordinate systems for aviation applications is the North, East, Down (NED) coordinate system [1], where its axes are point to the North, East, and down directions. It can be transformed from the ENU coordinate system by rotations about Y-axis and Z-axis as follows:

$$\mathbf{x}_{\text{NED}} = \mathbf{C}_{\text{ENU}}^{\text{NED}}\mathbf{x}_{\text{ENU}} = \begin{bmatrix} \cos\frac{\pi}{2} & \sin\frac{\pi}{2} & 0 \\ -\sin\frac{\pi}{2} & \cos\frac{\pi}{2} & 0 \\ 0 & 0 & 1 \end{bmatrix} \begin{bmatrix} \cos\pi & 0 & -\sin\pi \\ 0 & 1 & 0 \\ \sin\pi & 0 & \cos\pi \end{bmatrix} \mathbf{x}_{\text{ENU}}$$
$$= \begin{bmatrix} 0 & 1 & 0 \\ 1 & 0 & 0 \\ 0 & 0 & -1 \end{bmatrix} \mathbf{x}_{\text{ENU}} \tag{2.5}$$

In coordinate transformations, especially when aligning the local body coordinate system with the local tangent plane coordinate system (ENU), it is often necessary to describe how an object is oriented relative to the chosen reference frame. This orientation, known as attitude, can be parameterized in various ways. In the ENU coordinate system, the attitude of an object can be represented by the deviations of its local body coordinate system axes from the ENU axes. This is typically expressed in terms of Euler angles, describing 3 degrees of freedom (3-DOF): pitch, roll, and yaw. For example, as shown in Figure 2.5(a), the yaw angle ψ (positive in the counterclockwise direction following the right-hand rule) describes the rotation of the local X- and Y-axes about the Z-axis away from the E- and N-axes of the ENU system. Similarly, pitch and roll angles describe the rotation of the local body coordinate system about its X- and Y-axes relative to the ENU frame.

It is important to note that attitude (orientation) is different from bearing, which describes the direction toward which an object is located. Bearing in the ENU system is often represented by the elevation angle, the angle from the ground plane to the object's line of sight (positive when above the ground), and the azimuth angle, the angle from the North direction to the object's projection on the ground plane (positive counterclockwise). Figure 2.5(b) shows an example of bearing represented by azimuth angle. For applications requiring different reference planes or axes, bearing can also be defined relative to other coordinate system elements. It is worth noting that, in the NED coordinate system, the azimuth angle is defined as the down axis and has the opposite convention compared to the ENU system. A more detailed discussion of attitude representation, including alternative parameterizations such as rotation matrices and quaternions, is provided in Section 2.2.

For the position of object A, the transformation between the ENU coordinate system and the local body coordinate system can be achieved by

$$\mathbf{x}_{\text{BODY},A} = \mathbf{C}_{\text{ENU}}^{\text{BODY}}\left(\mathbf{x}_{\text{ENU},A} - \mathbf{x}_{\text{ENU},B}\right) \tag{2.6}$$

$$\mathbf{x}_{\text{ENU},A} = \mathbf{C}_{\text{BODY}}^{\text{ENU}}\left(\mathbf{x}_{\text{BODY},A} - \mathbf{x}_{\text{BODY},R}\right) \tag{2.7}$$

where $\mathbf{C}_{\text{ENU}}^{\text{BODY}}$ is the rotation matrix based on the attitude pitch, roll, and yaw angle of the body coordinate system from the ENU coordinate system. $\mathbf{x}_{\text{ENU},B}$ is the position of the origin of the local body coordinate system in the ENU coordinate system. $\mathbf{x}_{\text{BODY},R}$ is the position of the origin of the ENU coordinate system in the local body coordinate system.

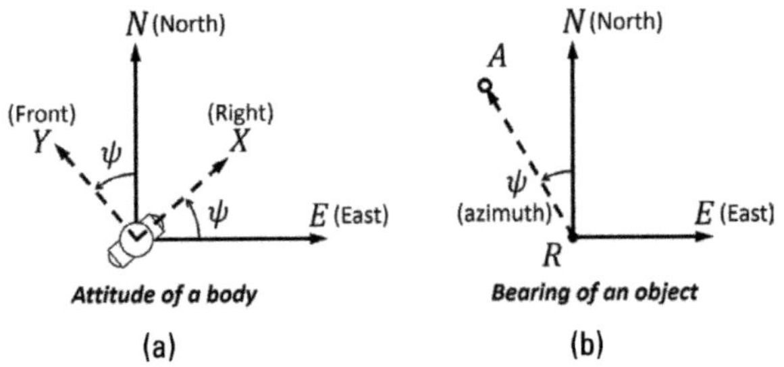

Figure 2.5 (a) Attitude representation of a body by yaw angle. (b) Bearing representation of an object by azimuth angle.

2.1.4 Geodetic Coordinate System: Latitude, Longitude, Height Coordinate System

Another popular coordinate system for positioning in a large-scale area is the geodetic coordinate system [1], in which the ground curvature cannot be neglected. Then the Earth is modeled by an ellipsoid, as Figure 2.6 shows. One of the popular standards is the World Geodetic System 1984 (WGS-84), which defines necessary parameters for the coordinate system, such as the equatorial radius, the flattening, and the geoid [3]. Based on such a standard, the position of the concerning object on the Earth can be described by the geodetic coordinates: latitude φ, longitude λ, and ellipsoid height h. For an object (or a user) A, the unique line $\overline{AI}$ passing through the object normal to the ellipsoid can be obtained, which intersects the ellipsoid by point P. The distance from point P to point A denotes the shortest distance between the object and the Earth, namely the ellipsoid height h. Then the angle between the normal line $\overline{AI}$ and the equatorial plane is the geodetic latitude φ, which is positive for the Northern Hemisphere and is negative for the Southern Hemisphere. Note that the intersection of this normal line and the equatorial plane, I, may not coincide with the ellipsoid center O. Finally, the longitude λ is the angle from the plane of Prime Meridian to the projection of the normal line $\overline{AI}$ on the equatorial plane. The longitude λ is positive for the East of the Prime Meridian and is negative for the West of the Prime Meridian. Then the position of object A can be represented by

$$\mathbf{x}_{LLH} = \begin{bmatrix} \varphi & \lambda & h \end{bmatrix}^{\mathrm{T}} \tag{2.8}$$

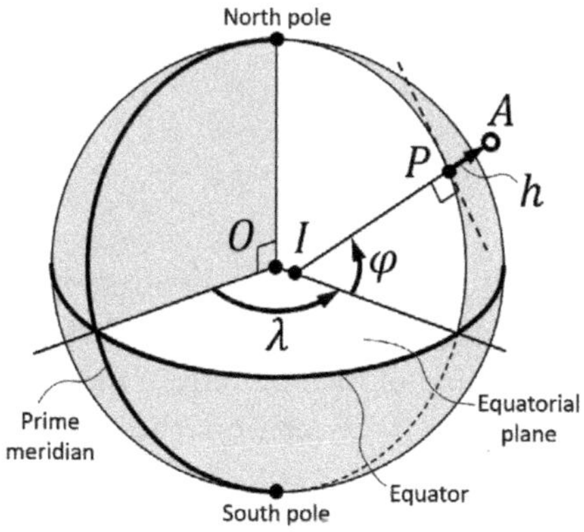

Figure 2.6 Geodetic coordinate system.

Meanwhile, the actual surface of the Earth is fluctuated compared to the ellipsoid surface. An alternative measure is the height above the geoid (or the mean sea level), which considers the natural curvature of the Earth's surface. This is a global coordinate system that describes the position of an object with respect to the Earth.

The position expression can be transformed between the global geodetic coordinate system and the local ENU coordinate system. Note that the global geodetic coordinate system is not a Cartesian coordinate system. Thus, a global Cartesian coordinate system, the Earth-centered Earth-fixed (ECEF) coordinate system, is needed during the transformation in between. A detailed introduction of the ECEF coordinate system can be found in [2]. The position transformation from the geodetic coordinate system in (2.1) to the ECEF coordinate system is achieved by

$$\mathbf{x}_{\text{ECEF}} = \begin{bmatrix} \dfrac{a\cos\lambda}{\sqrt{1+\left(1-e^2\right)\tan^2\varphi}} + h\cos\lambda\cos\varphi \\ \dfrac{a\sin\lambda}{\sqrt{1+\left(1-e^2\right)\tan^2\varphi}} + h\sin\lambda\cos\varphi \\ \dfrac{a\left(1-e^2\sin\varphi\right)}{\sqrt{1-e^2\sin^2\varphi}} + h\sin\varphi \end{bmatrix} \tag{2.9}$$

where a and e are the Earth's ellipsoid semimajor axis and eccentricity, respectively. Then the position of an object A in the ENU coordinate system with a selected reference point R as the origin is obtained by

$$\begin{aligned} \mathbf{x}_{\text{ENU},A} &= \mathbf{C}_{\text{ECEF}}^{\text{ENU}}\left(\mathbf{x}_{\text{ECEF},A} - \mathbf{x}_{\text{ECEF},R}\right) \\ &= \begin{bmatrix} -\sin\lambda & \cos\lambda & 0 \\ -\sin\varphi\cos\lambda & -\sin\varphi\sin\lambda & \cos\varphi \\ \cos\varphi\cos\lambda & \cos\varphi\sin\lambda & \sin\varphi \end{bmatrix}\left(\mathbf{x}_{\text{ECEF},A} - \mathbf{x}_{\text{ECEF},R}\right) \end{aligned} \tag{2.10}$$

where $\mathbf{x}_{\text{ECEF},A}$ and $\mathbf{x}_{\text{ECEF},R}$ are the positions of object A and reference R in the ECEF coordinate system, respectively, based on (2.9). ($\mathbf{x}_{\text{ECEF},A} - \mathbf{x}_{\text{ECEF},R}$) can be regarded as the translation of the origin from the Earth ellipsoid center to the reference point R. $\mathbf{C}_{\text{ECEF}}^{\text{ENU}}$ is the rotation matrix, which rotates the axes in the ECEF coordinate system to align with the East, North, and up directions. The position transformation from the local ENU coordinate system to the global geodetic coordinate system can be achieved via the ECEF coordinate system with an iteration method, which can be found in [2]. Note that the global geodetic coordinate system is less popular for IPIN applications, as the coordinates are not intuitive for demonstrating the relative distance information we care about.

2.2 Attitude: Definition and Representation

2.2.1 Definition

In the context of IPIN applications, attitude refers to the orientation of an object (or its local body coordinate frame) relative to a chosen reference coordinate system (e.g., a local map or tangent plane coordinate system). Attitude is a purely rotational concept, distinct from position, and it describes how the object is oriented in 3-D space. For a rigid body in 3-D space, attitude is characterized by 3-DOF, representing rotations about three orthogonal axes of the local body coordinate system.

These three rotations are commonly described by roll, pitch, and yaw angles. The precise assignment of axes depends on the coordinate system convention adopted. For example, in a right-forward-up (*X*-right, *Y*-forward, *Z*-up) coordinate system, roll is typically defined as rotation about the *Y*-axis, pitch about the *X*-axis, and yaw about the *Z*-axis. The exact rotation sequence and axis assignment may vary by convention (e.g., aerospace versus robotics). Nonetheless, these three angles collectively describe the orientation of the object relative to the reference frame. Figure 2.7 illustrates these three principal rotations defining attitude (roll, pitch, yaw), showing the object's rotations about its *Y*, *X*, and *Z*-axes. Figure 2.7 also shows an example of how a yaw rotation transforms the reference coordinate system.

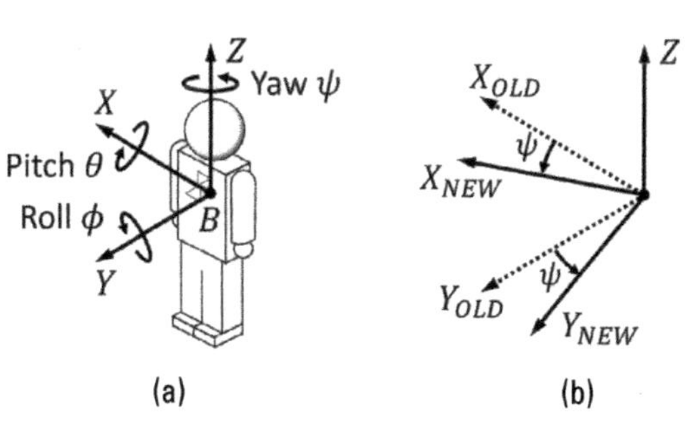

Figure 2.7 (a) Attitude movement. (b) Example of transformation from old to new coordinate systems by the yaw angle.

2.2.2 Representation

There are multiple ways to represent a 3-D orientation (attitude) in mathematical terms. The three major representation methods discussed here are Euler angles, rotation matrices, and quaternions [4]. Each representation has a different form, respectively: a set of three angles, a 3×3 matrix, and a 4-parameter unit quaternion, and each comes with its own advantages, limitations, and theoretical underpinnings. All these methods are equivalent in the sense that they can describe any possible orientation in 3-D space, and one can convert from one representation to another. However, they differ in interpretability, computational convenience, presence or absence of singularities, and compactness. We will describe each in turn, providing definitions and the underlying theory, followed by a comparative summary.

2.2.2.1 Euler Angles

The attitude movement of an object can also be described in terms of a local body coordinate system, as Figure 2.7(a) shows. Following the right-hand rule, rotations on X, Y, and Z-axes are denoted by the pitch angle θ, roll angle ϕ, and yaw angle ψ, respectively. Thus, the 3-DOF angular movement of an object can be represented correspondingly. Besides, the attitude movement of an object can be regarded as a transformation between the old and the new coordinate systems, as Figure 2.7(b) shows.

However, Euler angles have important limitations, notably the issue of singularities. Since the space of orientations is not topologically equivalent to a 3-D Euclidean space, no single continuous set of three parameters can cover all possible orientations without discontinuities. In practice, this results in a gimbal lock, a phenomenon where, at certain critical orientations, two of the rotation axes align and the system loses 1 DOF. In the standard yaw-pitch-roll sequence, this singularity occurs at $\theta = \pm 90°$; at that attitude, the yaw and roll axes coincide, and an increment of yaw produces the same effect as an increment of roll. Consequently, the system effectively "locks up" with only two independent rotation directions instead of three. As an example, at $\theta = 90°$, changing yaw or roll results in the same motion, making those angles indistinguishable. As a result, Euler angle representations become unreliable near these singular configurations, small changes in orientation can lead to large, abrupt changes in the angle values or even make them undefined. Additionally, Euler angles complicate certain calculations: rotation composition requires applying the appropriate rotation matrices rather than simply adding angles, and the noncommutative nature of rotations necessitates strict adherence to the specified rotation order. Nonetheless, Euler angles remain popular due to their intuitive interpretability and minimal parameter count. In summary, while Euler angles are highly interpretable, their main drawbacks include the presence of singularities (gimbal lock) and the complexity of composing rotations.

2.2.2.2 Rotation Matrix

Many applications may prefer a coordinate system with a customized orientation for better position expression, which requires rotation adjustments on both three axes from a standard coordinate system. This adjustment can be regarded as changing the basis of a vector space in linear algebra while preserving the distance of a point, as multiplying a transformation matrix:

$$\mathbf{x}_{\mathrm{NEW},A} = \mathbf{C}_{\mathrm{OLD}}^{\mathrm{NEW}} \mathbf{x}_{\mathrm{OLD},A} \tag{2.11}$$

where $\mathbf{x}_{\mathrm{OLD},A}$ and $\mathbf{x}_{\mathrm{NEW},A}$ denote the position of object A before and after the transformation. $\mathbf{C}_{\mathrm{OLD}}^{\mathrm{NEW}}$ is the transformation matrix where the subscript and superscript indicate the coordinate systems before and after the transformation, namely the rotation matrix. As the rotation matrix is orthogonal, it has the following property:

$$\mathbf{C}_{\mathrm{OLD}}^{\mathrm{NEW}} = \left(\mathbf{C}_{\mathrm{NEW}}^{\mathrm{OLD}}\right)^{-1} = \left(\mathbf{C}_{\mathrm{NEW}}^{\mathrm{OLD}}\right)^{\mathrm{T}} \tag{2.12}$$

where the superscript T denotes the transpose of a matrix. The transformation can be applied backwards by multiplying the transpose of the rotation matrix. For a Cartesian coordinate system, such as the local body coordinate system, its rotation about each axis with the corresponding Euler angle can be achieved by multiplying a transformation matrix. As shown in the example in Figure 2.8, the position of object A in the new coordinate system XYZ^* after a yaw rotation ψ can be represented by the coordinate values from the old coordinate system XYZ as follows:

$$\mathbf{x}^* = \begin{bmatrix} x^* \\ y^* \\ z^* \end{bmatrix} = \begin{bmatrix} x\cos\psi + y\sin\psi \\ -x\sin\psi + y\cos\psi \\ z \end{bmatrix} = \begin{bmatrix} \cos\psi & \sin\psi & 0 \\ -\sin\psi & \cos\psi & 0 \\ 0 & 0 & 1 \end{bmatrix} \begin{bmatrix} x \\ y \\ z \end{bmatrix} \tag{2.13}$$

where the superscript $*$ denotes the variable in the new coordinate system. Note that the coordinate value on the Z-axis remained unchanged after the rotation. Thus, the rotation matrix in (2.11) representing the rotation of XY-plane to new XY^*-plane about the Z-axis with a yaw angle ψ is

$$\mathbf{C}_{XY}^{XY^*} = \begin{bmatrix} \cos\psi & \sin\psi & 0 \\ -\sin\psi & \cos\psi & 0 \\ 0 & 0 & 1 \end{bmatrix} \tag{2.14}$$

Similarly, the rotation matrix of the XZ-plane to a new XZ^*-plane about the Y-axis by a roll angle ϕ is:

$$\mathbf{C}_{XZ}^{XZ^*} = \begin{bmatrix} \cos\phi & 0 & -\sin\phi \\ 0 & 1 & 0 \\ \sin\phi & 0 & \cos\phi \end{bmatrix} \tag{2.15}$$

The rotation matrix of the YZ-plane to a new YZ^*-plane about the X-axis by a pitch angle θ is:

$$\mathbf{C}_{YZ}^{YZ^*} = \begin{bmatrix} 1 & 0 & 0 \\ 0 & \cos\theta & \sin\theta \\ 0 & -\sin\theta & \cos\theta \end{bmatrix} \tag{2.16}$$

Then the rotation matrix with all three axes from the old XYZ coordinate to the new XYZ^* is represented by Euler angles as below, following the sequence of yaw rotation, row rotation, and pitch rotation (i.e., about the Z-, Y-, and X-axes):

$$\begin{aligned} \mathbf{C}_{\text{OLD}}^{\text{NEW}} &= \mathbf{C}_{YZ}^{YZ^*}\mathbf{C}_{XZ}^{XZ^*}\mathbf{C}_{XY}^{XY^*} \\ &= \begin{bmatrix} \cos\phi\cos\psi & \cos\phi\sin\psi & -\sin\phi \\ -\cos\theta\sin\psi + \sin\theta\sin\phi\cos\psi & \cos\theta\cos\psi + \sin\theta\sin\phi\sin\psi & \sin\theta\cos\phi \\ \sin\theta\sin\psi + \cos\theta\sin\phi\cos\psi & -\sin\theta\cos\psi + \cos\theta\sin\phi\sin\psi & \cos\theta\cos\phi \end{bmatrix} \end{aligned} \tag{2.17}$$

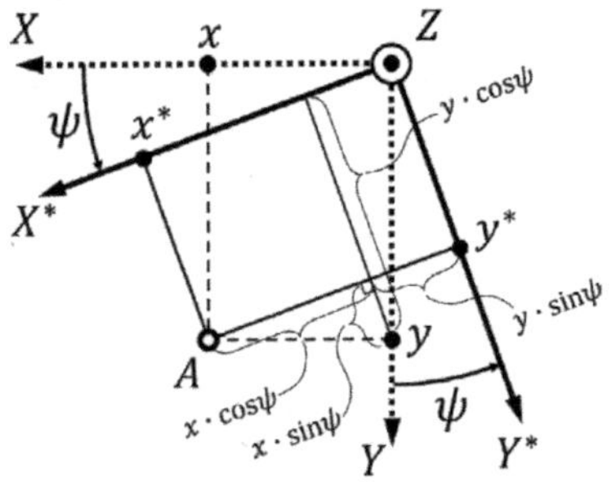

Figure 2.8 Example of position representation after yaw rotation. The superscript * denotes the variable in the new coordinate system.

Note that the rotations based on Euler angles (as matrix multiplication) are not commutative. For a general case, two coordinate systems are not necessary to have the same origin, which needs a translation of the origin to perform coordinate transformation. Thus, the position representation of object A in a new coordinate system can be converted from its position $\mathbf{x}_{\mathrm{OLD},A}$ in the old coordinate system by

$$\mathbf{x}_{\mathrm{NEW},A} = \mathbf{C}_{\mathrm{OLD}}^{\mathrm{NEW}}\left(\mathbf{x}_{\mathrm{OLD},A} - \mathbf{x}_{\mathrm{OLD},O_{\mathrm{NEW}}}\right) \tag{2.18}$$

where $\mathbf{x}_{\mathrm{OLD},O_{\mathrm{NEW}}}$ is the origin position of the new coordinate system in the old coordinate system.

Rotation matrices are also straightforward to use in transformations. The downside is that rotation matrices are less compact than minimal representations (since 9 numbers are stored, or 3×3 matrix, although only 3 are independent). They also may require maintaining orthonormality in a computational system; for example, if one integrates many small rotations (from gyroscope readings, say), numerical errors might cause the matrix to drift away from perfect orthogonality and require re-orthonormalization. In terms of interpretability, rotation matrices are not as immediately intuitive as Euler angles; one can interpret the columns or rows as basis vectors directions, but it is not as simple as "an angle about an axis" for a layperson. However, they are very convenient mathematically: combining rotations is just matrix multiplication, and many algorithms in navigation and robotics use rotation matrices internally due to their lack of singularity and direct applicability in coordinate transformation calculations. They are also the standard representation in many scientific and engineering libraries. To summarize, a rotation matrix provides a complete and nonsingular description of attitude and is well-suited for computations, at the expense of using more parameters (with constraints) and being less human-readable than Euler angles.

2.2.2.3 Quaternion

There are conditions that the rotations cannot be represented by means of Euler angles. For example, when the roll angle ϕ equals $\pi/2$, the rotation matrix in (2.17) is represented as

$$\mathbf{C}_{\mathrm{OLD}}^{\mathrm{NEW}} = \begin{bmatrix} 0 & 0 & -1 \\ \sin(\theta-\psi) & \cos(\theta-\psi) & 0 \\ \cos(\theta-\psi) & -\sin(\theta-\psi) & 0 \end{bmatrix} \tag{2.19}$$

The Euler angles of pitch and yaw have the same effect on the final rotation representation, in which 1 DOF is lost. Thus, the attitude related to the lost DOF cannot be represented. Note that this gimbal lock is more likely to occur for IPIN with a portable device, which is expected to have all possible orientations during a practical application. To avoid it, the quaternion is usually employed. Quaternion is a four-dimensional (4-D) complex number $\mathbf{q}_{\text{OLD}}^{\text{NEW}} = q_0 + q_1\mathbf{i} + q_2\mathbf{j} + q_3\mathbf{k}$, which is usually denoted by only the coefficients using

$$\mathbf{q}_{\text{OLD}}^{\text{NEW}} = \left[q_0, q_1, q_2, q_3\right] \tag{2.20}$$

It is popular for describing 3-D rotations. A comprehensive introduction can be found in [5]. The rotation matrix corresponding to a set of quaternions and its inverse representation are given by

$$\mathbf{C}_{\text{OLD}}^{\text{NEW}} = \begin{bmatrix} 1-2q_2^2-2q_3^2 & 2q_1q_2+2q_0q_3 & 2q_1q_3-2q_0q_2 \\ 2q_1q_2-2q_0q_3 & 1-2q_1{}^2-2q_3{}^2 & 2q_2q_3+2q_0q_1 \\ 2q_1q_3+2q_0q_2 & 2q_2q_3-2q_0q_1 & 1-2q_1^2-2q_2^2 \end{bmatrix} \tag{2.21}$$

The Euler angles corresponding to a set of quaternions and its inverse representation are given by

$$\begin{aligned} \theta &= \tan^{-1}\frac{2\left(q_0q_1+q_2q_3\right)}{1-2\left(q_1^2+q_2^2\right)} \\ \phi &= \sin^{-1}2\left(q_0q_2-q_3q_1\right) \\ \psi &= \tan^{-1}\frac{2\left(q_0q_3+q_1q_2\right)}{1-2\left(q_2^2+q_3^2\right)} \end{aligned} \tag{2.22}$$

$$\begin{aligned} q_0 &= \cos\left(\frac{\theta}{2}\right)\cos\left(\frac{\phi}{2}\right)\cos\left(\frac{\psi}{2}\right)+\sin\left(\frac{\theta}{2}\right)\sin\left(\frac{\phi}{2}\right)\sin\left(\frac{\psi}{2}\right) \\ q_1 &= \sin\left(\frac{\theta}{2}\right)\cos\left(\frac{\phi}{2}\right)\cos\left(\frac{\psi}{2}\right)-\cos\left(\frac{\theta}{2}\right)\sin\left(\frac{\phi}{2}\right)\sin\left(\frac{\psi}{2}\right) \\ q_2 &= \cos\left(\frac{\theta}{2}\right)\sin\left(\frac{\phi}{2}\right)\cos\left(\frac{\psi}{2}\right)+\sin\left(\frac{\theta}{2}\right)\cos\left(\frac{\phi}{2}\right)\sin\left(\frac{\psi}{2}\right) \\ q_3 &= \cos\left(\frac{\theta}{2}\right)\cos\left(\frac{\phi}{2}\right)\sin\left(\frac{\psi}{2}\right)-\sin\left(\frac{\theta}{2}\right)\sin\left(\frac{\phi}{2}\right)\cos\left(\frac{\psi}{2}\right) \end{aligned} \tag{2.23}$$

A key advantage of the quaternion representation is that it avoids gimbal lock singularities. By using four parameters instead of three, quaternions provide smooth and continuous coverage of all possible orientations in 3-D space. The only caveat is the double-cover property represent the same physical orientation, so one must account for sign ambiguity when comparing quaternions. Additionally, quaternions must remain unit-length, so algorithms often renormalize them periodically to mitigate drift from numerical integration. Compared to rotation matrices, quaternions are more compact (4 numbers versus 9) and computationally efficient, especially for combining rotations and performing interpolation. However, they lack intuitive interpretability: their components do not directly correspond to roll, pitch, and yaw angles, requiring conversion for human-understandable descriptions. Despite this, quaternions are widely used in navigation systems due to their stability, efficiency, and freedom from singularities, even if outputs are often ultimately presented in Euler angles for readability.

Table 2.1 compares the three attitude representation methods. For IPIN applications, quaternions representation is the most popular option due to its computational efficiency and its nature on nonsingularity issues. In the rest of the book, we use rotation matrix representation to show the derivation for clarity.

Table 2.1
Comparison of Attitude Representation Methods

Representation	**Interpretability**	**Computational Efficiency**	**Numerical Stability**	**Singularity Issues**
Euler angles	Highly interpretable; it directly corresponds to yaw, pitch, roll rotations that are easy to visualize and understand	Moderate for computation; it has only 3 parameters, but combining rotations requires careful use of rotation formulas (not a simple vector addition); efficient for small rotations but composition can be cumbersome	Moderate for stability; numerical drift can occur in long computations, especially near singular configurations, requiring careful numerical treatment	Yes; it is susceptible to gimbal lock (loss of a DOF) at specific orientations (e.g., pitch = ±90° in a typical sequence)

Table 2.1
(Cont.)

Representation	Interpretability	Computational Efficiency	Numerical Stability	Singularity Issues
Rotation matrix	Moderate interpretability; it contains direction cosines of axes; harder to read at a glance, but rows/columns indicate orientation of basis vectors	Good for many operations; applying rotation to vectors is a simple matrix multiply; composition of rotations is matrix multiplication (costlier than quaternion multiplication, but still straightforward)	Moderate for stability; numerical drift can occur over long computations, requiring periodic orthonormalization to maintain valid rotations	No singularities; it can represent all orientations with no gimbal lock
Quaternions	Low interpretability; the components are not directly meaningful to humans (must be converted to axis-angle or Euler angles to interpret physically)	High efficiency; the rotation composition via quaternion multiplication is faster and more stable than using matrices; suitable for real-time systems; easy to normalize	High for stability; it requires special attention to maintain unit norm (which is essential to avoid numerical drift)	No singularities; global coverage without gimbal lock

2.3 Conclusions

This chapter introduced the foundational concepts of coordinate systems and attitude representation essential for IPIN applications. These fundamental topics provide the basis for the estimation methods and algorithms discussed in the following chapters.

References

[1] Groves, P., *Principles of GNSS, Inertial, and Multisensor Integrated Navigation Systems*, Second Edition, Norwood, MA: Artech House, 2013.

[2] Kaplan, E. D., and C. Hegarty, (eds.), *Understanding GPS/GNSS: Principles and Applications*, Norwood, MA: Artech House, 2017.

[3] Mularie, W. M., *Department of Defense World Geodetic System 1984, Its Definition and Relationships with Local Geodetic Systems*, National Geospatial-Intelligence Agency, Tech. Rep 152, 2000.

[4] Diebel, J., "Representing Attitude: Euler Angles, Unit Quaternions, and Rotation Vectors." *Matrix*, Vol. 58, No. 15-16, 2006, pp. 1–35.

[5] Barfoot, T. D., *State Estimation for Robotics*, Cambridge, UK: Cambridge University Press, 2024.

3

Estimators and Filters for Indoor Positioning

Indoor positioning and navigation rely on robust state estimation to determine location from sensor data. This chapter reviews key estimators and filters used in indoor positioning, ranging from classical least squares methods to modern probabilistic filters and optimization frameworks. We discuss both the theoretical foundations and practical considerations of each approach, including their background, mathematical derivations, and applications in indoor positioning. Emphasis is placed on understanding how each method works, along with its advantages and limitations. These methods have evolved from early analytical solutions, such as least squares estimation, to optimal recursive estimators like the Kalman filter (introduced in 1960), as well as more advanced approaches such as particle filters and factor graph optimization (FGO), which leverage modern computing. The remainder of this chapter is organized as follows. Section 3.1 introduces least squares estimation. Section 3.2 discusses the Kalman filter and its nonlinear extensions, including the extended Kalman filter (EKF), iterated EKF (IEKF), and unscented Kalman filter (UKF). Section 3.3 presents particle filtering methods. Section 3.4 covers FGO for smoothing. Section 3.5 provides a comparative analysis of these approaches, while Section 3.6 concludes the chapter with final insights and future research directions.

3.1 Least Squares Estimation

Least-squares estimation is a fundamental technique for determining an unknown parameter (or set of parameters) by minimizing the sum of squared differences between observed values and the values predicted by a model. Suppose we have m measurements $y_1, \ldots, y_m$ related to an unknown parameter vector $\mathbf{x} \in \mathbb{R}^n$ through the model $y_i = h_i(\mathbf{x}) + \epsilon_i$, where ϵ_i represents measurement noise. In the least-squares estimation, we seek the estimate $\hat{\mathbf{x}}$ that minimizes the cost function shown below.

$$J(\mathbf{x}) = \sum_{i=1}^{m} \left(y_i - h_i(\mathbf{x})\right)^2 \tag{3.1}$$

In other words, the total squared error between the model predictions $h_i(\mathbf{x})$ and the observations y_i. This method originated from the work of Gauss and Legendre on fitting astronomical data and remains a staple due to its simplicity and optimality under Gaussian noise assumptions.

In many indoor positioning problems, least-squares serves as a fundamental starting point. For example, determining a user's 2-D position from distance measurements to Wi-Fi APs or UWB beacons can be formulated as a least-squares problem: the unknown position $\mathbf{x} = (x,y)$ is estimated by minimizing the squared errors between the measured distances and the distances predicted based on an assumed position. Least-squares is also widely used as an underlying solver for calibration tasks, such as determining sensor biases, and frequently functions as a subroutine within more complex estimation frameworks.

If the ith measurement model $h_i(\mathbf{x})$ is linear in $\mathbf{x}$, that is, $h_i(\mathbf{x}) = \mathbf{h}_i^T\mathbf{x}$ for some known vector $\mathbf{h}_i$, then the least-squares solution can be obtained in a closed form. All vectors $\mathbf{h}_i^T$ can be stacked to form an $m \times n$ matrix $\mathbf{H}$. Let $\mathbf{y} = [y_1, \ldots ,y_m]^T$. The cost function then becomes $J(\mathbf{x}) = \|\mathbf{y} - \mathbf{H}\mathbf{x}\|^2$. Setting the gradient to 0 yields the normal equations

$$\mathbf{H}^T\mathbf{H}\hat{\mathbf{x}} = \mathbf{H}^T\mathbf{y} \tag{3.2}$$

assuming $\mathbf{H}$ has full column rank. The solution is

$$\hat{\mathbf{x}} = \left(\mathbf{H}^T\mathbf{H}\right)^{-1}\mathbf{H}^T\mathbf{y} \tag{3.3}$$

which is the well-known ordinary least-squares estimator. This solution minimizes $J(\mathbf{x})$ and represents the maximum likelihood estimate if the noise ϵ_i is independent, zero-mean Gaussian. Moreover, it is an unbiased estimator and achieves the lowest possible variance among all linear unbiased estimators, as stated by the Gauss-Markov theorem.

When the model $h_i(\mathbf{x})$ is nonlinear (as is common in positioning, e.g., $h_i(\mathbf{x}) = \sqrt{(x - x_i)^2 + (y - y_i)^2}$ for distance to a known anchor at (x_i, y_i)), no closed-form solution exists. Instead, one must solve

$$\sum_{i=1}^{m}\left(y_i - h_i(\mathbf{x})\right)\nabla h_i(\mathbf{x}) = 0 \tag{3.4}$$

iteratively. $\nabla h_i(\mathbf{x})$ is the gradient of $h_i(\mathbf{x})$. A common approach is the Gauss-Newton method: start from an initial guess $\mathbf{x}_0$, and iteratively update $\mathbf{x}$ by linearizing $h_i(\mathbf{x})$ around the current estimate. At iteration t, one computes the Jacobian $J_t = \partial h/\partial x|_{\mathbf{x}_t}$ and residual $\mathbf{r}_t = \boldsymbol{y} - h(\mathbf{x}_t)$, then solves the linear least-squares $J_t\Delta\mathbf{x} \approx \mathbf{r}_t$. The estimate is updated as $\mathbf{x}_{t+1} = \mathbf{x}_t + \Delta\mathbf{x}$. This process repeats until convergence. The Gauss-Newton iteration effectively solves a sequence of linear approximations to find the minimum of $J(\mathbf{x})$. In practice, convergence to the global minimum is only guaranteed if the initial guess is close to the true solution and the cost function is relatively well-behaved (convex in a large neighborhood). Chapter 4 will introduce point positioning based on this iterative least squares in detail.

3.1.1 Robust Estimators

There are several extensions of basic least squares that improve their robustness and applicability. Weighted least squares assigns a weight w_i to each residual in the cost: $J(\mathbf{x}) = \sum_i w_i(y_i - h_i(\mathbf{x}))^2$. This is useful when measurements have different noise variances; setting $w_i = 1/\sigma_i^2$ (the inverse of noise variance) yields the best linear unbiased estimate in the linear case. Robust least squares modifies the cost function to reduce the influence of outliers by using alternative loss functions. This is particularly important in indoor environments, where some measurements may be outliers due to nonline-of-sight (NLOS) conditions. To avoid the influence of these large-biased measurements, a robust estimator is a popular means. Table 3.1 shows a few popular robust estimators. Their pros and cons are discussed in [1].

If there is a measurement with a large bias, it means that its residual is big. As mentioned earlier, the estimation is achieved when the "error function" is minimized. A robust estimator is a function to bridge the residual function to a cost function that can mitigate the impact of the measurements with a large bias. Figure 3.1 shows the idea. The x-axis denotes the

Table 3.1
Popular Robust Estimators

Robust Estimator	**Error Function**
L2	$e(\mathbf{x}) = \frac{1}{2}\lVert \mathbf{r}(\mathbf{x}) \rVert^2$
Cauchy	$e(\mathbf{x}) = \frac{1}{2}\ln\left(1+\lVert \mathbf{r}(\mathbf{x}) \rVert^2\right)$
Huber	$e(\mathbf{x}) = \begin{cases} \frac{1}{2}\lVert \mathbf{r}(\mathbf{x}) \rVert^2, & \lvert \mathbf{r}(\mathbf{x}) \rvert \le \delta \\ \delta\left(\lvert \mathbf{r}(\mathbf{x}) \rvert - \frac{1}{2}\delta\right), & \text{otherwise} \end{cases}$
G-M	$e(\mathbf{x}) = \frac{1}{2}\frac{\lVert \mathbf{r}(\mathbf{x}) \rVert^2}{1+\lVert \mathbf{r}(\mathbf{x}) \rVert^2}$

measurement error, which is the error between the actual measurement and expected measurement from converting the true state into the measurement domain. The y-axis denotes the costs from the measurement errors based on different robust estimators, which is expected to be minimized during the estimation. For example, the measurement with a higher cost has a higher priority to minimize its corresponding residual (i.e., has more impact during

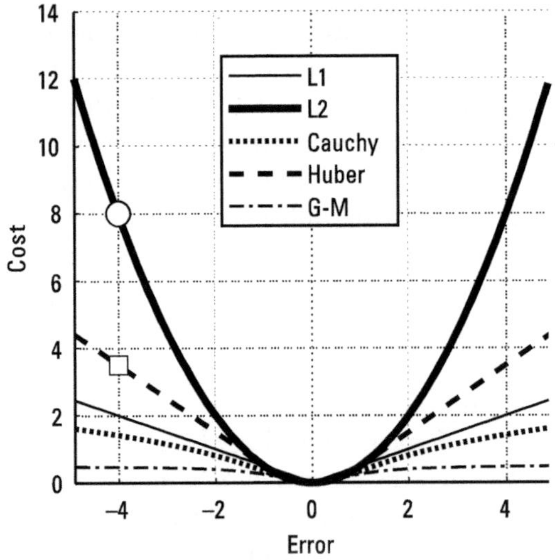

Figure 3.1 Cost of different robust functions.

the estimation). The circle and square markers denote the cost calculated from a large-biased measurement using the L2 estimator and Huber estimator, respectively. The corresponding cost is large for the L2 estimator, resulting in a significant influence from this large-biased measurement during estimation. However, the cost when using the Huber estimator (a popular robust estimator) is much lower, in which the large-biased measurement will not affect the estimation too much (i.e., more robust against outliers). However, the robust estimator must be carefully tuned since the definition of a healthy measurement might not be the same for different IPIN applications. A state-of-the-art research proposed to adaptively tune the parameters of G-M can be found at [2].

Another extension is regularized least squares, which adds a penalty term ($\lambda \|\mathbf{x}\|^2$) to encourage smaller parameter norms. This approach is commonly used in ill-posed problems to improve numerical stability and prevent overfitting.

The appeal of least squares lies in its simplicity and optimality under ideal conditions. It provides a closed-form solution for linear models and can be efficiently computed even with large numbers of measurements. Least-squares estimates are statistically optimal, achieving minimum-variance unbiased estimates when errors are Gaussian and independent. Furthermore, least-squares solutions are unique and globally optimal if the problem is convex, which holds for linear models. However, least-squares methods also have significant limitations. They are highly sensitive to outliers, as a single erroneous measurement can substantially skew the result due to the squared error penalty. Nonlinear least squares can converge to a local minimum if the initial guess is poor or if the cost surface has multiple minima, a common issue in complex indoor environments, where ambiguous distances or symmetries may arise. Additionally, least squares assume that the model form is correct; systematic errors, such as measurement bias or misrepresented physics, can lead to biased estimates. In dynamic scenarios, a simple least-squares approach that processes all data in a batch may not be suitable for real-time updates. Despite these limitations, least-squares remains a foundational technique and often serves as a starting point for more advanced filtering approaches, which are discussed later in this chapter.

3.2 Kalman Filters and Extensions

The Kalman filter is a recursive algorithm that provides optimal state estimates for linear dynamical systems with Gaussian noise. Originally formulated by

R. E. Kalman in 1960 [3], it has become a cornerstone of tracking and sensor fusion in navigation. We first describe the standard linear Kalman filter and then introduce several extensions designed for nonlinear systems.

3.2.1 Linear Kalman Filter

The Kalman filter (KF) is the most popular sensor fusion algorithm applied in navigation, particularly because it can achieve optimal performance, in this case, a maximum a posterior (MAP) estimation, under certain assumptions. During the early days of computing, such as in the 1960s, when computational power was very limited, directly applying MAP was impractical due to the exponential increase in computational load with an increasing number of time frames. An ideal solution is to estimate the state of the current time frame using a recursive approach, which fully uses the information from all previous time frames. The MAP estimation for this example can be represented as follows:

$$\hat{\chi}_{\text{MAP}} = \arg\max_{\chi}\left(P\left(\chi|\mathbf{Z},\mathbf{U}\right)\right) \tag{3.5}$$

where $P(\chi|\mathbf{Z},\mathbf{U})$ is the classic posterior probability of $\chi = \{\mathbf{x}_0, \mathbf{x}_1, \mathbf{x}_2, \cdots, \mathbf{x}_k\}$ given the measurement $\mathbf{Z} = \{\mathbf{z}_0, \mathbf{z}_1, \mathbf{z}_2, \cdots, \mathbf{z}_k\}$ and control input $\mathbf{U} = \{\mathbf{u}_0, \mathbf{u}_1, \mathbf{u}_2, \cdots, \mathbf{u}_k\}$. The control input describes how the system moves from one state to the next. Equation (3.5) implies that MAP is achieved when a set of state vectors χ (across all time frames) can be found that satisfies the sets of conditions defined by the control input and measurement vectors (again, across all time frames). To decouple (3.5), the Bayes rule is employed, leading to the following equation.

$$P\left(\chi|\mathbf{Z},\mathbf{U}\right) \propto \prod_k P\left(\mathbf{z}_k|\mathbf{x}_k\right)P\left(\mathbf{x}_0\right)\prod_k P\left(\mathbf{x}_k|\mathbf{x}_{k-1},\mathbf{u}_{k-1}\right) \tag{3.6}$$

where $\prod_k P(\mathbf{z}_k|\mathbf{x}_k)$ represents the joint probability of the likelihoods $P(\mathbf{z}_k|\mathbf{x}_k)$, and $\prod_k P(\mathbf{x}_k|\mathbf{x}_{k-1},\mathbf{u}_{k-1})$ denotes the joint probability of the propagation probabilities $P(\mathbf{x}_k|\mathbf{x}_{k-1},\mathbf{u}_{k-1})$. Additionally, $P(\mathbf{x}_0)$ is the prior probability. It is evident that MAP estimation accounts for the states, measurements, and propagation across all time frames, which can result in a significant computational load. The MAP estimate can then be obtained using the following equation.

$$P\left(\chi|\mathbf{Z},\mathbf{U}\right) \xleftrightarrow{\text{with assumptions}} P\left(\mathbf{x}_k|\mathbf{z}_k,\mathbf{u}_k\right) \propto P\left(\mathbf{z}_k|\mathbf{x}_k\right) P\left(\mathbf{x}_k|\mathbf{x}_{k-1},\mathbf{u}_{k-1}\right) \tag{3.7}$$

where $P(\mathbf{z}_k|\mathbf{x}_k)$ represents the likelihood of the current kth time frame and $P(\mathbf{x}_k|\mathbf{x}_{k-1},\mathbf{u}_{k-1})$ is the prior probability of the state at the current time frame, based on the current propagation and the previous measurement. This conceptualization is realized through the Kalman filter.

To achieve (3.7), it is necessary to represent the information from all previous time frames using only the most recent time frame. The Kalman filter accomplishes this by leveraging the concept of uncertainty, represented by the state noise covariance matrix **P**. However, this is only possible under the following three essential assumptions. Any violation of these assumptions compromises the optimality of the Kalman filter:

- *Assumption 3.1:* The first essential assumption of the Kalman filter is linearity. This assumption states that both the system being modeled and the measurement process exhibit linear behavior. In other words, the relationship between the system's state and its measurements can be accurately described using linear equations. This implies that the system dynamics and the measurement process can be represented by matrices and vectors that undergo linear transformations. By assuming linearity, the Kalman filter leverages the principles of linear algebra and estimation theory to estimate the system's state and its uncertainty. It uses a linear model to propagate the state forward in time and update it based on new measurements. However, if the system or measurement process deviates significantly from linearity, the Kalman filter may produce suboptimal results or fail to provide accurate estimates.
- *Assumption 3.2:* The second crucial assumption of the Kalman filter is that the noise in both the system dynamics and measurements follows a Gaussian (normal) distribution. This assumption enables the Kalman filter to make probabilistic inferences and compute optimal estimates. By assuming Gaussian noise, the Kalman filter effectively manages uncertainty and refines state estimates using measurement data. It leverages the properties of the Gaussian distribution, particularly its mean and covariance, to determine the optimal weights for integrating predicted states with actual measurements. This Gaussian assumption is essential because deviations from Gaussian characteristics, such as non-Gaussian or heavy-tailed noise distributions, can introduce inaccuracies into the estimation process. Such deviations may significantly impact the accuracy and optimality of the Kalman filter's estimates.
- *Assumption 3.3:* The third crucial assumption of the Kalman filter is stationarity. This implies that both the system dynamics and the statistical properties of the noise remain consistent over time. In essence, the system being modeled is assumed to be time-invariant. Under the

stationarity assumption, the Kalman filter operates with the premise that the system's behavior remains relatively unchanged during the estimation process. This allows the filter to use a consistent set of parameters and matrices to characterize the system dynamics. Additionally, it assumes that the statistical properties of the noise, including its mean and covariance, remain constant throughout the estimation process. The first assumption of linearity enables the use of matrices to describe uncertainty, which can be calculated digitally. The second assumption, involving Gaussian noise, facilitates the application of statistical tools to handle the mean and covariance in describing independent probability density functions. With these two assumptions in place, the measurement model can be expressed as:

$$\mathbf{z}_k^{\text{measurement}} = \mathbf{H}_k \mathbf{x}_k + \mathbf{w}_{z,k},\ \mathbf{w}_{z,k} \in N\left(0, \sigma_{z,k}\right) \tag{3.8}$$

where $N(0, \sigma)$ represents a zero-mean Gaussian distribution with a standard deviation σ. In (3.8), $\mathbf{H}_k$ represents the measurement model, and $\mathbf{w}_{z,k}$ represents the Gaussian random noise associated with the measurement. The mean and covariance of the likelihood can be calculated as follows:

$$\begin{aligned} \mathrm{E}\left[P\left(\mathbf{z}_k | \mathbf{x}_k\right)\right] &= \mathbf{H}_k \mathbf{x}_k \\ \mathrm{Cov}\left[P\left(\mathbf{z}_k | \mathbf{x}_k\right)\right] &= \Sigma_{\mathbf{w}_{z,k}} \end{aligned} \tag{3.9}$$

Consequently, the likelihood can be described by the following Gaussian distribution,

$$P\left(\mathbf{z}_k | \mathbf{x}_k\right) \propto -\exp\left(\left(\mathbf{z}_k - \mathbf{H}_k \mathbf{x}_k\right)^T \Sigma_{\mathbf{w}_{z,k}}^{-1} \left(\mathbf{z}_k - \mathbf{H}_k \mathbf{x}_k\right)\right) \tag{3.10}$$

Similarly, the propagation model can be expressed as:

$$\begin{aligned} &\mathbf{x}_{k,k-1}^{\text{propagation}} = \mathbf{F}_k \mathbf{x}_{k-1} + \mathbf{w}_{x_{k-1}} + \mathbf{u}_k + \mathbf{w}_{u,k} \\ &\mathbf{w}_{x_{k-1}} \in N\left(0, \sigma_{x_{k-1}}\right) \text{ and } \mathbf{w}_{u,k} \in N\left(0, \sigma_{u,k}\right) \end{aligned} \tag{3.11}$$

where $\mathbf{F}_k$ denotes the state propagation model and $\mathbf{u}_k$ is the control input vector. The mean and covariance of the prior probability are

$$\begin{aligned} \mathrm{E}\left[P\left(\mathbf{x}_k | \mathbf{z}_{k-1}, \mathbf{u}_k\right)\right] &= \mathbf{F}_k \mathbf{x}_{k-1} + \mathbf{u}_k = \mathbf{x}_{k,k-1} \\ \mathrm{Cov}\left[P\left(\mathbf{x}_k | \mathbf{z}_{k-1}, \mathbf{u}_k\right)\right] &= \mathbf{F}_k \Sigma_{\mathbf{x}_{k-1}} \mathbf{F}_k^T + \Sigma_{\mathbf{w}_{u,k}} = \mathbf{P}_{k,k-1} \end{aligned} \tag{3.12}$$

Thus, the prior probability can also be described as a Gaussian distribution,

$$P\left(\mathbf{x}_k|\mathbf{z}_{k-1},\mathbf{u}_k\right) \propto -\exp\left(\left(\mathbf{x}_k - \mathbf{x}_{k,k-1}\right)^T \mathbf{P}_{k,k-1}^{-1}\left(\mathbf{x}_k - \mathbf{x}_{k,k-1}\right)\right) \quad (3.13)$$

Substituting (3.10) and (3.13) into (3.7) yields the posterior probability:

$$P\left(\mathbf{x}_k|\mathbf{z}_k,\mathbf{u}_k\right) = -\mathrm{C}\exp\begin{pmatrix}\left(\mathbf{z}_k - \mathbf{H}_k\mathbf{x}_k\right)^T \boldsymbol{\Sigma}_{\mathbf{w}_{z,k}}^{-1}\left(\mathbf{z}_k - \mathbf{H}_k\mathbf{x}_k\right) \\ \cdot\left(\mathbf{x}_k - \mathbf{x}_{k,k-1}\right)^T \mathbf{P}_{k,k-1}^{-1}\left(\mathbf{x}_k - \mathbf{x}_{k,k-1}\right)\end{pmatrix} \quad (3.14)$$

where C is a normalizing constant.

As outlined in the assumptions, an accurate representation of $P(\mathbf{x}_0)$ is available, meaning that the uncertainty of the initial state $\mathbf{x}_0$ can be presented by $\boldsymbol{\Sigma}_0$ If the MAP of $\mathbf{x}_1$ (denoted as $\hat{\mathbf{x}}_{1,\mathrm{MAP}}$) and its associated uncertainty $\boldsymbol{\Sigma}_1$ can be estimated using $\mathbf{x}_0$ and $\boldsymbol{\Sigma}_0$, then $\hat{\mathbf{x}}_{k,\mathrm{MAP}}$ can be estimated recursively from the time frame 0, 1, …, $k-1$. In essence, with the accurate representation of $P(\mathbf{x}_0)$, the first-order Markov chain property is utilized in the Kalman filter. Therefore, a typical two-slice Kalman filter suffices to continuously provide MAP estimation for the current state in each time frame, as shown in Figure 3.2. The key lies in determining $\hat{\mathbf{x}}_{k,\mathrm{MAP}}$ using $\mathbf{x}_{k-1}$ and $\boldsymbol{\Sigma}_{\mathbf{x}_{k-1}}$. It is intuitive to apply the first-order derivative rule to find a state that maximizes its posterior probability. The MAP estimation can be derived using the following equation:

$$\left.\frac{\partial P\left(\mathbf{x}_k|\mathbf{z}_k,\mathbf{u}_k\right)}{\partial \mathbf{x}_k}\right|_{\mathbf{x}_k=\hat{\mathbf{x}}_{k,\mathrm{MAP}}} = 0 \quad (3.15)$$

As a result, the MAP of the state at the current time frame is given by:

$$\hat{\mathbf{x}}_{k,\mathrm{MAP}} = \left(\mathbf{H}_k^T \boldsymbol{\Sigma}_{\mathbf{w}_{z,k}}^{-1} \mathbf{H}_k + \mathbf{P}_{k,k-1}^{-1}\right)^{-1} \times \left(\mathbf{P}_{k,k-1}^{-1}\mathbf{x}_{k,k-1} + \mathbf{H}_k^T \boldsymbol{\Sigma}_{\mathbf{w}_{z,k}}^{-1}\mathbf{z}_k\right) \quad (3.16)$$

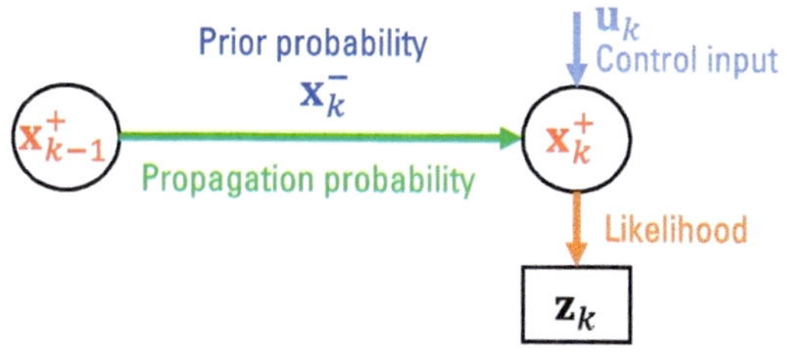

Figure 3.2 A typical example of a two slices Kalman filtering problem.

Equation (3.16) can be reorganized as:

$$\hat{\mathbf{x}}_{k,\text{MAP}} = \mathbf{x}_{k,k-1} + \mathbf{K}_k\left(\mathbf{z}_k - \mathbf{H}_k\mathbf{x}_{k,k-1}\right) \tag{3.17}$$

$$\mathbf{K}_k = \mathbf{P}_{k,k-1}\mathbf{H}_k^T\left(\mathbf{H}_k\mathbf{P}_{k,k-1}\mathbf{H}_k^T + \Sigma_{\mathbf{w}_{z,k}}\right)^{-1} \tag{3.18}$$

where **K** represents the Kalman filter gain, which balances the weightings between the prior probability and likelihood. As illustrated in Figure 3.3, the term inside the inverse operation in the Kalman filter gain calculation reflects the uncertainty associated with the likelihood. It takes into account both the measurement uncertainty in the current time frame and the propagated state uncertainty, along with the observability of the measurement model. Conversely, the other terms represent the uncertainty of the prior probability, taking into account both the uncertainty in the propagation input at the current time frame and the observability of the propagation model. Let's explore some extreme cases to understand this better:

1. $\Sigma_{\mathbf{w}_{u,k}} \ll \Sigma_{\mathbf{w}_{z,k}}$: The uncertainty of the system propagation is much smaller than that of the measurement update. Thus, $\mathbf{K}_k \to 0$, $\mathbf{x}_{k,\text{MAP}} \approx \mathbf{x}_{k,k-1}$.
2. $\Sigma_{\mathbf{w}_{u,k}} \gg \Sigma_{\mathbf{w}_{z,k}}$: The uncertainty of the measurement update is much smaller than that of the system propagation. Thus, $\mathbf{K}_k \to \infty$, $\hat{\mathbf{x}}_{k,\text{MAP}} \approx (\mathbf{H}_k^T\mathbf{H}_k)^{-1}\mathbf{H}_k^T\mathbf{z}_k$.

Intuitively speaking, if the system uncertainty is much smaller, it will rely more on the prior estimate. Otherwise, the posterior estimate will place greater

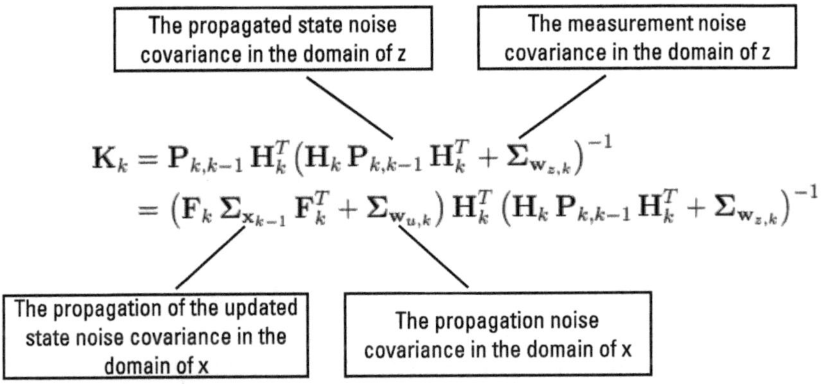

Figure 3.3 The computation of Kalman filter gain.

trust in the measurement-updated result. If the uncertainties fall between these two extremes, the Kalman gain is capable of optimally adjusting its weightings. This adjustment considers the observability of the model, ensuring that the gain is appropriately balanced for the given conditions. This ability to fine-tune its response based on varying levels of uncertainty is the core beauty of the Kalman filter, demonstrating its versatility and effectiveness in diverse scenarios.

Due to the recursive nature of the Kalman filter, the noise covariance of the current state, denoted as $\boldsymbol{\Sigma}_{\mathbf{x}_k}$, is indispensable for estimating the state in the subsequent time frame. The derivation of this covariance can be represented as follows:

$$\boldsymbol{\Sigma}_{\mathbf{x}_k} = \mathrm{Cov}\left[\mathbf{x}_k - \hat{\mathbf{x}}_{k,\mathrm{MAP}}\right] = \mathbf{P}_{k,k-1} - \mathbf{F}_k\mathbf{K}_k\mathbf{H}_k\mathbf{P}_{k,k-1} \tag{3.19}$$

To summarize, based on the above derivation, the renowned five equations of the Kalman filter are sufficient to achieve MAP, provided that the three key assumptions are met. In practical scenarios, a certain degree of nonlinearity, non-Gaussian properties, and inaccuracies in the initial prior probability can be tolerated, allowing the Kalman filter to be effectively applied in many engineering contexts. However, implementing the Kalman filter for sensor fusion does necessitate careful calibration $\boldsymbol{\Sigma}_0$, $\boldsymbol{\Sigma}_{\mathbf{w}_u,k}$, and $\boldsymbol{\Sigma}_{\mathbf{w}_z,k}$ (also well-known as $\mathbf{P}_0$, $\mathbf{Q}$, and $\mathbf{R}$) to approximate the MAP. These parameters are crucial because they underpin the fundamental assumptions ensuring the optimality of the Kalman filter.

$$\begin{aligned}
\mathbf{x}_{k,k-1} &= \mathbf{F}_k\mathbf{x}_{k-1} + \mathbf{u}_k \\
\mathbf{P}_{k,k-1} &= \mathbf{F}_k\boldsymbol{\Sigma}_{\mathbf{x}_{k-1}}\mathbf{F}_k^T + \boldsymbol{\Sigma}_{\mathbf{w}_{u,k}} \\
\mathbf{K}_k &= \mathbf{P}_{k,k-1}\mathbf{H}_k^T\left(\mathbf{H}_k\mathbf{P}_{k,k-1}\mathbf{H}_k^T + \boldsymbol{\Sigma}_{\mathbf{w}_{z,k}}\right)^{-1} \\
\hat{\mathbf{x}}_{k,\mathrm{MAP}} &= \mathbf{x}_{k,k-1} + \mathbf{K}_k\left(\mathbf{z}_k - \mathbf{H}_k\mathbf{x}_{k,k-1}\right) \\
\boldsymbol{\Sigma}_{\mathbf{x}_k} &= \mathbf{P}_{k,k-1} - \mathbf{F}_k\mathbf{K}_k\mathbf{H}_k\mathbf{P}_{k,k-1}
\end{aligned} \tag{3.20}$$

For linear-Gaussian problems, the Kalman filter is unmatched in both performance and elegance.

3.2.2 Extended Kalman Filter

The extended Kalman filter (EKF) is the most commonly adapted version of the Kalman filter for handling nonlinear systems. In an EKF, either the state transition model, the observation model, or both can be nonlinear functions:

$$\mathbf{x}_k = f\left(\mathbf{x}_{k-1}, \mathbf{u}_k\right) + \mathbf{w}_k, \mathbf{z}_k = h\left(\mathbf{x}_k\right) + \mathbf{v}_k \tag{3.21}$$

where $f(\cdot)$ and $h(\cdot)$ are nonlinear mappings (not necessarily matrices), such models are commonly found in indoor environments. For example, f might represent the motion equations of a robot, which are nonlinear due to rotations, while h could describe a distance measurement as a nonlinear function of position. The EKF applies the logic of the Kalman filter by linearizing these nonlinear functions around the current estimate. Essentially, it performs a local linear approximation, allowing the standard Kalman update equations to be used.

The EKF maintains a similar recursive structure: at each step, it predicts the state using f and then updates it using the new measurement and the linearized h. While it is not an optimal estimator in the general nonlinear case, it often performs well when the system's nonlinearity is mild or when the estimation error remains small, ensuring the validity of linearization. The EKF has been a reliable tool in indoor navigation. For instance, in pedestrian tracking, where inertial sensors provide a nonlinear motion model and radio beacons contribute nonlinear range measurements, the EKF has been widely adopted due to its relative simplicity and efficiency.

Start with mathematical derivation. At time $k-1$, assume that we have the state estimate $\hat{\mathbf{x}}_{k-1}$ and covariance $\mathbf{P}_{k-1}$. The EKF then proceeds as follows:

Prediction: Apply the nonlinear process model:

$$\begin{aligned} \hat{\mathbf{x}}_k^- &= f\left(\hat{\mathbf{x}}_{k-1}^-, \mathbf{u}_k\right) \\ \mathbf{P}_k^- &= \mathbf{F}_{k-1}\mathbf{P}_{k-1}\mathbf{F}_{k-1}^T + \mathbf{Q} \end{aligned} \tag{3.22}$$

where $\mathbf{F}_{k-1} = \partial f / \partial \mathbf{x}|_{\hat{\mathbf{x}}_{k-1}}$ is the Jacobian of f with respect to the state, evaluated at the current estimate. This Jacobian, $\mathbf{F}_{k-1}$, plays the role of the state transition matrix in propagating uncertainty. Equation (3.22) is essentially a first-order approximation of how covariance evolves. The linearization of process model in EKF is shown in Figure 3.4.

Update: When $\mathbf{z}_k$ arrives, we linearize the observation model around the predicted state $\hat{\mathbf{x}}_k^-$. We compute $\mathbf{H}_k = \partial h / \partial \mathbf{x}|_{\hat{\mathbf{x}}_k^-}$, the Jacobian of h evaluated at the predicted state. Next, we form the innovation $\mathbf{y}_k = \mathbf{z}_k - h(\hat{\mathbf{x}}_k^-)$ and compute the innovation covariance as $\mathbf{S}_k = \mathbf{H}_k\mathbf{P}_k^-\mathbf{H}_k^T + \mathbf{R}$. The Kalman gain and update formulas then follow the same structure as in the linear case:

$$\begin{aligned} \mathbf{K}_k &= \mathbf{P}_k^-\mathbf{H}_k^T\mathbf{S}_k^{-1} \\ \hat{\mathbf{x}}_k &= \hat{\mathbf{x}}_k^- + \mathbf{K}_k\mathbf{y}_k \\ \mathbf{P}_k &= \left(\mathbf{I} - \mathbf{K}_k\mathbf{H}_k\right)\mathbf{P}_k^- \end{aligned} \tag{3.23}$$

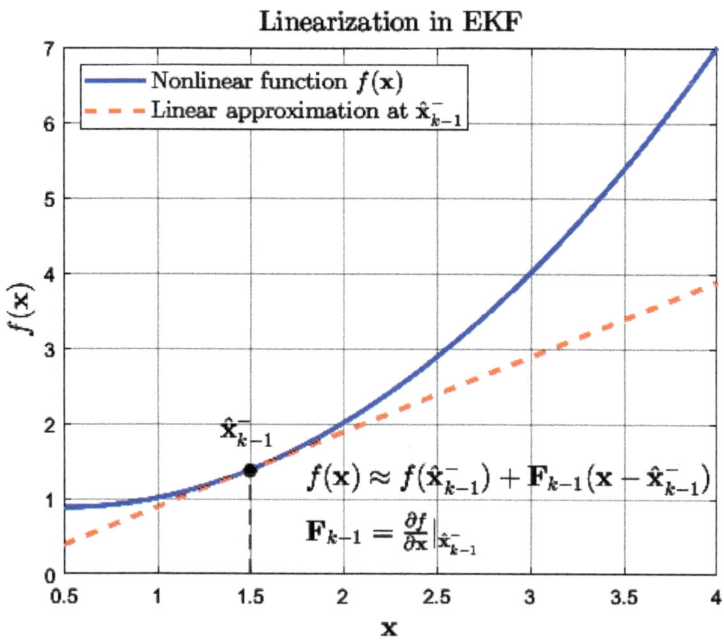

Figure 3.4 The linearization of process model in EKF.

These EKF equations demonstrate that, apart from the Jacobian calculations, the filter operations remain the same as those in the linear Kalman filter. The success of the EKF depends on the validity of the linearization as an approximation of the system's behavior. If $\hat{\mathbf{x}}_k^-$ is close to the true state, $\mathbf{H}_k$ will accurately capture how changes in the state affect the measurement, allowing the update step to correct the estimate appropriately. However, if the estimate is poor or the system is highly nonlinear, $\mathbf{H}_k$ may be an inadequate approximation, leading to incorrect updates that can potentially destabilize the filter.

The EKF inherits many strengths of the linear Kalman filter: it is relatively easy to implement, is computationally efficient (often $O(n^3)$ per step due to matrix operations, which is fine for moderate state sizes), and provides state estimates in real time with an associated covariance. It works well for a vast array of mild nonlinear problems and has been the default choice for many navigation systems.

3.2.3 Iterated Extended Kalman Filter

When measurement nonlinearities are severe, one way to improve the EKF's accuracy is to iterate the update step multiple times per measurement. The iterated extended Kalman filter (IEKF) applies the measurement update repeatedly in a single time step, relinearizing at the updated state each time. This effectively performs a Newton-like refinement to solve the measurement equation more accurately. Bell and Cathey (1993) showed that the IEKF update is equivalent to one or more Gauss-Newton iterations applied to the measurement residual [4].

Followed by the EKF equations, after obtaining the predicted state $(\hat{\mathbf{x}}_k^-, \mathbf{P}_k^-)$, the IEKF performs:

1. Initialize $\mathbf{x}_k^{(0)} = \hat{\mathbf{x}}_k^-$.
2. For $j = 0, 1 \ldots, N$ (a small number of iterations, until convergence):
   - Compute $\mathbf{H}_k^{(j)} = \partial h / \partial \mathbf{x}|_{\mathbf{x}_k^{(j)}}$.
   - Compute the innovation $\mathbf{y}_k^{(j)} = \mathbf{z}_k - h(\mathbf{x}_k^{(j)}) + \mathbf{H}_k^{(j)}(\mathbf{x}_k^{(j)} - \hat{\mathbf{x}}_k^-)$. This is a modified residual that accounts for the shift in linearization point (so that $\mathbf{y}_k^{(j)}$ would be 0 if $\mathbf{x}_k^{(j)}$ were the correct state).
   - Compute the Kalman gain $\mathbf{K}_k^{(j)} = \mathbf{P}_k^-[\mathbf{H}_k^{(j)}]^T(\mathbf{H}_k^{(j)}\mathbf{P}_k^-[\mathbf{H}_k^{(j)}]^T + \mathbf{R})^{-1}$.
   - Update the state estimate: $\mathbf{x}_k^{(j+1)} = \hat{\mathbf{x}}_k^- + \mathbf{K}_k^{(j)}\mathbf{y}_k^{(j)}$.
3. Set $\hat{\mathbf{x}}_k = \mathbf{x}_k^{(N+1)}$ as the final estimate, and update covariance $\mathbf{P}_k = (\mathbf{I} - \mathbf{K}_k^{(N)}\mathbf{H}_k^{(N)})\mathbf{P}_k^-$.

In essence, the IEKF repeatedly linearizes the measurement model at the latest state estimate and applies a correction. The modified residual $\mathbf{y}_k^{(j)}$ ensures that, upon convergence, $\mathbf{x}_k^{(j)}$ closely satisfies the measurement equation. Mathematically, the IEKF is equivalent to performing a Gauss-Newton optimization of the measurement likelihood at each time step, starting from the predicted state. By reducing linearization error, it enhances filter stability, particularly in highly nonlinear scenarios.

3.2.4 Unscented Kalman Filter

The unscented Kalman filter (UKF), developed by Julier and Uhlmann [5], was introduced as an alternative to the EKF to overcome the limitations of linearization. The UKF belongs to the family of sigma-point filters. Its key idea is to statistically propagate a Gaussian distribution through nonlinear functions without relying on linearization. Instead of using derivatives, the UKF selects a set of carefully chosen sample points (sigma points) that capture the mean and covariance of the prior distribution. These points are then passed through the nonlinear function, and a new Gaussian approximation is computed from the transformed points. As a result, the UKF often achieves second-order accuracy for the mean and covariance in nonlinear transformations, whereas the EKF is only first-order accurate. Notably, the UKF does not require analytic Jacobians, simplifying implementation for complex models. This feature is particularly valuable in indoor positioning, where it can handle intricate motion models (potentially learned from data) or measurement models that are difficult to differentiate analytically. An example of mean and covariance propagation among actual (sampling), EKF, and UKF is shown in Figure 3.5.

We describe one time step of the UKF, which follows the standard predict-update structure:

- *Sigma point generation:* At time step $k-1$, given the mean $\hat{\mathbf{x}}_{k-1}$ and covariance $\mathbf{P}_{k-1}$, we select $2n+1$ sigma points in the state space, where n is the state dimension:

$$\chi_0 = \hat{\mathbf{x}}_{k-1}, \chi_i = \hat{\mathbf{x}}_{k-1} + \delta_i, \chi_{i+n} = \hat{\mathbf{x}}_{k-1} - \delta_i, i = 1,\ldots,n, \tag{3.24}$$

where δ_i is the ith column of $\mathbf{L} = \sqrt{(n+\lambda)\mathbf{P}_{k-1}}$, the matrix square root of $(n+\lambda)\mathbf{P}_{k-1}$ (commonly computed via the Cholesky decomposition). The parameter λ is a scaling factor (usually $\lambda = n(\alpha^2 - 1)$ for some small α). Intuitively, χ_i are spread around the mean so that their sample mean and covariance match $\hat{\mathbf{x}}_{k-1}$ and $\mathbf{P}_{k-1}$ up to the third order of smallness. Associated weights $W_i^{(\text{state})}$ and $W_i^{(\text{covariance})}$ are set for

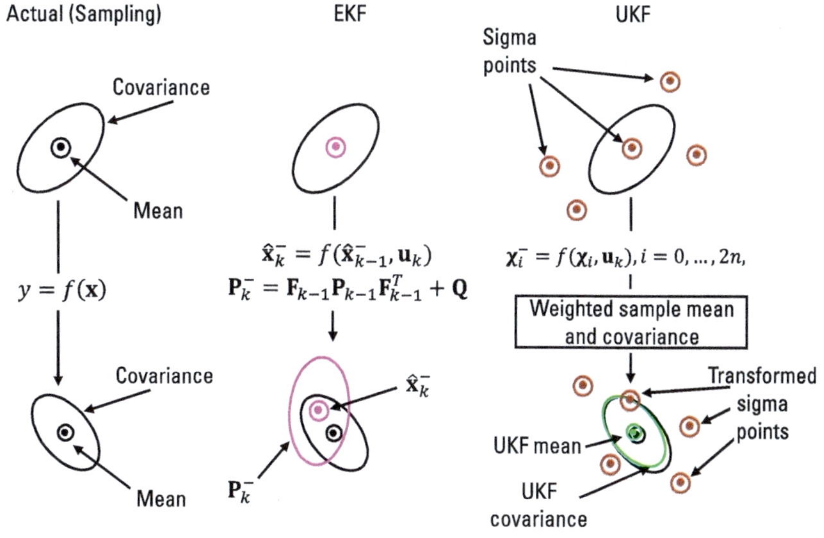

Figure 3.5 Example of mean and covariance propagation: actual (sampling), EKF, and UKF [6].

computing mean and covariance (with $W_0^{(\text{State})} = \lambda/(n+\lambda)$, $W_0^{(\text{Covariance})} = \lambda/(n+\lambda) + (1-\alpha^2+\beta)$, and $W_i^{(\text{State})} = W_i^{(\text{Covariance})} = 1/[2(n+\lambda)]$ for $i > 0$; β is an extra parameter often set to 2 for Gaussian priors) [6].

- *Prediction:* Each sigma point is propagated through the potentially nonlinear motion model:

$$\chi_i^- = f\left(\chi_i, \mathbf{u}_k\right),\ i = 0,\ldots,2n \tag{3.25}$$

This yields a set of predicted sigma points, χ_i^-, for the state at time k. The prior state mean and covariance are then reconstructed as weighted averages:

$$\begin{gathered}\hat{\mathbf{x}}_k^- = \sum_{i=0}^{2n} W_i^{(\text{State})} \chi_i^- \\ \mathbf{P}_k^- = \sum_{i=0}^{2n} W_i^{(\text{Covariance})} \left(\chi_i^- - \hat{\mathbf{x}}_k^-\right)\left(\chi_i^- - \hat{\mathbf{x}}_k^-\right)^T + \mathbf{Q}\end{gathered} \tag{3.26}$$

Equation (3.26) adds the process noise covariance $\mathbf{Q}$ after computing the spread of the transformed sigma points. At this stage, the predicted mean and covariance are obtained without ever computing the Jacobian of f.

- *Measurement update:* Next, we propagate the sigma points through the observation model:

$$\mathcal{Z}_i = h\left(\chi_i^-\right),\ i = 0,\ldots,2n \tag{3.27}$$

This generates a set of predicted measurements, from which we compute the predicted measurement mean and covariance:

$$\begin{gathered}\hat{\mathbf{z}}_k = \sum_{i=0}^{2n} W_i^{(\text{State})} \mathcal{Z}_i, \\ \mathbf{P}_{zz} = \sum_{i=0}^{2n} W_i^{(\text{Covariance})} \left(\mathcal{Z}_i - \hat{\mathbf{z}}_k\right)\left(\mathcal{Z}_i - \hat{\mathbf{z}}_k\right)^T + \mathbf{R}\end{gathered} \tag{3.28}$$

We also compute the cross-covariance between the state and the measurement:

$$\mathbf{P}_{xz} = \sum_{i=0}^{2n} W_i^{(\text{Covariance})} \left(\chi_i^- - \hat{\mathbf{x}}_k^-\right)\left(\mathcal{Z}_i - \hat{\mathbf{z}}_k\right)^T \tag{3.29}$$

Finally, we compute the Kalman gain and perform the update, just as in the Kalman filter:

$$\begin{aligned} \mathbf{K}_k &= \mathbf{P}_{xz}\mathbf{P}_{zz}^{-1}, \\ \hat{\mathbf{x}}_k &= \hat{\mathbf{x}}_k^- + \mathbf{K}_k\left(\mathbf{z}_k - \hat{\mathbf{z}}_k\right), \\ \mathbf{P}_k &= \mathbf{P}_k^- - \mathbf{K}_k\mathbf{P}_{zz}\mathbf{K}_k^T. \end{aligned} \tag{3.30}$$

This yields the posterior state estimate $\hat{\mathbf{x}}_k$ and covariance $\mathbf{P}_k$.

The UKF requires carrying and transforming $2n + 1$ points, making its computational complexity per step approximately $O(n^3)$, due to the covariance updates and matrix inversions involving matrices of size n and the measurement dimension. For moderate dimensions, this complexity remains manageable. The key advantage of the UKF is its improved accuracy. With a proper choice of parameters (α, β, κ), it can capture the effects of nonlinearities more effectively than an EKF linearization. For instance, if the true posterior after a prediction or update is mildly non-Gaussian, the sigma points can approximate the mean and covariance more accurately than an EKF, which only propagates means and linearized variances.

3.2.5 Comparison Between the Kalman Filter and Its Variants

Each of these variants has its place. EKF is commonly the default choice in many engineering applications due to its familiarity and simplicity. IEKF iteratively refines EKF updates for improved linearization accuracy at the cost of additional computation. The UKF is preferred when model fidelity is critical or when avoiding Jacobians simplifies implementation. To select which Kalman filter to use, the following questions should be answered: When does each filter work best? Which filter is the most efficient and scalable in high-dimensional problems? What are the challenges in deriving and tuning parameters? How does each filter handle uncertainty and outliers? For the ease of the reader, Table 3.2 compares these aspects.

In the next section, we move beyond Kalman filters to particle filtering, which takes a fundamentally different, sampling-based approach to state estimation.

3.3 Particle Filters

When assumptions of Kalman filters, such as Gaussian noise, unimodal distributions, and near-linear dynamics, do not hold, particle filters provide a

Table 3.2
Comparison Between the Kalman Filter and Its Variants

Aspect	**Kalman Filter**	**EKF**	**IEKF**	**UKF**
Accuracy and applicability [8–10]	Optimal for linear Gaussian systems: achieves minimum-variance, unbiased estimates when both system dynamics and observations are perfectly linear with Gaussian noise. Limitations with nonlinear systems: performance deteriorates significantly in strongly nonlinear environments, which is why it is rarely used as a standalone solution for such cases.	Handles mild nonlinearities: uses a first-order Taylor approximation to linearize around the current state; performs best when nonlinearities are moderate, and the operating point changes slowly between steps. Limitations with high nonlinearity: struggles with highly nonlinear problems or poor initial estimates, as unmodeled higher-order effects may introduce bias or lead to divergence.	Enhanced accuracy via iteration: improves accuracy in strongly nonlinear systems by iteratively refining the measurement update (Gauss-Newton method), providing a more optimal linearization point at each step. Effective for significant nonlinearity: particularly beneficial when a single EKF linearization is insufficient. IEKF achieves closer-to-optimal estimates by converging locally to the nonlinear update solution, provided the initial prediction is reasonably accurate.	Designed for highly nonlinear systems: utilizes sigma points to propagate Gaussian distributions through nonlinear transformations, accurately capturing mean and covariance without explicit linearization or analytic Jacobians, making it ideal for complex or black-box models. Effective in complex scenarios: particularly suitable for models with significant nonlinearities or complexities, such as angle wrapping or range measurements, where EKF's first-order approximations would introduce substantial errors.
Computational complexity [8, 9, 11]	Computational efficiency: each update involves matrix operations, making it efficient and suitable for real-time use.	Computational cost: per-step computational cost is comparable to a standard Kalman filter, involving similar matrix multiplications	Iterative computational cost: IEKF requires multiple iterations per update, increasing computation	Higher computational cost than EKF: Requires generating and propagating $(2n+1)$ sigma points (for

Table 3.2
(Cont.)

Aspect	Kalman Filter	EKF	IEKF	UKF
	Baseline computational load: provides the lowest computational effort among Kalman filter variants, as it requires neither iterative refinements nor sampling-based computations.	and inversions, plus the additional but typically minor; cost of computing Jacobians. Real-time feasibility: remains suitable for real-time applications, especially when Jacobian computations represent only a small fraction of the total update cost; sparsity optimizations further enable its use in large-scale problems such as SLAM.	proportionally to the number of iterations. Typically, 2–5 iterations are sufficient, resulting in roughly 2–5 times the cost of a single EKF update. Practical efficiency: Although iterations increase computational load, IEKF remains computationally efficient in practice. In most indoor applications, 1 to 2 iterations provide an effective balance between accuracy and computational complexity.	state dimension (*n*)), increasing computational load and memory usage linearly with state size. Although per-step complexity remains polynomial, the constant factors are higher compared to EKF. Practical performance and scalability: generally, fast enough for moderate-sized state dimensions but can become computationally challenging in high-dimensional scenarios.
Ease of implementation [11–13]	Simple implementation: easy to implement using well-known standard formulas, requiring only the specification of linear system matrices (**F**, **H**) and tuning of noise covariances (**Q**, **R**).	Moderate implementation difficulty: requires analytic Jacobians of motion and measurement models, which can be challenging to derive manually, especially in complex scenarios involving trigonometric relationships; these linearization steps increase initial development effort, debugging complexity, and maintenance if models change.	Simple extension of EKF: IEKF builds directly upon the EKF framework by introducing iterative updates, requiring minimal adjustments to existing EKF implementations. However, careful coding is needed to implement convergence checks or fixed-iteration loops clearly and correctly.	No Jacobian needed, direct model usage: directly utilizes the model's forward function without explicit linearization, simplifying initial setup and enabling easy application to complex or black-box systems. Implementation complexity: implementation involves unscented transform, including generating sigma points, propagating them

Table 3.2
(Cont.)

Aspect	Kalman Filter	EKF	IEKF	UKF
				through nonlinear models, and recombining results.
Robustness to noise and model errors [8, 9, 11, 14]	Sensitive to assumption violations: Kalman filter heavily relies on model accuracy; if the model is incorrect or noise is non-Gaussian, estimation errors can increase, potentially leading to divergence. Highly susceptible to outliers: Kalman filter treats unexpected sensor jumps as genuine observations, which can significantly distort state estimates unless additional measures such as outlier rejection or covariance inflation are applied.	Same assumption sensitivity as Kalman filter: the EKF relies on accurate models and Gaussian noise; if the model is significantly incorrect or the noise is highly non-Gaussian, EKF has no built-in resilience and may fail. Underestimates true uncertainty: EKF often provides overly optimistic covariance estimates, making it less adaptable to unexpected changes and exacerbating the impact of model errors. Requires a good initial estimate: if initialized too far from the true state, the linearization may be invalid, preventing the filter from recovering.	Reduces linearization error: by refining the estimate iteratively during the update, the IEKF reduces linearization errors, leading to more consistent estimates in highly nonlinear scenarios. More robust to poor initial linearization: if the initial linearization at a given time step is inaccurate, IEKF iterates to correct it. Still relies on a correct model: like EKF, IEKF assumes the overall model is correct; it does not inherently handle measurement outliers or bad data, which can still mislead the filter unless detected.	More robust to modeling errors and nonlinear effects: UKF does not rely on linearization, reducing the risk of divergence due to a poorly chosen linearization point. It also tracks state uncertainty more accurately, making it more adaptable when reality deviates from the assumed model. Less dependent on an accurate prior estimate: by propagating sigma points through the actual nonlinear system, UKF relies more on incoming measurements and is less sensitive to errors accumulated in the model; this allows it to correct deviations more effectively than EKF.

Table 3.2
(Cont.)

Aspect	Kalman Filter	EKF	IEKF	UKF
				Still assumes Gaussian noise: like Kalman filter and EKF, UKF does not inherently reject outliers or handle non-Gaussian noise; without additional robust techniques or careful noise covariance tuning, it will incorporate outliers just as the other filters would.
Best use cases for IPIN applications [10, 14–18]	Direct position measurements Barometric altimeter smoothing Single-axis drift correction Baseline for other filters	Default choice for moderate nonlinearities IPIN applications Indoor sensor fusion combining IMU data with radio measurements (e.g., Wi-Fi/BLE RSSI or UWB range) Smartphone navigation	Lidar-inertial odometry for robots Visual-inertial odometry	RF-based (Wi-Fi, BLE, or UWB positioning) localization Robust navigation for autonomous robots

more general framework for state estimation. Particle filters belong to the broader class of sequential Monte Carlo methods and represent the state's probability distribution using a set of random samples (particles) rather than a closed-form expression (mean and covariance). This approach allows particle filters to handle multimodal distributions, non-Gaussian noise, and highly nonlinear processes, albeit at the cost of increased computational complexity. Particle filter implements Bayes filtering (recursive Bayesian estimation) through sampling. The goal is to estimate $p(\mathbf{x}_k|\mathbf{z}_{1:k})$, the posterior distribution of the state $\mathbf{x}_k$ given all measurements up to time k. The Bayes filter update consists of two steps:

$$p(\mathbf{x}_k|\mathbf{z}_{1:k-1}) = \int p(\mathbf{x}_k|\mathbf{x}_{k-1})\, p(\mathbf{x}_{k-1}|\mathbf{z}_{1:k-1})\, d\mathbf{x}_{k-1} \tag{3.31}$$

$$p(\mathbf{x}_k|\mathbf{z}_{1:k}) \propto p(\mathbf{z}_k|\mathbf{x}_k)\, p(\mathbf{x}_k|\mathbf{z}_{1:k-1}) \tag{3.32}$$

In closed form, these integrals and multiplications are only tractable for simple cases—one such example is the Kalman filter, where everything remains Gaussian. Particle filters approximate these steps using a set of weighted samples $\{\mathbf{x}_k^{(i)}, w_k^{(i)}\}_{i=1}^N$ to represent $p(\mathbf{x}_k|\mathbf{z}_{1:k})$, where each $\mathbf{x}_k^{(i)}$ is a possible state (particle) and $w_k^{(i)}$ is its corresponding weight (importance probability).

The most commonly used algorithm is the sequential importance resampling (SIR) filter, also known as the Bootstrap filter [19]. The process of particle filter iteration is shown in Figure 3.6.

It follows these steps:

1. *Initialization:* Draw N samples $\mathbf{x}_0^{(i)}$ from the initial distribution $p(\mathbf{x}_0)$ and assign equal weights $w_0^{(i)} = 1/N$.

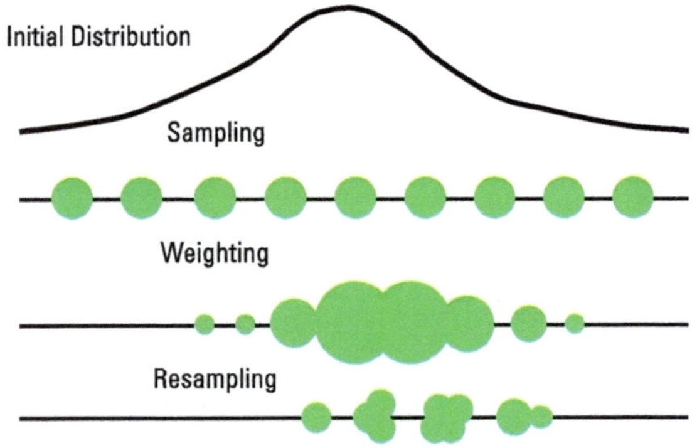

Figure 3.6 The process of particle filter iteration.

2. For each time step $k = 1, 2, \ldots$:
   (a) *Sampling (prediction):* For each particle i, generate a new sample from the process model:

$$\mathbf{x}_k^{(i)} \sim p\left(\mathbf{x}_k | \mathbf{x}_{k-1}^{(i)}\right) \tag{3.33}$$

   This is analogous to propagating each sample through the motion model with added noise.

   (b) *Importance weighting (update):* For each particle, compute the likelihood of the new measurement given the particle's state:

$$\tilde{w}_k^{(i)} = w_{k-1}^{(i)} p\left(\mathbf{z}_k | \mathbf{x}_k^{(i)}\right) \tag{3.34}$$

   Since, after resampling in the previous step, we typically have $w_{k-1}^{(i)} = 1/N$, this simplifies to $\tilde{w}_k^{(i)} \propto p(\mathbf{z}_k | \mathbf{x}_k^{(i)})$. Intuitively, if a particle's predicted measurement closely matches the actual observation $\mathbf{z}_k$, its weight increases; otherwise, its weight decreases.

   (c) *Normalization:* Normalize the weights $\{\tilde{w}_k^{(i)}\}$ so that $\Sigma_i w_k^{(i)} = 1$.

   (d) *Resampling:* Draw N particles from the current set $\{\mathbf{x}_k^{(i)}\}$ with probabilities proportional to $w_k^{(i)}$. This results in a new set $\{\mathbf{x}_k^{(i)}\}$ (reindexed), where particles with higher weights are more likely to be replicated multiple times, while low-weight particles may be eliminated. After resampling, set $w_k^{(i)} = 1/N$ for all i. Resampling helps prevent degeneracy, a phenomenon in which most particles acquire negligible weights after multiple updates.

3. The state estimate can be computed from the particles, for example, as the weighted mean $\hat{\mathbf{x}}_k = \Sigma_i w_k^{(i)} \mathbf{x}_k^{(i)}$. Alternatively, the particle with the highest weight may be selected as the representative estimate. The full posterior distribution is captured by the ensemble of particles.

This algorithm can represent virtually any distribution given a sufficient number of particles. For example, if there are two likely positions for a user, perhaps two symmetric locations relative to certain beacons, the particle set can maintain two distinct clusters. In contrast, a Kalman filter would instead approximate the distribution with a single Gaussian, resulting in a compromise solution that places the mean between the two modes, which is incorrect if the distribution is truly bimodal.

Particle filters can be computationally intensive, especially when the state dimension is high or when high accuracy (low variance) is required, necessitating a large number of particles. The number of particles N needed increases with the complexity of the distribution. For a 2-D position, a few hundred to a few thousand particles may be sufficient for room-level accuracy. However, for a high-dimensional state, such as a full simultaneous localization and mapping (SLAM) problem that includes both the robot's pose and multiple landmarks, basic particle filters become impractical. This challenge led to specialized methods like the Rao-Blackwellized particle filter (RBPF) for SLAM, where some variables are tracked using analytical filters while only part of the state is sampled.

Resampling can also introduce issues, such as particle impoverishment, where diversity is lost if all but a few particles have negligible weight. To mitigate this, techniques like roughening (adding random noise after resampling) or systematic resampling (which ensures better diversity) are commonly used.

In indoor scenarios, the primary state variables of interest, position, and possibly heading are typically low-dimensional, which makes particle filters feasible. However, if additional variables such as velocity or sensor biases are included in the state, the dimensionality increases, requiring more particles. A practical strategy to address this is hybrid estimation, where some state components are estimated using a Kalman filter within each particle, the core idea behind Rao-Blackwellized particle filters. For instance, in SLAM, FastSLAM 0 [20] utilizes a particle filter to estimate the robot's pose while embedding a Kalman filter in each particle to estimate the map. Similarly, in indoor positioning, a particle filter could sample a user's position, while a small Kalman filter within each particle estimates step length or phone orientation. This hybrid approach improves efficiency by reducing the number of required particles while maintaining accurate state estimation.

Particle filters offer several advantages that make it a powerful and flexible estimation technique. Unlike Kalman filter-based methods, particle filters can represent arbitrary probability distributions, including multimodal and skewed distributions, allowing it to handle complex and highly nonlinear systems. Its adaptability extends to any system model without requiring linearization or Gaussian assumptions, making it suitable for a wide range of applications. Additionally, particle filters can seamlessly integrate nontraditional information sources, such as map data or logical constraints, enhancing estimation accuracy in real-world scenarios. Its conceptual simplicity, combined with its ability to solve difficult localization problems—such as the kidnapped robot problem and global localization—further demonstrates its robustness in cases where traditional filtering techniques struggle.

Despite its strengths, the particle filter has several limitations that can pose challenges in practical applications. One major drawback is its computational intensity, as achieving stable results often requires a large number of particles, making real-time implementation difficult. However, a particle filter is well-suited for parallel processing since each particle independently undergoes prediction and weighting. Additionally, a particle filter suffers from approximation errors due to the finite number of particles; if the sample size is too small, the estimated distribution may contain significant sampling noise, leading to inconsistent or inaccurate state estimates. Another key issue is degeneracy, where insufficient particle spread or low process noise causes most particles to concentrate on an incorrect hypothesis, leading to divergence. While resampling helps mitigate degeneracy, it can also inadvertently discard useful particles due to randomness. Finally, a particle filter requires careful tuning of parameters such as the number of particles, process noise models, and measurement likelihoods. An inadequate number of particles may fail to capture the true uncertainty, while excessive particles increase computational costs, making parameter selection a critical but challenging aspect of particle filter implementation.

Particle filters have been successfully implemented on mobile and embedded devices for indoor localization. They are often combined with other methods, for example, using a particle filter for coarse global localization and then switching to a Kalman filter for fine tracking once a reliable hypothesis has been established.

3.4 Factor Graph Optimization

The methods discussed so far, Kalman filters and particle filters, perform filtering, estimating the current state sequentially in an online manner. In contrast, factor graph optimization (FGO) typically addresses smoothing or batch estimation, considering a collection of states over time and solving for the entire trajectory that best fits all measurements. This approach is closely related to least-squares estimation but is generally formulated within a probabilistic graphical model framework. The section begins with the derivation of FGO. Typical algorithm models are presented to illustrate the flexibility of the factor graph in solving problems related to multisensor integration.

Similar to the previous section, we are interested in the unknown state variables $\mathbf{X}$, such as position and orientation, given the measurements $\mathbf{Z}$. The MAP estimation is so named because it tries to maximize the posterior density $p(\mathbf{X}|\mathbf{Z})$ of the states $\mathbf{X}$ given the measurements $\mathbf{Z}$:

$$\begin{aligned}\hat{\mathbf{X}}_{\text{MAP}} &= \arg\max_{\mathbf{X}} p(\mathbf{X}|\mathbf{Z}) \\ &= \arg\max_{\mathbf{X}} \frac{p(\mathbf{Z}|\mathbf{X})\,p(\mathbf{X})}{p(\mathbf{Z})}\end{aligned} \tag{3.35}$$

The second equation above is obtained based on Bayes' law [21] and expresses the posterior as the product of the measurement density $p(\mathbf{Z}|\mathbf{X})$ and the prior $p(\mathbf{X})$ over the states, normalized by the factor $p(\mathbf{Z})$.

A distinct expression of Bayes' law is crucial to describe the actual computation underlying MAP estimation. Indeed, all the quantities in Bayes' law can theoretically be calculated from the Bayesian net [21]. In particular, the measurements $\mathbf{Z}$ are provided, and the normalization factor $p(\mathbf{Z})$ becomes irrelevant to the maximization and can be disregarded or assumed as $p(\mathbf{Z}) = 1$. Furthermore, while the conditional density $p(\mathbf{Z}|\mathbf{X})$ is a properly normalized Gaussian density in (3.35), our concern lies only with it as a function in the unknown states $\mathbf{X}$. Hence, the simplified form of Bayes' law is as follows:

$$\hat{\mathbf{X}}_{\text{MAP}} = \underset{\mathbf{X}}{\arg\max}\, l(\mathbf{X};\mathbf{Z})\, p(\mathbf{X}) \tag{3.36}$$

Here $l(\mathbf{X}; \mathbf{Z})$ is the likelihood of the states $\mathbf{X}$ given the measurements $\mathbf{Z}$, and is defined as a function that is proportional to $p(\mathbf{Z}|\mathbf{X})$:

$$l(\mathbf{X};\mathbf{Z}) \propto p(\mathbf{Z}|\mathbf{X}) \tag{3.37}$$

Factor graphs are graphical models [22–24] that are well suited to modeling complex estimation problems, such as SLAM or a structure from motion (SFM) [25]. In particular, there are variables and factors in factor graphs. The variables represent unknown quantities in the problem, and the factors represent functions on subsets of the variables. Edges in the factor graph are always between factors and variables and indicate that a particular factor depends on a particular variable.

A Wi-Fi inertial system is a widely used navigation system that combines the capabilities of both a Wi-Fi and an IMU. Let's begin with a classic example of Wi-Fi inertial navigation using FGO as illustrated in Figure 3.7. As can be seen, a factor graph is a bipartite graph with variable nodes (states at different times, or other parameters) and factor nodes (representing constraints or measurements relating variables). The factor graph captures the structure of the estimation problem.

In this derivation, $\mathbf{x}_k$, $\mathbf{u}_k$, and $\mathbf{z}_k$ represent the state vector, the control (or propagation) vector (e.g., from IMU), and the measurement vector in the kth time frame, respectively. In the application of Wi-Fi inertial navigation, $\mathbf{u}_k$ represents the acceleration and angular velocity from the IMU. We define that $\mathbf{Z} = \{\mathbf{z}_0, \mathbf{z}_1, \cdots, \mathbf{z}_k\}$, $\mathbf{X} = \{\mathbf{x}_0, \mathbf{x}_1, \cdots, \mathbf{x}_k\}$ and $\mathbf{U} = \{\mathbf{u}_0, \mathbf{u}_1, \cdots, \mathbf{u}_{k-1}\}$. In addition, state propagation is $\mathbf{x}_k = \mathrm{f}(\mathbf{x}_{k-1}, \mathbf{u}_{k-1})$, which can be derived based on IMU-based preintegration [26].

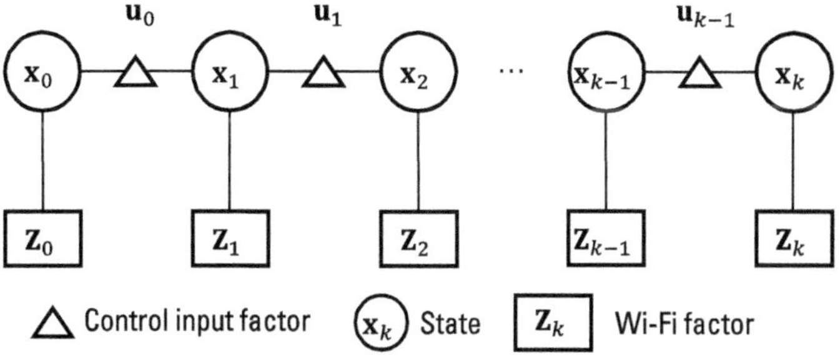

Figure 3.7 A typical example of the Wi-Fi inertial navigation problem. The triangle icon denotes the control input factor (e.g., from IMU). The rectangle icon denotes the Wi-Fi factor.

This proof does not delve into the details of the state propagation formula. More details and examples of IMU preintegration will be provided next. The measurement model is represented as $\mathbf{z}_k = \mathrm{h}(\mathbf{x}_k)$, which depends on the physical model of the sensor, such as Wi-Fi. Unlike the model for the EKF, the factor graph model is represented as an undirected graphical model, allowing the relationship between the states and the factors to be propagated both forward and backward.

Similar to EKF, which discusses the formulation of the EKF, the Gaussian noise assumption is adopted for both the state and all measurement models. Therefore, the state propagation conditional probability density, the measurement conditional probability density, and the prior probability density can be formulated as follows:

$$
\begin{aligned}
p\left(\mathbf{x}_k|\mathbf{u}_{k-1},\mathbf{x}_{k-1}\right) &= \frac{1}{\sqrt{|2\pi\mathbf{R}|}}\exp\left\{-\frac{1}{2}\left\|\mathrm{f}\left(\mathbf{x}_{k-1},\mathbf{u}_{k-1}\right)-\mathbf{x}_k\right\|_{\mathbf{R}}^2\right\} \\
p\left(\mathbf{z}_k|\mathbf{x}_k\right) &= \frac{1}{\sqrt{|2\pi\mathbf{P}|}}\exp\left\{-\frac{1}{2}\left\|\mathrm{h}\left(\mathbf{x}_k\right)-\mathbf{z}_k\right\|_{\mathbf{P}}^2\right\} \\
p\left(\mathbf{x}_0\right) &= \frac{1}{\sqrt{|2\pi\mathbf{Q}|}}\exp\left\{-\frac{1}{2}\left\|\mathbf{x}_0-\mu\right\|_{\mathbf{Q}}^2\right\}
\end{aligned}
\tag{3.38}
$$

where **R**, **P**, and **Q** represent the covariance matrix of state propagation noise, measurement noise, and initial state noise, respectively. In particular, all these noises are subjected to Gaussian noise. In addition, the covariance matrix is applied as $\|\mathbf{F}\|_{\Sigma}^2 = \mathbf{F}^{\mathrm{T}}\Sigma^{-1}\mathbf{F}$. The μ denotes the prior information of the state. However, all these noises are determined based on the properties of the sensors applied. For example, the **R** for the IMU should be calibrated in advance.

According to (3.38), $p(\mathbf{Z}|\mathbf{X})p(\mathbf{X})$ can be transformed as follows:

$$\begin{aligned}
p(\mathbf{Z}|\mathbf{X})p(\mathbf{X}) &= \left\{\prod_{k}^{0} p(\mathbf{z}_k|\mathbf{x}_k)\right\}\left\{p(\mathbf{x}_0)\prod_{k}^{1} p(\mathbf{x}_k|\mathbf{u}_{k-1},\mathbf{x}_{k-1})\right\} \\
&= \left\{\prod_{k}^{0}\frac{1}{\sqrt{|2\pi\mathbf{P}|}}\exp\left\{-\frac{1}{2}\left\|\mathrm{h}(\mathbf{x}_k)-\mathbf{z}_k\right\|_{\mathbf{P}}^{2}\right\}\right\} \\
&\quad * \left\{\begin{array}{c}\frac{1}{\sqrt{|2\pi\mathbf{Q}|}}\exp\left\{-\frac{1}{2}\left\|\mathbf{x}_0-\mu\right\|_{\mathbf{Q}}^{2}\right\} \\ \prod_{k}^{1}\frac{1}{\sqrt{|2\pi\mathbf{R}|}}\exp\left\{-\frac{1}{2}\left\|\mathrm{f}(\mathbf{x}_{k-1},\mathbf{u}_{k-1})-\mathbf{x}_k\right\|_{\mathbf{R}}^{2}\right\}\end{array}\right\} \\
&= \mathbf{K} * \exp\left\{\begin{array}{c}-\frac{1}{2}\left\|\mathbf{x}_0-\mu\right\|_{\mathbf{Q}}^{2}-\sum_{k}^{1}\frac{1}{2}\left\|\mathrm{f}(\mathbf{x}_{k-1},\mathbf{u}_{k-1})-\mathbf{x}_k\right\|_{\mathbf{R}}^{2} \\ -\sum_{k}^{0}\frac{1}{2}\left\|\mathrm{h}(\mathbf{x}_k)-\mathbf{z}_k\right\|_{\mathbf{P}}^{2}\end{array}\right\}
\end{aligned} \tag{3.39}$$

where $\mathbf{K}=\left(1/\sqrt{|2\pi\mathbf{Q}|}\right)\left(1/\sqrt{|2\pi\mathbf{R}|}\right)^{k}\left(1/\sqrt{|2\pi\mathbf{P}|}\right)^{k+1}$.

Calculating the logarithm of both sides of (3.39) yields:

$$\begin{aligned}
\log\left(p(\mathbf{Z}|\mathbf{X})p(\mathbf{X})\right) &= \log(\mathbf{K})-\frac{1}{2}\left\|\mathbf{x}_0-\mu\right\|_{\mathbf{Q}}^{2}-\sum_{k}^{1}\frac{1}{2}\left\|\mathrm{f}(\mathbf{x}_{k-1},\mathbf{u}_{k-1})-\mathbf{x}_k\right\|_{\mathbf{R}}^{2} \\
&\quad -\sum_{k}^{0}\frac{1}{2}\left\|\mathrm{h}(\mathbf{x}_k)-\mathbf{z}_k\right\|_{\mathbf{P}}^{2}
\end{aligned} \tag{3.40}$$

Then (3.36) can be derived as follows:

$$\begin{aligned}
\hat{\mathbf{X}}_{\mathrm{MAP}} &= \arg\max_{\mathbf{X}} l(\mathbf{X};\mathbf{Z})p(\mathbf{X}) = \arg\max_{\mathbf{X}} p(\mathbf{Z}|\mathbf{X})p(\mathbf{X}) \\
&= \arg\max_{\mathbf{X}}\log\left(p(\mathbf{Z}|\mathbf{X})p(\mathbf{X})\right) \\
&= \arg\max_{\mathbf{X}}\left\{\begin{array}{c}-\frac{1}{2}\left\|\mathbf{x}_0-\mu\right\|_{\mathbf{Q}}^{2}-\sum_{k}^{1}\frac{1}{2}\left\|\mathrm{f}(\mathbf{x}_{k-1},\mathbf{u}_{k-1})-\mathbf{x}_k\right\|_{\mathbf{R}}^{2} \\ -\sum_{k}^{0}\frac{1}{2}\left\|\mathrm{h}(\mathbf{x}_k)-\mathbf{z}_k\right\|_{\mathbf{P}}^{2}\end{array}\right\}
\end{aligned}$$

$$
= \arg\min_{\mathbf{x}} \left\{ \begin{array}{c} \frac{1}{2}\left\|\mathbf{x}_0 - \boldsymbol{\mu}\right\|_{\mathbf{Q}}^2 + \sum_{k}^{1} \frac{1}{2}\left\| \mathrm{f}\left(\mathbf{x}_{k-1}, \mathbf{u}_{k-1}\right) - \mathbf{x}_k \right\|_{\mathbf{R}}^2 \\ + \sum_{k}^{0} \frac{1}{2}\left\| \mathrm{h}\left(\mathbf{x}_k\right) - \mathbf{z}_k \right\|_{\mathbf{P}}^2 \end{array} \right\} \tag{3.41}
$$

In this case, the conditional probability problem is transformed into a quadratic problem. The minimum quadratic problem (3.41) can be solved using the Gaussian-Newton method, as referenced in [27]. In summary, FGO is derived based on the MAP estimation mentioned above. In practice, several popular open-source solvers are available to the community, including GTSAM [22] and the Ceres Solver [28]. For instance, GTSAM is open-source software that facilitates defining factors and solving FGO problems. It is one of the most widely used open-source tools for solving FGO in robotics and navigation. GTSAM comes with many commonly used factors already implemented and provided, such as the visual factor, IMU preintegration factor, and GNSS factor. Unlike GTSAM, the Ceres Solver is another open-source solver for nonlinear optimization, extensively used in robotics for state estimation, offering more foundational tools for users. Specifically, it allows users to customize the formulation of factors from scratch, which makes it a handy tool for users.

3.4.1 Numerical Optimization Methods

In optimization-based methods, three popular techniques are commonly employed to estimate the states by minimizing error functions: Gradient Descent, Gauss-Newton, and Levenberg-Marquardt, which are introduced.

3.4.1.1 Gradient Descent Method

A popular strategy to find the solution to minimize the cost function is the line search strategy, which iteratively updates the state by a certain step α along a chosen direction $\mathbf{d}$:

$$
\mathbf{x}_{k+1} = \mathbf{x}_k + \alpha\mathbf{d} \tag{3.42}
$$

where subscripts k and $k + 1$ denotes the current and next iteration indexes, respectively. To ensure that the state update approaches the solution with the minimal cost function value, the cost function value is required to be decreased after each iteration:

$$
f\left(\mathbf{x}_{k+1}\right) = f\left(\mathbf{x}_k + \alpha\mathbf{d}\right) < f\left(\mathbf{x}_k\right) \tag{3.43}
$$

Thus, the line search strategy follows the procedures [29]: (1) find a descent direction **d** for the cost function in the state space; and (2) find a step length α achieving sufficient decrement on the cost function value. By repeating such procedures, the state converges to the solution when no sufficient decrement can be found. In practice, the step length α can be determined by generating a set of candidate values and selecting the one that achieves the minimum cost. A standard pseudocode is shown in Algorithm 3.1.

A straightforward way to guarantee a decrease in the cost function is to update the state along the direction of the steepest descent, which is known as the gradient descent method. The first-order Taylor expansion of cost function after an update can be expressed by

$$f\left(\mathbf{x}_k + \alpha \mathbf{d}\right) \approx f\left(\mathbf{x}_k\right) + \alpha \mathbf{d}^{\mathrm{T}} \mathbf{f}'\left(\mathbf{x}_k\right) \tag{3.44}$$

Assuming that the direction term **d** is normalized as a unit vector, the change rate of the objective function f due to an update with step size α in the gradient descent method can be expressed as

$$\mathbf{d}^{\mathrm{T}} \mathbf{f}'\left(\mathbf{x}_k\right) = \|\mathbf{d}\| \left\|\mathbf{f}'\left(\mathbf{x}_k\right)\right\| \cos\theta = \left\|\mathbf{f}'\left(\mathbf{x}_k\right)\right\| \cos\theta \tag{3.45}$$

where θ denotes the angle between **d** and the gradient $\mathbf{f}'(\mathbf{x}_k)$. The gradient descent method aims to decrease the objective function most effectively by

Algorithm 3.1

Standard Line Search Strategy for Optimization

```
Input:   Initial state x_0, cost function f(x), maximum iteration threshold k_max
Output:  Solution state x̂

Initialize iteration step k = 0
While f(x_k) is not the lowest in the vicinity of x_k and k < k_max
    Find a descent direction d on x_k reducing f(x)
    Find a step length α that f(x_k + αd) is sufficiently lower than f(x_k)
    Update state x_{k+1} = x_k + αd
    Update iteration step k = k + 1
End while
Obtain solution state x̂ = x_k
```

aligning the update direction **d** with the negative gradient, that is, $\theta = \pi$ (or $\cos\theta = -1$). Hence, the update direction of the gradient descent method is given by

$$\mathbf{d}_{\mathrm{gd}} = -\mathbf{f}'(\mathbf{x}_k) \tag{3.46}$$

This ensures that each update step is taken in the direction of the steepest decrease of the function.

Detailed analysis of its convergence can be found in [30]. An example of state optimization using the gradient descent method is shown in Figure 3.8. The gradient descent method is easy to implement and only requires computing the first derivative of the cost function. However, it typically converges linearly, which can make it very slow during the final stage of optimization.

3.4.1.2 Gauss-Newton Method

Another popular optimization method is the Newton's method, which has a higher convergence rate compared to the gradient descent method. Instead of finding the gradient direction, the Newton method optimizes the state by finding a direction that the first derivative of cost function approaches 0, which is the necessary condition of a solution [29]. The first-order Taylor expansion of the first derivatives of the cost function after an update can be expressed by

$$\mathbf{f}'(\mathbf{x}_k + \mathbf{d}_n) \approx \mathbf{f}'(\mathbf{x}_k) + \mathbf{f}''(\mathbf{x}_k)\mathbf{d}_n \tag{3.47}$$

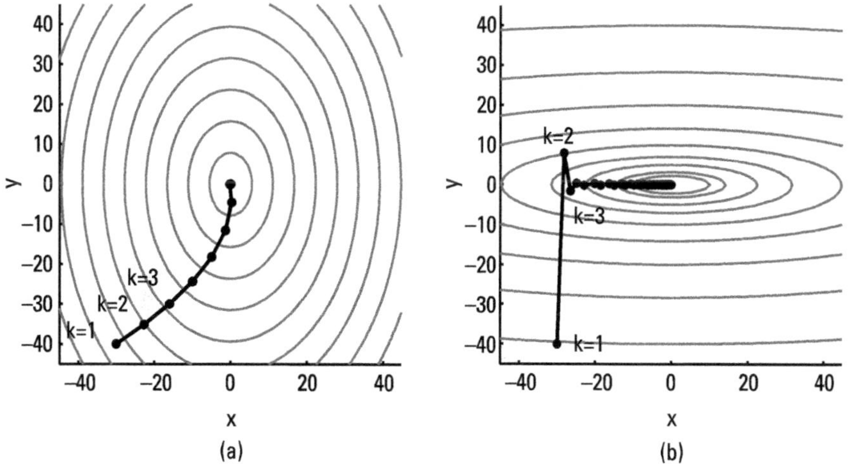

Figure 3.8 Example of gradient descent method. (a) Good convergence. (b) Slow convergence.

where $\mathbf{f}''(\mathbf{x}_k)$ denotes the second derivative (or Hessian) of the cost function with state $\mathbf{x}_k$. Thus, by assuming $\mathbf{f}'(\mathbf{x}_k + \mathbf{d}_n)$ approaches 0 after the update, the state update direction $\mathbf{d}_n$ can be computed by solving

$$\mathbf{f}''(\mathbf{x}_k)\mathbf{d}_n = -\mathbf{f}'(\mathbf{x}_k) \tag{3.48}$$

Then the state is updated by $\mathbf{x}_{k+1} = \mathbf{x}_k + \mathbf{d}_n$. Note that the classic Newton method updates the state with a unit step length $\alpha = 1$. The Newton method can achieve a quadratic convergence rate and good for the final stage of optimization [30]. However, the Hessian $\mathbf{f}''(\mathbf{x}_k)$ needs to be positive definite to guarantee $\mathbf{d}_n$ is a descent direction for optimization. Moreover, $\mathbf{f}''(\mathbf{x}_k)$ may hard to be computed or not even available for complicated problems.

To improve the efficiency and practicability, the Newton method is modified to approximate the calculation of the Hessian $\mathbf{f}''(\mathbf{x}_k)$ by the Jacobian of the residual function $\mathbf{r}(\mathbf{x}_k)$, namely the Gauss-Newton method. The first and second derivatives of the cost function can be derived by [30]:

$$\mathbf{f}'(\mathbf{x}_k) = \mathbf{r}'(\mathbf{x}_k)^{\mathrm{T}}\mathbf{r}(\mathbf{x}_k) \tag{3.49}$$

$$\mathbf{f}''(\mathbf{x}_k) = \mathbf{r}'(\mathbf{x}_k)^{\mathrm{T}}\mathbf{r}'(\mathbf{x}_k) + \sum_{m=1}^{M} r_m(\mathbf{x}_k) r_m''(\mathbf{x}_k) \tag{3.50}$$

For small residuals $r_m(\mathbf{x}_k)$, $\mathbf{f}''(\mathbf{x}_k)$ is dominated by the first term, such that

$$\mathbf{f}''(\mathbf{x}_k) \approx \mathbf{r}'(\mathbf{x}_k)^{\mathrm{T}}\mathbf{r}'(\mathbf{x}_k) \tag{3.51}$$

With this approximation, the Gauss-Newton method obtains the update direction $\mathbf{d}_{gn}$ by solving (3.48) under (3.49) and (3.51), as follows:

$$\left(\mathbf{r}'(\mathbf{x}_k)^{\mathrm{T}}\mathbf{r}'(\mathbf{x}_k)\right)\mathbf{d}_{gn} = -\mathbf{r}'(\mathbf{x}_k)^{\mathrm{T}}\mathbf{r}(\mathbf{x}_k) \tag{3.52}$$

Then the state is updated by $\mathbf{x}_{k+1} = \mathbf{x}_k + \mathbf{d}_{gn}$. A comparison of Gauss-Newton method with the gradient descent method is shown in Figure 3.9. The Gauss-Newton method avoids the computation of $r''_m(\mathbf{x}_k)$ and maintains a quadratic convergence rate in the final stage of optimization. Meanwhile, a descent direction can be guaranteed when $\mathbf{r}'(\mathbf{x}_k)$ has full rank and $\mathbf{f}'(\mathbf{x}_k)$ is nonzero. However, its convergence rate is still expected to be linear for general situations [29]. Besides, the Gauss-Newton method is not well behaved with infinite number of solutions of $\mathbf{d}_{gn}$ when $\mathbf{r}'(\mathbf{x}_k)$ is rank-deficient. It may fail to converge when having large residuals (i.e., a poor initial guess in the beginning stage of optimization [31]).

3.4.1.3 Levenberg-Marquardt Method

The Levenberg-Marquardt method [32, 33] is developed by adding a damping term on the Gauss-Newton method, which can avoid its drawbacks under rank-deficient and large residuals. One can regard the Levenberg-Marquardt method as an integration of the gradient descent method and the Gauss-Newton method [34]. The update direction of the Levenberg-Marquardt method is obtained by solving

$$\left(\mathbf{r}'(\mathbf{x}_k)^{\mathrm{T}}\mathbf{r}'(\mathbf{x}_k)+\mu\mathbf{I}\right)\mathbf{d}_{lm}=-\mathbf{r}'(\mathbf{x}_k)^{\mathrm{T}}\mathbf{r}(\mathbf{x}_k) \tag{3.53}$$

where $\mu \geq 0$ is the damping parameter and $\mathbf{I}$ is the identity matrix. The value of μ controls the update behavior during optimization. A large value of μ makes (3.53) dominated by the damping term, resulting in an update direction in the form of the gradient descent method, as follows:

$$\mathbf{d}_{lm}=-\left(\mathbf{r}'(\mathbf{x}_k)^{\mathrm{T}}\mathbf{r}'(\mathbf{x}_k)+\mu\mathbf{I}\right)^{-1}\mathbf{r}'(\mathbf{x}_k)^{\mathrm{T}}\mathbf{r}(\mathbf{x}_k)\approx-\frac{\mathbf{f}'(\mathbf{x}_k)}{\mu} \tag{3.54}$$

On the contrary, a small value of μ makes (3.53) the damping term becomes little, resulting in a dominating update direction in the form of the Gauss-Newton method, as follows:

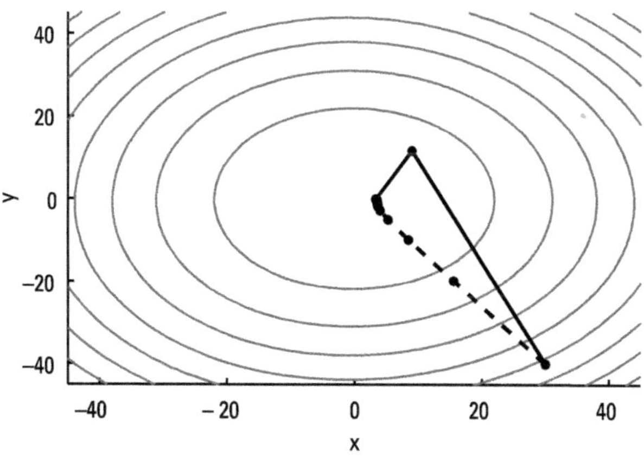

Figure 3.9 Example of optimization using the Gauss-Newton method (dashed line) and the gradient descent method (solid line).

$$\mathbf{d}_{lm} = -\left(\mathbf{r}'(\mathbf{x}_k)^{\mathrm{T}}\mathbf{r}'(\mathbf{x}_k)+\mu\mathbf{I}\right)^{-1}\mathbf{r}'(\mathbf{x}_k)^{\mathrm{T}}\mathbf{r}(\mathbf{x}_k)$$
$$\approx -\left(\mathbf{r}'(\mathbf{x}_k)^{\mathrm{T}}\mathbf{r}'(\mathbf{x}_k)\right)^{-1}\mathbf{r}'(\mathbf{x}_k)^{\mathrm{T}}\mathbf{r}(\mathbf{x}_k) \tag{3.55}$$

By appropriate adjustments of μ, the Levenberg-Marquardt method can behave like the gradient descent method in the beginning stage to achieve good convergence under large residuals and can behave like the Gauss-Newton method in the final stage to achieve quadratic convergence under small residuals. Note that the convergence rate of the Levenberg-Marquardt method is still linear under large residuals.

In practice, the value of μ is initialized by user experiences or the elements in $\mathbf{r}'(\mathbf{x}_k)^{\mathrm{T}}\mathbf{r}'(\mathbf{x}_k)$ [34]. Then the Levenberg-Marquardt method iteratively updates the state by $\mathbf{x}_{k+1} = \mathbf{x}_k + \mathbf{d}_{lm}$ with $\mathbf{d}_{lm}$ solved from (3.53) and adjusts μ for the next update based on the gain ratio g derived from [29]:

$$g = \frac{f(\mathbf{x}_k) - f(\mathbf{x}_k + \mathbf{d}_{lm})}{\frac{1}{2}\mathbf{d}_{lm}^{\mathrm{T}}\left(\mu\mathbf{d}_{lm} - \mathbf{f}'(\mathbf{x}_k)\right)} \tag{3.56}$$

The damping parameter μ is reduced for a large g, indicating that the update fits with the approximation (3.51) by the Gauss-Newton method, and is increased for a small g, indicating a poor approximation. The iterative update will be terminated either if the first derivative approaches 0 (i.e., $\|\mathbf{f}'(\mathbf{x}_{k+1})\|_\infty$ is sufficiently small), or the change of state $\|\mathbf{x}_{k+1} - \mathbf{x}_k\|$ is sufficiently small. The example and outline of applying the Levenberg-Marquardt method are shown in Figure 3.10 and Algorithm 3.2 [29].

Finally, we summarize the key properties of the methods discussed above (see Table 3.3).

3.5 Comparison of Estimation Methods

3.5.1 Qualitative Comparison of Estimation Methods for Indoor Positioning

First, we compare the estimation methods presented in terms of typical accuracy, computational complexity, and convergence/stability properties, particularly in the context of indoor positioning. Table 3.4 provides a qualitative summary. In the Kalman filter family, EKF is selected to compare with the others due to its popularity in IPIN applications.

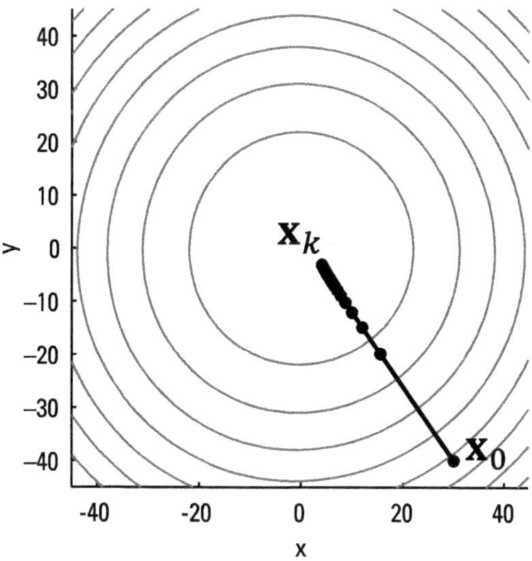

Figure 3.10 Example of optimization using the Levenberg-Marquardt method.

Algorithm 3.2

Example of Applying the Levenberg-Marquardt Method

Input: Initial state $\mathbf{x}_0$, cost function $f(\mathbf{x})$, maximum iteration threshold $k_{\max}$, initial damping parameter μ_0, terminate threshold ε_1 for the first derivative, and terminate threshold ε_2 for the change of state.

Output: Solution state $\hat{\mathbf{x}}$

Initialize iteration step $k = 0$, $\nu = 2$ (ν is chosen as a conservative initial value to control the adjustment speed of the damping parameter)

Solve $(\mathbf{r}'(\mathbf{x}_0)^{\mathrm{T}}\mathbf{r}'(\mathbf{x}_0) + \mu_0\mathbf{I})\mathbf{d}_{lm} = -\mathbf{r}'(\mathbf{x}_0)^{\mathrm{T}}\mathbf{r}(\mathbf{x}_0)$

While $\|\mathbf{f}'(\mathbf{x}_k)\|_\infty > \varepsilon_1$ **and** $\|\mathbf{d}_{lm}\| > \varepsilon_2(\|\mathbf{x}_k\| + \varepsilon_2)$ **and** $k < k_{\max}$

  Calculate gain ratio ϱ by (2.56)

  If $\varrho \le 0$

    Refuse poor update, $\mathbf{x}_{k+1} = \mathbf{x}_k$

    $\mu_{k+1} = \mu_k \cdot \nu$, $\nu = 2\nu$

  else

    Update state $\mathbf{x}_{k+1} = \mathbf{x}_k + \mathbf{d}_{lm}$

    $\mu_{k+1} = \mu_k \cdot \max\{1/3,\ 1 - (2\varrho - 1)^3\}$, $\nu = 2$

  End if

Algorithm 3.2
(Cont.)

Solve $(\mathbf{r}'(\mathbf{x}_{k+1})^{\mathrm{T}}\mathbf{r}'(\mathbf{x}_{k+1}) + \mu_{k+1}\mathbf{I})\mathbf{d}_{lm} = -\mathbf{r}'(\mathbf{x}_{k+1})^{\mathrm{T}}\mathbf{r}(\mathbf{x}_{k+1})$
Update iteration step $k = k + 1$
End while
Obtain solution state $\hat{\mathbf{x}} = \mathbf{x}_k$

Table 3.3
Comparison of Numerical Optimization Methods

Optimization Algorithm	Benefits	Drawbacks
Gradient descent method	Requires only the first derivative; simple and easy to implement	Linear convergence rate; can be slow in the final stages
Newton method	Quadratic convergence rate; performs well near the optimum	Second derivative may not be available; the update direction may not always be a descent direction
Gauss-Newton method	Approximates the second derivative using the first derivative term; nearly quadratic convergence rate near the solution	Poor performance if the Jacobian is rank-deficient; may fail with large residuals
Levenberg-Marquardt method	Initially behaves like the gradient descent method, enhancing global convergence; transitions to Gauss-Newton behavior near the solution, leading to faster local convergence	May exhibit linear convergence under large residuals

Ultimately, the choice among these estimation methods depends on the specific application requirements and constraints. For pedestrian tracking scenarios where the starting position is known and motion remains smooth, a well-tuned EKF offers a computationally efficient solution with sufficient accuracy. However, if the initial position is less known or if the system must incorporate map constraints, a particle filter becomes a more suitable option due to its ability to handle arbitrary probability distributions, despite its higher computational cost. For applications demanding maximum accuracy in complex multisensor fusion, such as robotic systems integrating lidar, cameras, and IMU, FGO provides the most effective data fusion framework. FGO enables global consistency and loop closure corrections, making it ideal for long-term navigation and mapping. In such cases, a hybrid approach often combines FGO with filtering methods to balance accuracy and real-time performance.

The discussion on selecting between EKF and FGO is becoming important due to the increasing available open-source code-based FGO. Next we detail their comparison.

3.5.2 Comparison of the EKF and FGO for Sensor Integration

We also focus on comparing EKF and FGO for sensor integration, as these are the most commonly used integration methods in modern navigation systems.

Accuracy and computational efficiency are the two most crucial factors for estimators. EKF has dominated estimation applications for decades due to its maturity and efficiency in sensor integration. However, with the increasing diversity of sensors for seamless positioning, the flexibility of the estimator becomes another key factor to consider before selecting the best estimator. For instance, the addition of new sensor measurements to the sensor integration problem often requires a redesign of the majority of the matrices for the EKF.

- *Computational accuracy:* EKF can achieve optimal state estimation accuracy if two key assumptions are perfectly met: the Gaussian noise and the first-order Markov assumptions. EKF mainly considers information from two consecutive epochs, assuming that the state at the current epoch is independent of past states and measurements. However, this first-order Markov assumption is challenging to satisfy in complex scenarios where measurements are highly time-correlated. For instance, NLOS effects in the UWB range measurements can introduce time-correlated errors due to environmental reflections. Consequently, violating this assumption can lead to accuracy degradation in sensor integration with the EKF. Unlike the EKF, the factor graph utilizes information from multiple epochs simultaneously to estimate a series of states, thereby eliminating the need for the first-order Markov assumption. This approach enables the exploration of time correlation within multiple epochs simultaneously. In other words, both forward and backward optimizations are feasible with the factor graph, an undirected graphical model, whereas only forward optimization is possible with the EKF. Additionally, the linearization error resulting from

Table 3.4
Qualitative Comparison of Estimation Methods for Indoor Positioning

Method	Accuracy	Computational Complexity	Convergence/Stability
LS	High for linear problems, achieving optimal accuracy in the least-squares sense. For nonlinear problems, accuracy can be high if the solution converges to the global minimum, which depends on the quality of the initial guess.	A one-time batch solve (using normal equations) typically has a complexity of $O(n^3)$ for n parameters or can be solved iteratively. Efficient for moderate-sized problems.	Guaranteed global optimum for linear models. For nonlinear problems, convergence is only local (as with Gauss-Newton or Newton methods) and may lead to an incorrect solution if the initial estimate is poor or the cost surface is nonconvex.
EKF	Provides good accuracy for moderate nonlinear systems but may exhibit bias in cases of strong nonlinearities. Often sufficient but not optimal for many indoor positioning scenarios.	Same complexity as a Kalman filter, plus the additional computation of Jacobians (either analytical or numerical). Still real-time for typical state sizes.	Locally stable if the initial error is small; however, it may diverge in highly nonlinear systems or with poor initialization, as it relies on linearization. Requires careful tuning but generally converges when operated within a valid regime.

Table 3.4
(Cont.)

Method	Accuracy	Computational Complexity	Convergence/Stability
Particle filter	Can achieve arbitrarily high accuracy given a sufficient number of particles, approaching the true Bayesian solution as $N \to \infty$. Effectively handles multimodal distributions, which EKF/UKF may fail to capture. However, a finite N introduces approximation errors.	Per step: $O(N \cdot n)$ complexity, where each particle update contributes to the overall cost. If N is large (hundreds or thousands of particles), computation can be significant. Best suited for lower-dimensional problems or scenarios where parallelization is feasible.	Converges in probability as N increases. With a limited number of particles, it may suffer from degeneracy (where the filter loses track if no particles represent the true state). Resampling helps maintain stability but introduces random fluctuations. Generally stable if N and noise models are appropriately chosen.
FGO	Typically achieves the highest accuracy, offering global least-squares optimality under Gaussian assumptions. Minimizes drift by incorporating all available data and can correct past errors.	A batch solve can be computationally expensive, with a naive complexity of $O(T^3)$ for T states. However, sparse solvers significantly improve efficiency. Sliding window or incremental methods further reduce the ongoing cost, often achieving near-linear complexity per step for new data. Still, FGO is heavier than filtering and may run at a lower frequency in real-time applications.	Converges to the global minimum if the problem is convex (or given a good initial guess and robust cost functions for nonconvex cases). Not prone to divergence, although it may converge to a local minimum if mis-initialized. Once converged, it is generally very stable and can adjust past estimates when new information (e.g., loop closures) becomes available.

the single iteration of the EKF can also limit computational accuracy. With increasing sensor diversity, the nonlinearity of the measurement model poses new challenges for the EKF, where a single iteration may not suffice to achieve global optimality. For example, the visual feature reprojection model [35] is highly nonlinear and requires several iterations to converge. In contrast, FGO employs multiple iterations to mitigate linearization errors. A comparison between the EKF and factor graph in GNSS/inertial integration, presented in [36], demonstrates the computational accuracy of the factor graph with several datasets collected in urban areas of Hong Kong.

- *Computational efficiency:* EKF efficiently solves the sensor integration problem due to the first-order Markov assumption, as it only considers information within two consecutive epochs. In contrast, the computational time of the factor graph increases significantly with the number of epochs considered. The complexity of the optimization is dominated by the variable k. Therefore, the computational load using the factor graph can increase significantly compared to EKF-based sensor integration. To address this challenge, sliding window-based FGO optimizes only the most recent epochs of states and measurements simultaneously, achieving a balance between computational accuracy and efficiency. Specifically, states and measurements outside the sliding window are marginalized as a prior factor to constrain the factor graph, using the Schur complement [37]. As shown in Figure 3.11, the dashed areas represent the marginalized information, while the remaining states are optimized simultaneously. The states within the dashed areas are not optimized but serve as a prior constraint. An example of sliding window FGO using the Schur complement will be provided next. Incremental smoothing [22] is another strategy to maintain the computational efficiency of the factor graph by applying a Bayes tree through incremental variable reordering and fluid relinearization, thus eliminating the need for periodic batch steps. This strategy presupposes that periodic batch relinearization is unnecessary, significantly enhancing computational efficiency, even with large-scale window sizes. Notably, the implementation of incremental smoothing is open-sourced in the GTSAM library. In summary, while the computational load increases with the factor graph compared to EKF-based sensor integration, computational efficiency can be achieved by employing either sliding window optimization with marginalization techniques [37] or incremental smoothing [22].

- *Estimator flexibility:* EKF is a recursive and sequential estimator that struggles with delayed measurements due to sensor synchronization issues. Furthermore, the core equations of the EKF require redesign whenever a new sensor is incorporated, limiting its flexibility in multisensor integration. Figure 3.12 illustrates an example of rich sensor integration, incorporating UWB, inertial, and Wi-Fi measurements. Compared to Figure 3.11, the UWB range measurement is a new addition, and its corresponding factor can be expressed as: $\left\| \left\| \mathbf{p}_k - \mathbf{p}^{\mathrm{UWB}} \right\|_2 - z_k^{\mathbf{UWB}} \right\|_{P_{\mathrm{UWB}}}^2$. Here, $\mathbf{p}_k$ and $\mathbf{p}^{\mathrm{UWB}}$ represent the position state and the UWB anchor position, respectively, while $z_k^{\mathbf{UWB}}$ and P_{UWB} denote the UWB range measurement and its associated variance, respectively. If EKF is employed for sensor integration, it necessitates a redesign of the measurement model of the estimator. However, with the inclusion of UWB, an additional cost function can be seamlessly added to (3.41), which can be efficiently solved using either the Ceres Solver or the GTSAM library. More importantly, the relevant Jacobian matrix for optimizing (3.41) can be automatically calculated using the Ceres Solver or the GTSAM library, further simplifying the implementation of sensor integration. The dashed pentagon in Figure 3.12 represents the delayed UWB measurements. These delayed UWB measurements can also be accommodated in FGO, a feat difficult to achieve with the EKF. In summary, the flexibility of the factor graph

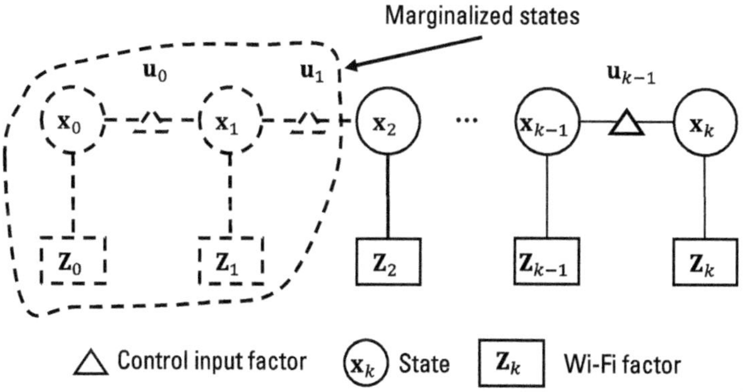

Figure 3.11 A typical example of the Wi-Fi inertial navigation problem. The triangle icon denotes the control input factor (e.g., from IMU). The rectangle icon denotes the Wi-Fi factor. The dashed areas denote the marginalized information (state and measurements).

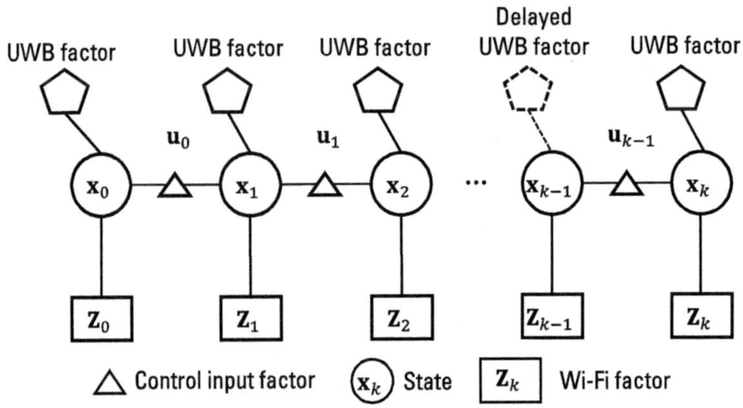

Figure 3.12 Illustration of the rich sensor integration problem, which includes the Wi-Fi, inertial, and UWB. The dashed pentagon icon denotes the delayed UWB measurements, for example, due to the sensor delay.

enhances the intuitiveness of sensor integration, providing plug-and-play capabilities.

3.6 Conclusions

This chapter reviewed key estimators and filters essential for indoor positioning. Starting with least-squares estimation for static problems, we discussed Kalman filters and variants (EKF, IEKF, and UKF), emphasizing their efficiency in real-time sensor fusion under uncertainty. These methods handle linear or linearized models and Gaussian assumptions, with UKF offering improved flexibility. Particle filters were introduced for managing non-Gaussian, multimodal uncertainties common in complex indoor environments, useful in ambiguous scenarios despite higher computational costs. FGO, performing batch or sliding-window optimization, was presented for its superior accuracy by revisiting past data, central to modern SLAM and visual-inertial navigation.

No single method dominates universally: EKF excels in speed and simplicity for small deviations; particle filters address ambiguity and complexity; and factor graphs achieve maximum accuracy given sufficient resources.

Emerging trends include automation through machine learning, distributed estimation across devices, and multimodal sensor fusion. Hybrid solutions integrating various filters and optimization techniques represent the forefront of indoor positioning, supporting critical applications such as emergency response, augmented reality, and smart buildings.

References

[1] MacTavish, K., and T. D. Barfoot, "At All Costs: A Comparison of Robust Cost Functions for Camera Correspondence Outliers," *2015 12th IEEE Conference on Computer and Robot Vision*, 2015, pp. 62–69.

[2] Yang, H., et al., "Graduated Non-Convexity for Robust Spatial Perception: From Non-Minimal Solvers to Global Outlier Rejection," *IEEE Robotics and Automation Letters*, Vol. 5, No. 2, 2020, pp. 1127–1134.

[3] Kalman, R. E., "A New Approach to Linear Filtering and Prediction Problems," *Journal of Basic Engineering*, Vol. 82, No. 1, March 1960, pp. 35–45.

[4] Bell, B. M., and F. W. Cathey, "The Iterated Kalman Filter Update as a Gauss-Newton Method," *IEEE Transactions on Automatic Control*, Vol. 38, No. 2, 1993, pp. 294–297.

[5] Julier, S. J., and J. K. Uhlmann, "Unscented Filtering and Nonlinear Estimation," *Proceedings of the IEEE*, Vol. 92, No. 3, 2004, pp. 401–422.

[6] Van Der Merwe, R., and E. A. Wan, "The Square-Root Unscented Kalman Filter for State and Parameter-Estimation," *2001 IEEE International Conference on Acoustics, Speech, and Signal Processing*, Proceedings (Cat. No. 01CH37221), Vol. 6, 2001, pp. 3461–3464.

[7] Mourikis, A. I., and S. I. Roumeliotis, "A Multi-State Constraint Kalman Filter for Vision-Aided Inertial Navigation," *Proceedings 2007 IEEE International Conference on Robotics and Automation*, 2007, pp. 3565–3572.

[8] Zanetti, R., "Adaptable Iterative and Recursive Kalman Filter Schemes," *AAS Spaceflight Mechanics Meeting*, No. JSC-CN-30315, 2014.

[9] Huang, G. P., A. I. Mourikis, and S. I. Roumeliotis, "On the Complexity and Consistency of UKF-Based SLAM," *2009 IEEE International Conference on Robotics and Automation*, 2009, pp. 4401–4408.

[10] Kolakowski, M., "Comparison of Extended and Unscented Kalman Filters Performance in a Hybrid BLE-UWB Localization System," *2020 23rd IEEE International Microwave and Radar Conference (MIKON)*, 2020, pp. 122–126.

[11] Yang, C., W. Shi, and W. Chen, "Comparison of Unscented and Extended Kalman Filters with Application in Vehicle Navigation," *The Journal of Navigation*, Vol. 70, No. 2, 2017, pp. 411–431.

[12] Shademan, A., and F. Janabi-Sharifi, "Sensitivity Analysis of EKF and Iterated EKF Pose Estimation for Position-Based Visual Servoing," *Proceedings of 2005 IEEE Conference on Control Applications*, 2005, pp. 755–760.

[13] LaViola, J. J., "A Comparison of Unscented and Extended Kalman Filtering for Estimating Quaternion Motion," *Proceedings of the 2003 IEEE American Control Conference*, Vol. 3, 2003, pp. 2435–2440.

[14] Bloesch, M., et al., "Iterated Extended Kalman Filter Based Visual-Inertial Odometry Using Direct Photometric Feedback," *The International Journal of Robotics Research*, Vol. 36, No. 10, 2017, pp. 1053–1072.

[15] Cruz, G. P. Jr., et al., "EKF-LOAM: An Adaptive Fusion of lidar SLAM with Wheel Odometry and Inertial Data for Confined Spaces with Few Geometric Features," *IEEE Transactions on Automation Science and Engineering*, Vol. 19, No. 3, 2022, pp. 1458–1471.

[16] Bai, S., et al., "Towards Persistent Spatial Awareness: A Review of Pedestrian Dead Reckoning-Centric Indoor Positioning with Smartphones," *IEEE Transactions on Instrumentation and Measurement*, 2024.

[17] Xu, W., et al., "Fast-lio2: Fast Direct Lidar-Inertial Odometry," *IEEE Transactions on Robotics*, Vol. 38, No. 4, 2022, pp. 2053–2073.

[18] You, W., et al., "Data Fusion of UWB and IMU Based on Unscented Kalman Filter for Indoor Localization of Quadrotor UAV," *IEEE Access*, Vol. 8, 2020, pp. 64971–64981.

[19] Arulampalam, M. S., et al., "A Tutorial on Particle Filters for Online Nonlinear/Non-Gaussian Bayesian Tracking," *IEEE Transactions on Signal Processing*, Vol. 50, No. 2, 2002, pp. 174–188.

[20] Montemerlo, M., et al., "FastSLAM: A Factored Solution to the Simultaneous Localization and Mapping Problem," *AAAI/IAAI*, 2002, pp. 593–598.

[21] Barfoot, T. D., *State Estimation for Robotics*, Cambridge, UK: Cambridge University Press, 2024.

[22] Kaess, M., et al., "iSAM2: Incremental Smoothing and Mapping Using the Bayes Tree," *The International Journal of Robotics Research*, Vol. 31, No. 2, 2012, pp. 216–235.

[23] Kaess, M., A. Ranganathan, and F. Dellaert, "iSAM: Incremental Smoothing and Mapping," *IEEE Transactions on Robotics*, Vol. 24, No. 6, 2008, pp. 1365–1378.

[24] Dellaert, F., and M. Kaess, "Factor Graphs for Robot Perception," *Foundations and Trends® in Robotics*, Vol. 6, No. 1-2, 2017, pp. 1–139.

[25] Zhao, L., Z. Mao, and S. Huang, "Feature-Based SLAM: Why Simultaneous Localisation and Mapping?" *Robotics: Science and Systems*, 2021.

[26] Forster, C., et al., "On-Manifold Preintegration for Real-Time Visual--Inertial Odometry," *IEEE Transactions on Robotics*, Vol. 33, No. 1, 2016, pp. 1–21.

[27] Grisetti, G., et al., "A Tutorial on Graph-Based SLAM," *IEEE Intelligent Transportation Systems Magazine*, Vol. 2, No. 4, 2010, pp. 31–43.

[28] Agarwal, S., and K. Mierle, "Ceres solver: Tutorial & Reference," Google Inc., No. 72, 2012, p. 8.

[29] Madsen, K., H. B. Nielsen, and O. Tingleff, *Methods for Non-Linear Least Squares Problems*, Second Edition, Informatics and Mathematical Modelling, Technical University of Denmark, 2004.

[30] Nocedal, J., and S. J. Wright, (eds.), *Numerical Optimization*, New York: Springer, 1999.

[31] Ipsen, I. C. F., C. T. Kelley, and S. R. Pope, "Rank-Deficient Nonlinear Least Squares Problems and Subset Selection," *SIAM Journal on Numerical Analysis*, Vol. 49, No. 3, 2011, pp. 1244–1266.

[32] Levenberg, K., "A Method for the Solution of Certain Non-Linear Problems in Least Squares," *Quarterly of Applied Mathematics*, Vol. 2, No. 2, 1944, pp. 164–168.

[33] Marquardt, D. W., "An Algorithm for Least-Squares Estimation of Nonlinear Parameters," *Journal of the Society for Industrial and Applied Mathematics*, Vol. 11, No. 2, 1963, pp. 431–441.

[34] Gavin, H. P., "The Levenberg-Marquardt Algorithm for Nonlinear Least Squares Curve-Fitting Problems," Department of Civil and Environmental Engineering, Duke University, August 3, 2019, pp. 1–23.

[35] Qin, T., P. Li, and S. Shen, "Vins-mono: A Robust and Versatile Monocular Visual-Inertial State Estimator," *IEEE Transactions on Robotics*, Vol. 34, No. 4, 2018, pp. 1004–1020.

[36] Wen, W., et al., "Factor Graph Optimization for GNSS/INS Integration: A Comparison with the Extended Kalman Filter," *NAVIGATION: Journal of the Institute of Navigation*, Vol. 68, No. 2, 2021, pp. 315–331.

[37] Leutenegger, S., et al., "Keyframe-Based Visual–Inertial Odometry Using Nonlinear Optimization," *The International Journal of Robotics Research*, Vol. 34, No. 3, 2015, pp. 314–334.

4

Point Positioning by Radio Signals

The indoor point positioning for an agent (such as a mobile device) is mostly determined by RF signal-based approaches. The beacons (e.g., reference points and anchors) are installed in a presurveyed location, usually described by a local coordinate system. Based on the distance and angle between the beacons and the agent, the agent's location can be determined. The popular fingerprint matching method, which requires a database of the signal features of the presurveyed locations, can also provide a point positioning result. It will be introduced in Chapter 5.

There are different sensors that make use of different wireless communication protocols to provide the ranging and angle measurements, including BLE, UWB, Wi-Fi, and 5G. Table 4.1 shows the available measurements of the popular radio signal sensors for indoor point positioning. Their expected performance in terms of the measurements is given in Table 4.1. The resolution and accuracy of their measurements are different according to their designed signal protocols.

This chapter will not cover the acquisition and the tracking parts of radio signals inside the receiver agent because different receiver algorithms are required for different communication protocols. Instead, this chapter focuses on the state and measurement models and the associated estimation theories and challenges for direct range, differential range, and angle-based indoor

Table 4.1
The Radio Signal Sensors and Their Measurements

	Direct Range			**Differential Range**	**Direct Angle**
Sensors	**TOA**	**Two-way TOA (Round Time Trip (RTT))**	**RSS-Ranging**	**TDOA**	**AOA**
BLE	Not supported	[1, 2]	[3–5]	Not supported	[6, 7]
UWB	[8–13]	[9, 14–16]	Not preferred	[8, 9, 17–20]	[8, 12, 13, 21, 22]
Wi-Fi	[23, 24]	[25–28]	[25, 27, 29]	[25, 29, 30]	[23, 24]
5G	[31–33]	[31, 34]	Not preferred	[31, 35]	[31, 33, 36]

positioning methods. Table 4.2 provides an overview of the states, measurements, and models for different point positioning methods.

There are three assumptions made for the positioning methods in this chapter:

1. There is an active system from the agent's perspective. The agent requests a positioning solution and then the beacons cooperate to provide radio signals for the agent's receiver to generate either timing or angle measurements between the agent and the beacons.
2. All the beacons' locations are known to the agent. It is not impossible to simultaneously determine the beacons' location at the same. This will require SLAM method to be incorporated in the state. The concept of this advanced method will be introduced in Chapter 9.
3. Both the states and measurements are in a local coordinate system (ENU).

The positioning methods introduced are based on weighted least squares (WLS) for easier comparison between direct ranging, differential ranging, and angle methods. The references of popular closed-form expression models that can use LS methods are included for the readers to perform further study.

The organization of this chapter is as follows. Section 4.1 briefly introduces popular time synchronization methods used in IPIN applications, and

Table 4.2
Comparison of Different Point Positioning Methods Using RF Signals in Terms of Measurement, Model, and State

Methods	**Measurement**	**Model**	**State**
Direct range positioning	d_a^i	$d_a^i = \left\lVert \mathbf{x}_a - \mathbf{x}^i \right\rVert$ in (4.16)	$\mathbf{x}_a = \begin{bmatrix} x_{e,a} \\ x_{n,a} \\ x_{u,a} \end{bmatrix}$
Differential range positioning (least squares)	$d_a^{(i,j)} = d_a^i - d_a^j$	$d_a^{(i,j)} = \left\lVert \mathbf{x}_a - \mathbf{x}^i \right\rVert - \left\lVert \mathbf{x}_a - \mathbf{x}^j \right\rVert$ in (4.36)	$\mathbf{x}_a = \begin{bmatrix} x_{e,a} \\ x_{n,a} \\ x_{u,a} \end{bmatrix}$
Differential range positioning (Fang's algorithm)	$d_a^{(i,j)} = d_a^i - d_a^j$	$\begin{aligned} &-2x_{e,a}\left(x_e^i - x_e^1\right) \\ &-2x_{n,a}\left(x_n^i - x_n^1\right) \\ &= {d_a^i}^2 - {d_a^1}^2 - {x_e^i}^2 \\ &+ {x_e^1}^2 - {x_n^i}^2 + {x_n^1}^2 \end{aligned}$ in (4.45)	$\mathbf{x}_a = \begin{bmatrix} x_{e,a} \\ x_{n,a} \end{bmatrix}$
Differential range positioning (Chan's algorithm)	$c\Delta t^i = d_a^i - d_a^1$	$\begin{aligned} &2\left(x_e^1 - x_e^i\right)x_{e,a} + 2\left(x_n^1 - x_n^i\right)x_{n,a} \\ &- 2d_a^1 c\Delta t^i = \left(c\Delta t^i\right)^2 \\ &-\left[\left({x_e^i}^2 + {x_n^i}^2\right) - \left({x_e^1}^2 + {x_n^1}^2\right)\right] \end{aligned}$ in (4.58)	$\mathbf{x}_a = \begin{bmatrix} x_{e,a} \\ x_{n,a} \\ d_a^1 \end{bmatrix}$
Angle-based positioning (least squares)	$\mathbf{z}_a^i = \begin{bmatrix} \sin\theta_a^i \\ \tan\psi_a^i \end{bmatrix}$	$\sin\theta_a^i = \dfrac{x_u^i - x_{u,a}}{\left\lVert \mathbf{x}_a - \mathbf{x}^i \right\rVert}$ $\tan\psi_a^i = \dfrac{x_e^i - x_{e,a}}{x_n^i - x_{n,a}}$ in (4.63) and (4.64)	$\mathbf{x}_a = \begin{bmatrix} x_{e,a} \\ x_{n,a} \\ x_{u,a} \end{bmatrix}$
Angle-based positioning (OVE)	θ_a^i and ψ_a^i	$\mathbf{S}_a \mathbf{x}_a = \mathbf{z}_a$ in (4.84)	$\mathbf{x}_a = \begin{bmatrix} x_{e,a} \\ x_{n,a} \\ x_{u,a} \end{bmatrix}$
Angle-based positioning (3-D PLE)	θ_a^i and ψ_a^i	$\mathbf{S}_a \begin{bmatrix} x_{e,a} \\ x_{n,a} \end{bmatrix} = \mathbf{z}_a$ in (4.90) $\tan\theta_a^i = \dfrac{x_u^i - x_{u,a}}{\left\lVert \mathbf{x}_a(1:2) - \mathbf{x}^i(1:2) \right\rVert}$ in (4.93)	$\begin{bmatrix} x_{e,a} \\ x_{n,a} \end{bmatrix}$ and $x_{u,a}$

Sections 4.2, 4.3, and 4.4 introduce the measurement model, state model, and the estimation for direct-ranging, differential-ranging, and AOA-based point positioning methods, respectively. Section 4.4 details the common challenges and issues faced by the radio signal-based point positioning methods. Section 4.5 gives coding examples for the three point-positioning methods introduced in the first three sections. The equations mentioned in this chapter will be used to program the positioning algorithms.

4.1 Time Synchronization Methods

High-precision time synchronization is fundamental to accurate point positioning in IPIN applications, where even small timing errors can lead to significant localization inaccuracies. Various synchronization methods are used depending on the required accuracy and system constraints, with performance ranging from subnanosecond to millisecond levels. Table 4.3 provides a comparison of different time synchronization methods.

For the high-end methods, IEEE 1588 Precision Time Protocol (PTP) is a widely used synchronization method that achieves nanosecond-level accuracy in networks supporting UWB TDOA, Wi-Fi RTT, and 5G NR positioning.

Table 4.3
Comparison of Different Time Synchronization Methods Usually Used in IPIN Applications

Synchronization Method	Common Use Cases	Accuracy
IEEE 1588 PTP [37]	UWB, TDOA, Wi-Fi FTM, 5G NR	Nanosecond
White Rabbit (WR) [38]	High-precision indoor systems	Subnanosecond
NTP [39]	Low-precision systems (BLE, Wi-Fi RSSI)	Millisecond
GPSDO [40]	Hybrid outdoor/indoor, UWB TDOA	Microsecond
SyncE [41]	Ethernet-connected anchor networks	Submicrosecond
Proprietary RF Sync [42]	UWB, BLE, Wi-Fi AOA/RTT	Nanosecond-microsecond
5G/LTE Sync [43]	Cellular-based positioning (TDOA, RTT)	Submicrosecond

It works by exchanging timestamped synchronization messages between a master clock and multiple slave clocks in a network, allowing devices to align their time to within nanosecond precision. The protocol uses sync and follow-up messages from the master, along with delay request and delay response messages from the slaves, to compensate for network delays and adjust local clocks accordingly. This allows distributed beacons or anchors to share a synchronized time reference, ensuring that time-based positioning methods can measure precise signal travel times. For example, in UWB-based TDOA, tightly synchronized anchors allow the accurate calculation of a device's position by analyzing differences in signal arrival times, reducing errors to the centimeter scale.

For ultrahigh precision applications, White Rabbit extends IEEE 1588 PTP with subnanosecond synchronization, making it the gold standard for scientific and industrial-grade, high-precision indoor positioning systems. White Rabbit ensures that clock discrepancies remain below a nanosecond, which is essential for laboratory environments, large-scale industrial automation, and sensor fusion applications where ultrafine timing control is required.

GPS Disciplined Oscillators (GPSDOs) are also used in hybrid indoor/outdoor positioning, particularly in UWB TDOA deployments where anchors need a common absolute time reference. GPSDOs ensure microsecond-level synchronization, which is sufficient for most indoor navigation applications where absolute positioning accuracy is required, such as in indoor-outdoor transition areas. However, the GPS receiver is heavily challenged in urban areas, making its accuracy compromised.

Synchronous Ethernet (SyncE) provides submicrosecond accuracy for Ethernet-connected anchor networks, stabilizing clock frequencies across wired infrastructure. SyncE is commonly used alongside IEEE 1588 PTP to eliminate frequency drift in positioning networks, ensuring stable and precise timestamping for Wi-Fi RTT, AOA, and UWB positioning.

At the lower end of the spectrum, Network Time Protocol (NTP) is a software-based synchronization method used in BLE and Wi-Fi RSSI-based positioning systems. However, with millisecond-level accuracy, NTP is unsuitable for high-precision applications. The large timing errors translate to meter-scale position errors, making it impractical for time-sensitive ranging techniques such as TOA or TDOA.

Also, there are proprietary RF synchronization methods. These are commonly found in UWB and BLE AOA-based positioning and offer a flexible alternative in which wired synchronization is infeasible. These systems can achieve nanosecond-to-microsecond accuracy, depending on radio propagation conditions and hardware capabilities. For example, UWB anchors can use wireless synchronization beacons to align their internal clocks within a few nanoseconds, ensuring robust performance in real-time tracking applications.

Finally, 5G and LTE synchronization provide submicrosecond accuracy for cellular-based positioning (TDOA and RTT methods). By leveraging GPS synchronization and network-wide timing protocols, 5G base stations can align timestamps to nanosecond precision. Again, this is similar to GPSDOs, the GPS-based synchronization will be compromised in urban areas.

4.2 Direct Range-Based Indoor Positioning

As mentioned in the assumption made in this chapter, the location of the beacons is presurveyed and is assumed to be known to the agent. If the ranges between the beacons and the agent can be measured by the receiver, the trilateration can be used to determine the agent's location, as shown in Figure 4.1. Taking the beacon's location as the center, the range between the beacon and the agent as the radius, the agent's location is in the circumference. As there could be more than one beacon, the number of the circle is more than 1 as well. If the ranging measurements and the beacons' location are ideal (perfect), the intersection of the circles is a point, indicating the agent's location.

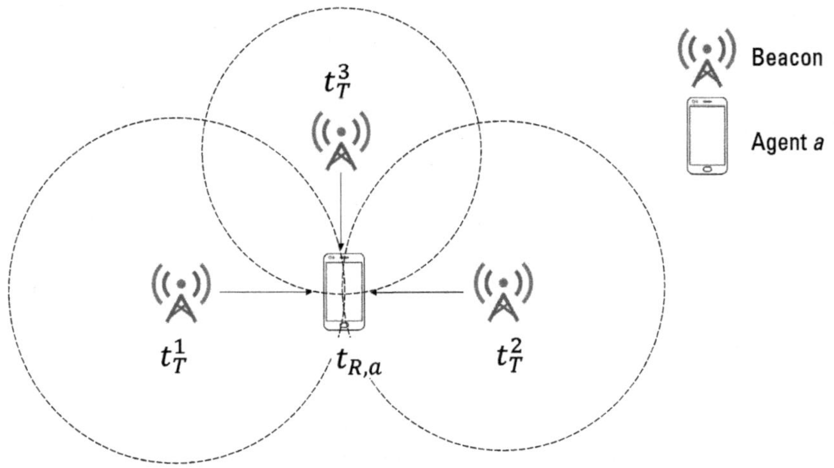

Figure 4.1 The trilateration based on the ranging measurements for indoor point positioning.

4.2.1 Direct Ranging

This section will introduce the three methods, TOA, two-way TOA, and RSS ranging, to generate the direct ranging measurement. More specifically, their models and the use of the measurements to estimate the agent's location will be detailed.

4.2.1.1 TOA Measurement

The direct range measurement based on TOA is generated using:

$$d_a^i = c\left(t_{R,a} - t_T^i\right) \tag{4.1}$$

where the superscript i denotes the index of the beacon, d_a^i denotes the range between beacon i and the target agent a, $t_{R,a}$, and t_T^i denote the time of reception on the agent a and transmission from the beacon i, respectively. Both $t_{R,a}$ and t_T^i are measured by the agent's receiver, and c is the speed of light. This equation holds under two major assumptions: (1) the time is synchronized between the beacons, and (2) the time is synchronized between the beacons and the agent. If we remove both assumptions, then

$$d_a^i = c\left(t_{R,a} - t_T^i + \Delta t_{R,a}^i\right) \tag{4.2}$$

where $\Delta t_{R,a}^i$ is the time-offset between the time of beacon i and the agent a. Obviously, (4.2) always adds one more unknown to the system for each additional beacon, which is not beneficial to the estimation of the agent's location. The first assumption is possible if the indoor beacons are synchronized to an acceptable accuracy by sophisticated synchronization methods, including IEEE 1588 PTP. For the second assumption, it is harder in the TOA measurement-based method because the oscillator used in the agent's receiver is usually low-cost and less stable compared to those used in beacons. Thus, if only the second assumption is removed, then

$$d_a^i = c\left(t_{R,a} - t_T^i + \Delta t_{R,a}\right) \tag{4.3}$$

where the $\Delta t_{R,a}$ is the time offset between the system time of beacons and the agent a, which is an additional unknown to the system. In general, (4.1) or (4.3) is used. However, it is important to consider the noise after applying the corrections of the time offset (usually generated by the beacon system provider and the receiver manufacturer) are used. As a result, the equations become

$$\tilde{d}_a^i = c\left(t_{R,a} - t_T^i\right) + \omega_{\Delta t^i_{R,a}} + \omega_{\Delta t_{R,a}} \tag{4.4}$$

$$\tilde{d}_a^i = c\left(t_{R,a} - t_T^i + \Delta t_{R,a}\right) + \omega_{\Delta t^i_{R,a}} \tag{4.5}$$

where $\omega_{\Delta t^i_{R,a}} \sim \mathcal{N}\left(0, \sigma^2_{\Delta t^i_{R,a}}\right)$ and $\omega_{\Delta t_{R,a}} \sim \mathcal{N}\left(0, \sigma^2_{\Delta t_{R,a}}\right)$ are the Gaussian random noise accounting for the noises after the time correction to synchronize the beacons (which is determined based on different time synchronization methods) and the time correction to synchronize the beacon and the agent, respectively. Note that the estimated time offsets are assumed to be unbiased in this book. If there is 1 ns of error (exemplified by IEEE 1588 PTP) in the time offset correction, then there is about 0.3m of error in the ranging measurement. Thus, it is utterly important to keep the beacons synchronized accurately. It is much harder to obtain the time offset between the agent and the beacons. Thus, the time offset between the system time of beacons and the agent is treated as a state to be estimated when using TOA-based measurement.

4.2.1.2 Two-Way TOA Measurement

The concept of two-way TOA aims to relax the strict time synchronization requirements in two aspects: (1) between beacons, and (2) between the beacon and the agent. In general, NTP, which achieves millisecond-level accuracy, is sufficient for most IPIN applications. This process is illustrated in Figure 4.2, where the signal travels in a round-trip path, making it commonly known

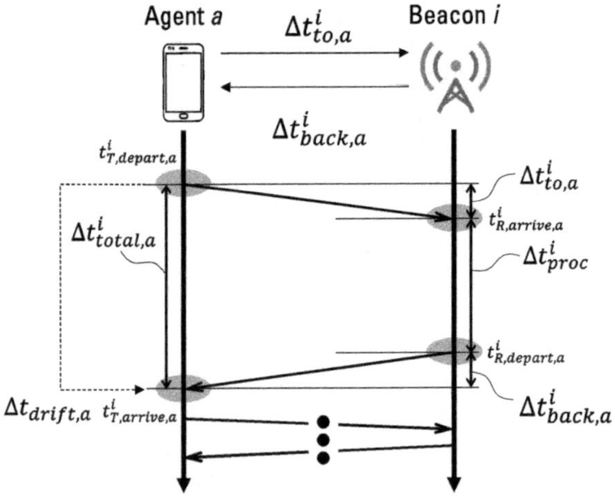

Figure 4.2 The time traveled to generate a two-way TOA ranging measurement.

as the round-trip time (RTT) method. In Wi-Fi Fine Time Measurement (FTM), this approach is explicitly defined as the RTT-based ranging method.

The distance between the agent a and the beacon i can be generated as:

$$d_a^i = c\frac{\left(\Delta t_{\text{to},a}^i + \Delta t_{\text{back},a}^i\right)}{2} \tag{4.6}$$

where $\Delta t_{\text{to},a}^i$ and $\Delta t_{\text{back},a}^i$ denote the time traveled from the agent a to the beacon i and from the beacon back to the agent, respectively. This equation assumes:

1. The beacon processes and retransmits the signal instantaneously (i.e., zero processing time).
2. The agent's clock is perfectly stable (i.e., no clock drift).

In real-world systems, beacons require finite time to process incoming signals before retransmitting them. To account for this processing delay (Δt_{proc}^i), the distance equation is updated to:

$$d_a^i = c\frac{\left(t_{T,\text{arrive},a}^i - t_{T,\text{depart},a}^i\right) - \Delta t_{\text{proc}}^i}{2} \tag{4.7}$$

where $t_{T,\text{depart},a}^i$ and $t_{T,\text{arrive},a}^i$ are the timestamps recorded by the agent when the signal departs and returns, respectively. Δt_{proc}^i is the processing time at the beacon, which must be estimated for accurate distance measurement. The accuracy of Δt_{proc}^i depends on the beacon's oscillator stability. This processing time is typically estimated using timestamps at the beacon upon receiving and transmitting the signal [28]. As processing time estimation significantly affects two-way TOA accuracy, manufacturers often treat it as a proprietary method, using custom calibration strategies to improve performance. Some manufacturers implement adaptive calibration algorithms, while others enforce a fixed processing delay to ensure consistency across beacons. This alternative approach is to define Δt_{proc}^i as a constant, where the beacon deliberately waits for a prespecified duration before retransmitting. This method, although less commonly used in practice, simplifies calibration and ensures consistency in signal processing time [14]. It also provides the beacon with sufficient time to modulate the necessary information into the response signal, which may be beneficial in certain applications.

The agent's oscillator also introduces timing errors due to clock drift, which accumulates over the total round-trip signal travel time and the beacon's processing time. If the agent's clock drifts by only 1 ns, it introduces a 0.3-m ranging error. To address this, the model is refined to:

$$d_a^i = c\frac{\left(t_{T,\text{arrive},a}^i - t_{T,\text{depart},a}^i\right) - \Delta t_{\text{proc}}^i - \Delta t_{\text{drift},a}}{2} \tag{4.8}$$

where $\Delta t_{\text{drift},a}$ is the drift-induced error in the agent's clock. In practical systems, temperature-compensated crystal oscillators (TCXOs), Kalman filtering, or external synchronization methods are commonly used to mitigate drift errors.

After accounting for noise introduced by both processing time estimation and clock drift compensation, the final range measurement is modeled as:

$$\tilde{d}_a^i = c\frac{\left(t_{T,\text{arrive},a}^i - t_{T,\text{depart},a}^i\right) - \Delta t_{\text{proc}}^i - \Delta t_{\text{drift},a}}{2} + \omega_{\Delta t_{\text{proc}}^i} + \omega_{\Delta t_{\text{drift},a}} \tag{4.9}$$

where $\omega_{\Delta t_{\text{proc}}^i} \sim \mathcal{N}\left(0,\sigma_{\Delta t_{\text{proc}}^i}^2\right)$ represents Gaussian noise from beacon processing time estimation, and $\omega_{\Delta t_{\text{drift},a}} \sim \mathcal{N}\left(0,\sigma_{\Delta t_{\text{drift},a}}^2\right)$ represents Gaussian noise from clock drift calibration at the agent. The standard deviations $\sigma_{\Delta t_{\text{proc}}^i}^2$ and $\sigma_{\Delta t_{\text{drift},a}}^2$ can typically be obtained from the fact sheets of the beacon and agent receiver hardware [44, 45]. These noise factors highlight the importance of high-quality oscillators and precise calibration in achieving accurate and reliable two-way ranging measurements.

4.2.1.3 RSS Ranging Measurement

Besides relying on the signal propagation time span, the range between a beacon and an agent can also be estimated based on the variation of the RSS, namely, RSS ranging. Following the path-loss model [46], the strength of a radio signal decays along its propagation path from the beacon (transmitter), as Figure 4.3 demonstrates.

The relationship in between is commonly described by a one-slope model, using

$$p_{R,a}^i = p_{\text{ref}} - 10\eta\log\left(\frac{d_a^i}{d_{\text{ref}}}\right) \tag{4.10}$$

where $P_{R,a}^i$ is the received signal power from the agent a with a distance d_a^i from the beacon i, p_{ref} is the measured power from a reference location with a distance d_{ref} from the beacon, and η is the path-loss propagation coefficient

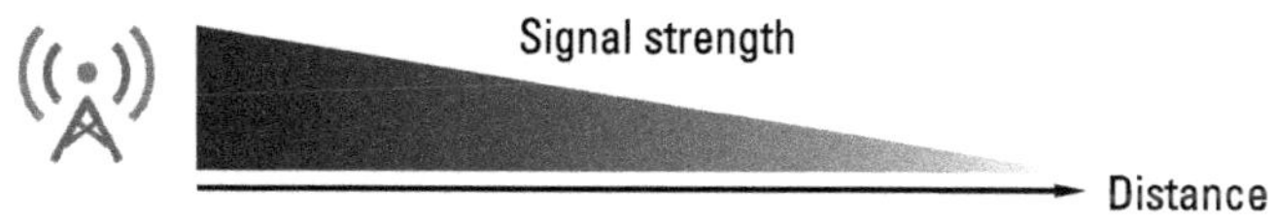

Figure 4.3 The radio signal strength decays along with its propagation distance.

indicating how the RSS decays along with propagation distance. Note that both the agent and the reference location have a certain distance from the beacon fulfilling the far-field assumption to apply the path-loss model. Therefore, it is feasible to inversely estimate the signal propagation distance based on the RSS measurement from the agent [47], as follows:

$$d_a^i = d_{\text{ref}} \cdot 10^{\frac{p_{\text{ref}} - p_{R,a}^i}{10\eta}} \tag{4.11}$$

Unlike the timing-based ranging methods, RSS ranging relies much less on time synchronization or oscillator quality. However, the RSS measurement still contains random variations in space and time [47], which can be modeled as

$$\tilde{p}_{R,a}^i = p_{R,a}^i + \omega_{\text{long},\mathbf{x}_a} + \omega_{\text{short},t} \tag{4.12}$$

where $\omega_{\text{long},\mathbf{x}_a}$ is the space-dependent error term, commonly known as the long-term fading process, $\omega_{\text{short},t}$ is the time-dependent error term, commonly known as the short-term fading process. The short-term fading process can be assumed as stationary and ergodic. Thus, its degradation can be mitigated by the averaging process on multiple samples over certain times, in which the mean value will be absorbed into the reference power term p_{ref} remaining constant [48]. However, the long-term fading process depends on the location of the agent and the surrounding environment, such as the severity of scattering effects. It cannot be mitigated via time averaging but is usually modeled as $\omega_{\text{long},\mathbf{x}_a} \sim \mathcal{N}(0, \sigma^2_{\text{long},\mathbf{x}_a})$ in the scale of dB [47]. Thus, the RSS ranging estimation yields

$$\begin{aligned} \tilde{d}_a^i &= d_{\text{ref}} \cdot 10^{\frac{p_{\text{ref}} - \tilde{p}_{R,a}^i}{10\eta}} = d_{\text{ref}} \cdot 10^{\frac{p_{\text{ref}} - p_{R,a}^i - \omega_{\text{long},\mathbf{x}_a}}{10\eta}} \\ &= d_{\text{ref}} \cdot 10^{\frac{p_{\text{ref}} - p_{R,a}^i}{10\eta}} \cdot 10^{\frac{-\omega_{\text{long},\mathbf{x}_a}}{10\eta}} \end{aligned} \tag{4.13}$$

where the long-term fading process degrades the RSS ranging as a multiplication factor that $\tilde{d}_a^i = d_a^i \cdot 10^{-\omega_{\text{long},\mathbf{x}_a}/10\eta}$.

In terms of the practicality, this method is practical because many of the radio signals do not include the data of time of transmission and the time of arrival (dated February 2025). For example, Wi-Fi can only provide fine-time measurement after the IEEE 802.11mc protocol, meaning that the earlier version of Wi-Fi cannot support FTM measurement methods.

4.2.2 Models and Estimation: Least Squares

In TOA positioning, the agent's location is estimated based on the absolute distances between the agent and multiple beacons. Because the TOA equations are nonlinear, an iterative weighted least squares (I-WLS) method is applied to iteratively refine the position estimate until convergence. This method is widely used in indoor positioning systems because it efficiently handles nonlinear ranging equations while minimizing the impact of measurement noise through a WLS approach.

The measurement vector and state vector are defined as:

$$\mathbf{z}_a = \begin{bmatrix} d_a^1 \\ \vdots \\ d_a^I \end{bmatrix} \tag{4.14}$$

$$\mathbf{x}_a = \begin{bmatrix} x_{e,a} \\ x_{n,a} \\ x_{u,a} \end{bmatrix} \tag{4.15}$$

where $x_{e,a}$, $x_{n,a}$, and $x_{u,a}$ denote the location of agent a on the East-axis, North-axis, and up-axis of the ENU coordinate system, respectively, and there are total I beacons. Note that this estimation algorithm is also applicable to other Cartesian coordinate systems.

Using trilateration, the TOA measurement model is:

$$d_a^i(\mathbf{x}) = \left\| \mathbf{x}_a - \mathbf{x}^i \right\| = \sqrt{\left(x_{e,a} - x_e^i\right)^2 + \left(x_{n,a} - x_n^i\right)^2 + \left(x_{u,a} - x_u^i\right)^2} \tag{4.16}$$

where $\mathbf{x}^i = (x_e^i, x_n^i, x_u^i)$ represents the known position of the ith beacon on the East-axis, North-axis, and up-axis, respectively. By taking the first order of the Taylor series expansion at initial guess $\mathbf{x}_{a(0)}$, the model is linearized:

$$d_a^i(\mathbf{x}_a) \approx d_a^i(\mathbf{x}_{a(0)}) + \mathbf{h}_{a(0)}^i\left(\mathbf{x}_a - \mathbf{x}_{a(0)}\right) \tag{4.17}$$

where $\mathbf{h}^i_{a(0)}$ is the Jacobian matrix, defined as:

$$\mathbf{h}^i_{a(0)} = \begin{bmatrix} \dfrac{x_{e,a(0)} - x^i_e}{d^i_a\left(\mathbf{x}_{a(0)}\right)} & \dfrac{x_{n,a(0)} - x^i_n}{d^i_a\left(\mathbf{x}_{a(0)}\right)} & \dfrac{x_{u,a(0)} - x^i_u}{d^i_a\left(\mathbf{x}_{a(0)}\right)} \end{bmatrix} \tag{4.18}$$

Thus, the linearized system for all beacons is:

$$\delta\mathbf{z}_a = \underbrace{\begin{bmatrix} \mathbf{h}^1_{a(0)} \\ \mathbf{h}^2_{a(0)} \\ \vdots \\ \mathbf{h}^I_{a(0)} \end{bmatrix}}_{\mathbf{H}_{a(0)}} \delta\mathbf{x}_a \tag{4.19}$$

where $\delta\mathbf{z}_a$ is the measurement residual vector, $\delta\mathbf{x}_a$ is the state correction vector representing the update to the agent's estimated position, and $\mathbf{H}_{a(0)}$ is the line-of-sight (LOS) matrix incorporating the $\mathbf{h}^i_{a(0)}$ of each beacon.

If at least four beacons are available ($I >= 4$), the system is overdetermined, allowing for an LS solution:

$$\delta\mathbf{x}_a = \left(\mathbf{H}_{a(0)}{}^{\mathrm{T}}\mathbf{H}_{a(0)}\right)^{-1}\mathbf{H}_{a(0)}{}^{\mathrm{T}}\delta\mathbf{z}_a \tag{4.20}$$

The estimated agent position is iteratively updated as:

$$\mathbf{x}_{a(1)} = \mathbf{x}_{a(0)} + \delta\mathbf{x}_a \tag{4.21}$$

The iteration stops when the update falls below a predefined threshold:

$$\left\|\mathbf{x}_{a(k)} - \mathbf{x}_{a(k-1)}\right\| \leq \text{iteration\_threshold} \tag{4.22}$$

where the threshold is selected based on the required localization accuracy.

If the measurement is unbiased and the associated noise standard deviations are available, the WLS estimation can be conducted to achieve a maximum likelihood estimation (MLE).

$$\begin{gathered} \delta\mathbf{x}_a = \left(\mathbf{H}_{a(0)}{}^{\mathrm{T}}\mathbf{W}_a\mathbf{H}_{a(0)}\right)^{-1}\mathbf{H}_{a(0)}{}^{\mathrm{T}}\mathbf{W}_a\delta\mathbf{z}_a \\ \mathbf{W}_a = \begin{bmatrix} \left(\sigma^2_{d^1_a}\right)^{-1} & \cdots & 0 \\ \vdots & \ddots & \vdots \\ 0 & \cdots & \left(\sigma^2_{d^I_a}\right)^{-1} \end{bmatrix} \end{gathered} \tag{4.23}$$

If the beacons and agent clocks are not synchronized, an additional clock bias term must be estimated. In this case, the state vector is expanded to include the unknown clock offset:

$$\mathbf{x}_a = \begin{bmatrix} x_{e,a} \\ x_{n,a} \\ x_{u,a} \\ c\Delta t_{R,a} \end{bmatrix} \tag{4.24}$$

$$d_a^i(\mathbf{x}_a) = \sqrt{\left(x_{e,a} - x_e^i\right)^2 + \left(x_{n,a} - x_n^i\right)^2 + \left(x_{u,a} - x_u^i\right)^2} + c\Delta t_{R,a} \tag{4.25}$$

By following the same linearization, the linearized LOS model based on an initial guess becomes

$$\mathbf{H}_{a(0)} = \begin{bmatrix} \frac{x_{e,a} - x_e^1}{d_a^1\left(\mathbf{x}_{a(0)}\right)} & \frac{x_{n,a} - x_n^1}{d_a^1\left(\mathbf{x}_{a(0)}\right)} & \frac{x_{u,a} - x_u^1}{d_a^1\left(\mathbf{x}_{a(0)}\right)} & 1 \\ & \vdots & & \vdots \\ \frac{x_{e,a} - x_e^I}{d_a^I\left(\mathbf{x}_{a(0)}\right)} & \frac{x_{n,a} - x_n^I}{d_a^I\left(\mathbf{x}_{a(0)}\right)} & \frac{x_{u,a} - x_u^I}{d_a^I\left(\mathbf{x}_{a(0)}\right)} & 1 \end{bmatrix} \tag{4.26}$$

In this case, the WLS estimation is performed to obtain the optimal position and clock bias estimate.

4.3 Differential Range-Based Indoor Positioning

TOA-based ranging requires precise time synchronization between beacons and the agent, which can be challenging and costly in practical deployments. To avoid this requirement, TDOA techniques are commonly used.

TDOA eliminates the need for absolute time synchronization by measuring relative time differences between signals received at multiple locations. This technique effectively removes common timing offsets, improving robustness against synchronization errors. It can be implemented in three ways, depending on how the time differences are measured:

1. *TDOA across transmitters:* Also known as uplink TDOA, this method is commonly used for snapshot positioning. Multiple beacons (receivers)

measure the time difference of a signal emitted by an agent. This is more popular for large-scale infrastructure-based tracking for the applications such as a real-time location system adapted in indoor logistic scenarios.

2. *TDOA across receivers:* This is also known as downlink TDOA, where a beacon measures the arrival time differences of signals from multiple agents. This approach is popularly used in cellular positioning, such as LTE/5G.

Their comparison is summarized in Table 4.4. The operations of uplink and download TDOAs are different, but the measurements model is actually the same, which is introduced in the following sections.

4.3.1 TDOA Measurement Across Beacons

To compute TDOA measurements, we first establish the direct range equations between the agent a and two beacons i and j:

$$\begin{aligned} d_a^i &= c\left(t_{R,a} - t_T^i + \Delta t_{R,a}\right) \\ d_a^j &= c\left(t_{R,a} - t_T^j + \Delta t_{R,a}\right) \end{aligned} \tag{4.27}$$

Table 4.4
Comparison Between Uplink and Downlink TDOA

Feature	Uplink TDOA	Downlink TDOA
Who transmits?	Agent (mobile device, tag, tracker)	Fixed beacon (Wi-Fi AP, base station for cell communication)
Who receives?	Fixed infrastructure (multiple beacons)	Mobile agents (phones, vehicles)
Who computes position?	Centralized server or infrastructure	The agent (self-localization)
Scalability	Limited (too many transmitting agents can cause congestion)	Higher (many devices can passively receive signals)
Best for	Large-scale tracking, RTLS, UWB-based indoor positioning	GNSS-like positioning, LTE/5G-based localization, multi-agent systems
Common applications	Industrial RTLS, logistics, emergency responder tracking	LTE/5G OTDOA, Wi-Fi FTM, autonomous navigation

where

- d_a^i and d_a^j are the distances between the agent and beacons i and j, respectively.
- $\Delta t_{R,a}$ is the receive timestamp at the agent.
- t_T^i and t_T^j are the transmit timestamps at beacons i and j.
- $t_{R,a}$ represents the receiver's clock offset.

By differencing the two direct range equations, we eliminate the receiver's clock offset:

$$d_a^{(i,j)} = d_a^i - d_a^j = c\left(t_{R,a} - t_T^i + \Delta t_{R,a}\right) - c\left(t_{R,a} - t_T^j + \Delta t_{R,a}\right) = c\left(t_T^j - t_T^i\right) \tag{4.28}$$

$d_a^{(i,j)}$ denotes the range difference between the direct ranges of d_a^i and d_a^j. This is a widely used method because receiver clock stability has less impact on range measurements compared to TOA-based methods. Equation (4.27) assumes that beacons are perfectly synchronized. However, in real-world implementations, beacons often operate on independent clocks, introducing an additional time offset $\Delta t_{R,a}^{(i,j)}$ between them. This modifies the equation:

$$\begin{aligned} d_a^i &= c\left(t_{R,a} - t_T^i + \Delta t_{R,a}^i\right) \\ d_u^j &= c\left(t_{R,a} - t_T^j + \Delta t_{R,a}^j\right) \\ d_a^{(i,j)} = d_a^i - d_a^j &= c\left(t_{R,a} - t_T^i + \Delta t_{R,a}^i\right) - c\left(t_{R,a} - t_T^j + \Delta t_{R,a}^j\right) \\ &= c\left(t_T^j - t_T^i + \Delta t_{R,a}^{(i,j)}\right) \end{aligned} \tag{4.29}$$

where $\Delta t_{R,a}^{(i,j)}$ represents the unknown time offset between beacons i and j. This introduces an additional unknown in the system, making direct estimation of the agent's position more difficult. Thus, beacon synchronization is generally required for TDOA-based positioning systems, similar to TOA methods. To account for remaining synchronization errors after time correction, Gaussian noise is introduced into the equation:

$$\tilde{d}_a^{i,j} = c\left(t_T^j - t_T^i\right) + \omega_{\Delta t_{R,a}} \tag{4.30}$$

where $\omega_{\Delta t_{R,a}} \sim \mathcal{N}\left(0, \sigma_{\Delta t_{R,a}}^2\right)$ represents random synchronization errors in the beacon clocks.

If beacon synchronization is impractical, an alternative approach is to derive TDOA from two-way TOA measurements. Assuming that the beacons process signals with identical delays (i.e., $t^i_{\text{proc}} = \Delta t^j_{\text{proc}} = \Delta t_{\text{proc}}$), the direct ranges become:

$$d^i_a = c\frac{\left(t^i_{T,\text{arrive},a} - t^i_{T,\text{depart},a}\right) - \Delta t_{\text{proc}}}{2}$$
$$d^j_a = c\frac{\left(t^j_{T,\text{arrive},a} - t^j_{T,\text{depart},a}\right) - \Delta t_{\text{proc}}}{2} \tag{4.31}$$

By differencing the two equations, we obtain:

$$d^{(i,j)}_a = d^i_a - d^j_a = c\frac{\left(t^i_{T,\text{arrive},a} - t^i_{T,\text{depart},a}\right) - \left(t^j_{T,\text{arrive},a} - t^j_{T,\text{depart},a}\right)}{2} \tag{4.32}$$

This equation benefits from:

1. *TDOA:* Eliminates receiver clock offsets.
2. *Two-way TOA:* Avoids requiring beacon synchronization.

As each TDOA measurement requires a pair of beacons, a system with I beacons provides only $I - 1$ independent TDOA measurements. Therefore, one beacon (e.g., beacon j) is selected as the reference, and all TDOA measurements are computed relative to it:

$$d^{(i,j)}_a = d^i_a - d^j_a, \ \forall i \neq j \tag{4.33}$$

The choice of reference beacon affects the system's geometry and localization accuracy. Typically, the reference beacon is:

- Placed centrally within the deployment area.
- Selected dynamically based on signal strength.

A poor choice of reference beacon can lead to a poor geometric dilution of precision (GDOP), reducing positioning accuracy, which is discussed later in the chapter.

4.3.2 Model and Estimation

TDOA positioning is widely used in wireless localization, radar, and sensor networks. It estimates an unknown signal source's position based on the time differences between signals arriving at multiple receivers. Various algorithms have been developed to solve this hyperbolic localization problem, with LS estimation, Fang's algorithm, and Chan's algorithm being the most commonly applied. Each method offers a trade-off between computational complexity, accuracy, and robustness. Their comparison is given in Table 4.5.

LS formulates the TDOA problem as a set of nonlinear equations that require iterative optimization. The LS approach is effective in minimizing residual errors but demands an initial position estimate and can be sensitive to poor initialization and convergence issues. When properly initialized, LS

Table 4.5
Comparison of the Models and Estimation Algorithms Used for TDOA-Based Positioning

Feature	LS Estimation [49]	Fang's Algorithm [50]	Chan's Algorithm [51]
Type of solution	Iterative	Closed-form	Two-step closed-form
Computational complexity	High (iterative updates)	Moderate (single LS step)	Low (sequential LS + quadratic constraint)
Handling of nonlinearity	Requires iterative linearization	Linearized via algebraic manipulation	Uses an auxiliary variable to linearize equations
Robustness to noise	Moderate (relies on noise model)	Sensitive (linearization amplifies errors)	High (WLS refines estimation)
Accuracy	High (converges to machine learning solution if noise is Gaussian)	Moderate (noise-dependent)	High (approaches Cramér-Rao bound for small errors)
Initial guess required?	Yes (iterative convergence)	No (direct LS solution)	No (direct LS + refinement)
Sensitivity to geometric configuration	High (bad initial guess affects convergence)	High (ill-conditioned geometry increases errors)	Lower (redundant measurements reduce GDOP effects)
WLS support	Yes	No	Yes (second step applies WLS)
Solution stability	Can diverge if poorly initialized	Can be unstable for poor anchor placement	More stable due to quadratic constraint correction
Application scenarios	High-precision GNSS, radar tracking	Real-time positioning with minimal computation	Robust localization in wireless networks, UWB, and sensor networks

estimation provides high accuracy and can handle complex error models through iterative refinements.

Fang's algorithm introduces a closed-form approach, transforming the hyperbolic TDOA equations into a linear system using algebraic manipulation. This method eliminates the need for iterative computation, making it computationally efficient. However, its accuracy can degrade in the presence of high measurement noise, and its performance is dependent on a favorable receiver geometry.

Chan's algorithm combines the advantages of LS and Fang's methods by employing a two-step process: first, an initial estimate is obtained through an unweighted LS solution, followed by a WLS refinement that accounts for measurement noise correlation. This approach improves robustness and accuracy, making it well-suited for real-world implementations in UWB, cellular positioning, and GNSS-like localization.

The following sections will provide a detailed mathematical representation of each method, highlighting their formulation for TDOA-based positioning systems.

4.3.3 Least Squares

TDOA positioning relies on solving a set of nonlinear hyperbolic equations, where the agent's position is estimated based on the range differences between pairs of beacons. To solve this problem, we apply the I-WLS method, which iteratively refines the position estimate by linearizing the nonlinear equations and minimizing the squared residuals.

This method is widely used in indoor positioning because it efficiently handles the nonlinear nature of TDOA measurements while providing accurate and computationally efficient solutions.

In this book, an iterative LS estimation is selected as the main point positioning estimation to introduce due to its popularity. Considering (4.27) and (4.30), the measurement vector $\mathbf{z}_a$ and the state vector $\mathbf{x}_a$ are given here:

$$\mathbf{z}_a = \begin{bmatrix} d_a^{1,j} \\ \vdots \\ d_a^{I-1,j} \end{bmatrix} \tag{4.34}$$

$$\mathbf{x}_a = \begin{bmatrix} x_{e,a} \\ x_{n,a} \\ x_{u,a} \end{bmatrix} \tag{4.35}$$

where $x_{e,a}$, $x_{n,a}$, and $x_{u,a}$ denote the agent's location on the East-axis, North-axis, and up-axis of the ENU coordinate system, respectively, and there are $I - 1$ beacons with beacon j as the reference beacon.

Using trilateration, the measurement model is:

$$d_a^{(i,j)} = \left\| \mathbf{x}_a - \mathbf{x}^i \right\| - \left\| \mathbf{x}_a - \mathbf{x}^j \right\| \tag{4.36}$$

where $\mathbf{x}^i$ represents the position of the ith beacon at (x_e^i, x_n^i, x_u^i). As shown in the equation above, each TDOA measurement corresponds to a hyperbola equation in 2-D (or a hyperboloid in 3-D). Figure 4.4 illustrates the TDOA model, where beacon 1 serves as the reference beacon. The solid lines are the hyperbola equations that represented the TDOA model of $d_a^{2,1}$ and $d_a^{3,1}$.

Since (4.36) is nonlinear, we apply a first-order Taylor series expansion around an initial position estimate $\mathbf{x}_{a(0)}$ to linearize the model:

$$d_a^{(i,j)}\left(\mathbf{x}_a\right) \approx d_a^{(i,j)}\left(\mathbf{x}_{a(0)}\right) + \mathbf{h}_{a(0)}^{(i,j)}\left(\mathbf{x}_a - \mathbf{x}_{a(0)}\right) \tag{4.37}$$

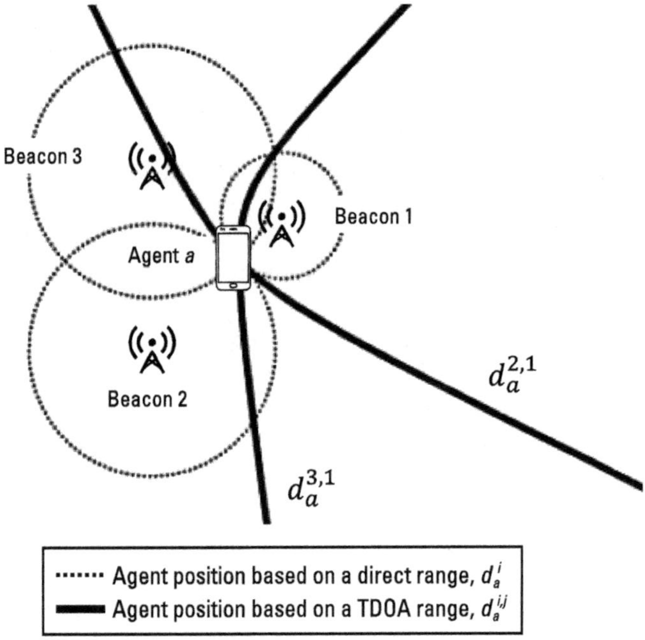

Figure 4.4 The measurement model of TDOA across the beacons. Beacon 1 acts as the reference beacon.

where $\mathbf{h}_{a(0)}^{(i,j)}$ is the Jacobian matrix, defined as:

$$\mathbf{h}_{a(0)}^{(i,j)} == \left[\frac{x_{e,a(0)} - x_e^i}{d_a^i\left(\mathbf{x}_{a(0)}\right)} \quad \frac{x_{n,a(0)} - x_n^i}{d_a^i\left(\mathbf{x}_{a(0)}\right)} \quad \frac{x_{u,a(0)} - x_u^i}{d_a^i\left(\mathbf{x}_{a(0)}\right)}\right] - \left[\frac{x_{e,a(0)} - x_e^j}{d_a^j\left(\mathbf{x}_{a(0)}\right)} \quad \frac{x_{n,a(0)} - x_n^j}{d_a^j\left(\mathbf{x}_{a(0)}\right)} \quad \frac{x_{u,a(0)} - x_u^j}{d_a^j\left(\mathbf{x}_{a(0)}\right)}\right] \tag{4.38}$$

Thus, the linearized system takes the form:

$$\delta\mathbf{z}_a = \mathbf{H}_{a(0)}\delta\mathbf{x}_a \tag{4.39}$$

where $\delta\mathbf{z}_a$ is the measurement residual vector, $\delta\mathbf{x}_a$ is the state correction vector representing the update to the agent's estimated position, and $\mathbf{H}_{a(0)}$ is differenced LOS matrix.

If at least four beacons are available, the system is overdetermined ($I - 1 > 3$) allowing for an LS solution:

$$\delta\mathbf{x}_a = \left(\mathbf{H}_{a(0)}{}^{\mathrm{T}}\mathbf{H}_{a(0)}\right)^{-1}\mathbf{H}_{a(0)}{}^{\mathrm{T}}\delta\mathbf{z}_a \tag{4.40}$$

If the noise statistics of the TDOA measurements are available, a WLS method can be used to an optimal solution.

$$\delta\mathbf{x}_a = \left(\mathbf{H}_{a(0)}{}^{\mathrm{T}}\mathbf{W}_a\mathbf{H}_{a(0)}\right)^{-1}\mathbf{H}_{a(0)}{}^{\mathrm{T}}\mathbf{W}_a\delta\mathbf{z}_a \tag{4.41}$$

where $\mathbf{W}_a$ is the weighting matrix, typically defined as $\mathbf{W}_a = \mathbf{\Sigma}_a^{-1}$, where $\mathbf{\Sigma}_a$ is the covariance matrix of measurement errors.

$$\mathbf{\Sigma}_a = \begin{bmatrix} \sigma_{d_a^{1,j}}^2 & \sigma_{\text{inter}}^2 & \cdots & \sigma_{\text{inter}}^2 \\ \sigma_{\text{inter}}^2 & \sigma_{d_a^{2,j}}^2 & \cdots & \sigma_{\text{inter}}^2 \\ \vdots & \vdots & \ddots & \vdots \\ \sigma_{\text{inter}}^2 & \sigma_{\text{inter}}^2 & \cdots & \sigma_{d_a^{I-1,j}}^2 \end{bmatrix} \tag{4.42}$$

It is important to note that the differenced measurements, meaning the measurements are actually correlated with each other. As a result, the off-diagonal terms are no longer zeros, denoted as the correction σ_{inter}^2 between each TDOA measurement.

4.3.4 Fang's Algorithms

Fang's algorithm is a closed-form LS-based method used for TDOA positioning. It provides a direct algebraic solution to the nonlinear TOA equations by reformulating them into a linear system. Unlike iterative approaches such as nonlinear LS, Fang's method avoids convergence issues and provides a computationally efficient localization solution.

Consider a mobile device at an unknown 2-D position $\mathbf{x}_a = \begin{bmatrix} x_{e,a} \\ x_{n,a} \end{bmatrix}$ with I known reference anchors located at (x_e^i, x_n^i) for $i = 1, 2, \cdots, I$. The TOA measurements provide the estimated distance to each anchor, using

$$d_a^i = \sqrt{\left(x_{e,a} - x_e^i\right)^2 + \left(x_{n,a} - x_n^i\right)^2} + \omega_i \tag{4.43}$$

where d_a^i is the estimated range from the mobile device a to anchor i, and ω_i is the measurement noise.

To derive a closed-form solution, Fang's algorithm linearizes these nonlinear range equations. Choose one of the anchors (e.g., the first anchor) as a reference and define the squared range difference between each anchor i and the reference anchor 1, by

$$\begin{aligned} d_a^{i\,2} - d_a^{1\,2} &= \left(x_{e,a} - x_e^i\right)^2 + \left(x_{n,a} - x_n^i\right)^2 - \left(x_{e,a} - x_e^1\right)^2 - \left(x_{n,a} - x_n^1\right)^2 \\ &= x_{e,a}^2 - 2x_{e,a}x_e^i + x_e^{i\,2} + x_{n,a}^{\;2} - 2x_{n,a}x_n^i + x_n^{i\,2} \\ &\quad - x_{e,a}^2 + 2x_{e,a}x_e^1 - x_e^{1\,2} - x_{n,a}^2 + 2x_{n,a}x_n^1 - x_n^{1\,2} \end{aligned} \tag{4.44}$$

The equation can be rearranged as

$$\begin{aligned} & d_a^{i\,2} - d_a^{1\,2} - x_e^{i\,2} + x_e^{1\,2} - x_n^{i\,2} + x_n^{1\,2} \\ & \quad = -2x_{e,a}\left(x_e^i - x_e^1\right) - -2x_{n,a}\left(x_n^i - x_n^1\right) \end{aligned} \tag{4.45}$$

Regarding the linear system corresponding to $\mathbf{x}_a$, the equation can be rewritten as

$$h_{e,a}^i x_{e,a} + h_{n,a}^i x_{n,a} = y_a^i \tag{4.46}$$

where

$$\begin{aligned} h_{e,a}^{i} &= -2\left(x_e^i - x_e^1\right) \\ h_{n,a}^{i} &= -2\left(x_n^i - x_n^1\right) \\ y_a^i = d_a^{i\,2} - d_a^{1\,2} &- x_e^{i\,2} + x_e^{1\,2} - x_n^{i\,2} + x_n^{1\,2} \end{aligned} \tag{4.47}$$

By incorporating all the anchors, the linear system of the agent's position can be written into a matrix form, by

$$\underbrace{\begin{bmatrix} h_{e,a}^2 & h_{n,a}^2 \\ h_{e,a}^3 & h_{n,a}^i \\ \vdots & \vdots \\ h_{e,a}^I & h_{n,a}^I \end{bmatrix}}_{\mathbf{H}_a} \begin{bmatrix} x_{e,a} \\ x_{n,a} \end{bmatrix} = \underbrace{\begin{bmatrix} y_a^2 \\ y_a^3 \\ \vdots \\ y_a^I \end{bmatrix}}_{\mathbf{y}_a} \tag{4.48}$$

This system can be solved using the LS estimation, as

$$\mathbf{x}_a = \left(\mathbf{H}_a^{\mathrm{T}}\mathbf{H}_a\right)^{-1}\mathbf{H}_a^{\mathrm{T}}\mathbf{y}_a \tag{4.49}$$

where $\mathbf{H}_a$ and $\mathbf{y}_a$ are the matrix and vector incorporating the terms from each anchor from index 2 to I by (4.48). The mobile device's location can be directly obtained by from the LS solution.

As can be seen, Fang's algorithm provides an efficient closed-form solution for TOA-based positioning by transforming nonlinear range equations into a linear system and solving using LS. This approach eliminates the need for iterative methods, making it computationally efficient and numerically stable. Unlike LS algorithms, Fang's method avoids convergence issues and is well-suited for real-time positioning applications in Wi-Fi FTM, UWB, and cellular TOA-based localization. However, Fang's algorithm has limitations. It is sensitive to measurement noise, particularly in multipath environments where TOA estimates may be biased. Additionally, at least three anchors are required for a valid 2-D solution, and poor anchor geometry can lead to GDOP, amplifying positioning errors. Although the LS solution is computationally fast, it does not inherently incorporate error modeling or filtering, making it less robust to noisy or biased TOA measurements compared to statistical estimation techniques such as Kalman filters. Overall, Fang's algorithm strikes a balance between computational simplicity and accuracy, making it a practical choice when computational efficiency is a priority. However, in high-noise environments, further refinement or hybrid approaches may be needed to enhance robustness.

4.3.5 Chan's Algorithms

Chan's algorithm is a well-known closed-form solution for source localization using TDOA measurements. It provides a noniterative method (via an auxiliary variable approach) to solve the hyperbolic location problem and is valued for its accuracy and efficiency [51]. Unlike earlier hyperbolic positioning methods (e.g., Fang's algorithm), Chan's method can incorporate redundant measurements from extra base stations, improving accuracy by using a two-step WLS approach.

For simplicity, we consider a 2-D plane with an unknown source (or mobile agent) located at $(x_{e,a}, x_{n,a})$. There are I known base stations (anchors) at fixed positions (x_e^i, x_n^i) for $i = 1, 2, \cdots, I$. One of these (e.g., the first anchor) is designated as the reference station. The goal of TDOA-based positioning is to estimate the unknown position of the agent using measured TDOAs between the reference station and the other stations.

The distance between the agent and each station i has the same form as (4.43), in which the noise is assumed to be negligible during estimation. In particular, the reference range for station 1 can be written as

$$d_a^1 = \sqrt{\left(x_{e,a} - x_e^1\right)^2 + \left(x_{n,a} - x_n^1\right)^2} \tag{4.50}$$

where the unknown terms can be used to form the state vector for estimation, as

$$\mathbf{x}_a = \begin{bmatrix} x_{e,a} \\ x_{n,a} \\ d_a^1 \end{bmatrix} \tag{4.51}$$

Here, the distance to the reference station, d_a^1, will be treated as an additional variable to be estimated during positioning.

The TDOA measurement between station i and the reference station (indexed by 1 here):

$$\Delta t^i = t^i - t^1 \tag{4.52}$$

where t^i is the signal arrival time at station i, and t^1 is the arrival time at the reference station. Δt^i is the observed difference in travel time from the source to station i and to station 1. It relates to the difference in distances by the signal propagation speed (e.g., the speed of radio, c), as follows:

$$c\Delta t^i = d_a^i - d_a^1 \tag{4.53}$$

Thus, the distance from the source to each station i can be written as

$$d_a^i = d_a^1 + c\Delta t^i \tag{4.54}$$

Geometrically, this equation represents a hyperbola: the locus of points for which the range difference, $d_a^i - d_a^1$, is constant. Each pair (i, 1) defines a hyperbolic curve on which the source must lie. With multiple stations, the source is at the intersection of multiple hyperbolae (similar to Figure 4.4).

The range difference equation (4.54) can be expanded in an explicit form as

$$\sqrt{\left(x_{e,a} - x_e^i\right)^2 + \left(x_{n,a} - x_n^i\right)^2} = \sqrt{\left(x_{e,a} - x_e^1\right)^2 + \left(x_{n,a} - x_n^1\right)^2} + c\Delta t^i \tag{4.55}$$

Squaring both sides and rearranging the equation yield

$$\begin{aligned} &\left(x_{e,a} - x_e^i\right)^2 + \left(x_{n,a} - x_n^i\right)^2 - \left(x_{e,a} - x_e^1\right)^2 - \left(x_{n,a} - x_n^1\right)^2 \\ &= 2c\Delta t^i \sqrt{\left(x_{e,a} - x_e^1\right)^2 + \left(x_{n,a} - x_n^1\right)^2} + \left(c\Delta t^i\right)^2 \end{aligned} \tag{4.56}$$

By substituting the square root term by (4.50), this becomes

$$\begin{aligned} &\left(x_{e,a} - x_e^i\right)^2 - \left(x_{e,a} - x_e^1\right)^2 + \left(x_{n,a} - x_n^i\right)^2 - \left(x_{n,a} - x_n^1\right)^2 \\ &= 2d_a^1 c\Delta t^i + \left(c\Delta t^i\right)^2 \end{aligned} \tag{4.57}$$

- Step 1: Solve a Linearized LS Problem

In this step, we assume that d_a^1 is an unknown constant and solve for the source position using a linearized LS approach. This step ignores the measurement error correlation (treating it as an LS problem) and, in the minimal case, provides a closed-form solution.

We rewrite the TDOA measurement equations in a linear form, introducing d_a^1 as an auxiliary unknown, as follows

$$\begin{aligned} &2\left(x_e^1 - x_e^i\right)x_{e,a} + 2\left(x_n^1 - x_n^i\right)x_{n,a} - 2d_a^1 c\Delta t^i \\ &= \left(c\Delta t^i\right)^2 - \left[\left(x_e^{i\,2} + x_n^{i\,2}\right) - \left(x_e^{1\,2} + x_n^{1\,2}\right)\right] \end{aligned} \tag{4.58}$$

Incorporating the TDOA measurement equations of all the stations (index from 2 to I), the system equation can be written in a matrix form, as

$$\underbrace{\begin{bmatrix} h_{e,a}^2 & h_{n,a}^2 & 2c\Delta t^2 \\ h_{e,a}^3 & h_{n,a}^3 & 2c\Delta t^3 \\ \vdots & \vdots & \vdots \\ h_{e,a}^I & h_{n,a}^I & 2c\Delta t^I \end{bmatrix}}_{\mathbf{H}_a} \begin{bmatrix} x_{e,a} \\ x_{n,a} \\ d_a^1 \end{bmatrix} = \underbrace{\begin{bmatrix} y_a^2 \\ y_a^3 \\ \vdots \\ y_a^I \end{bmatrix}}_{\mathbf{y}_a} \tag{4.59}$$

where

$$\begin{aligned} h_{e,a}^i &= 2\left(x_e^1 - x_e^i\right) \\ h_{n,a}^i &= 2\left(x_n^1 - x_n^i\right) \\ y_a^i = \left(c\Delta t^i\right)^2 &- \left[\left({x_e^i}^2 + {x_n^i}^2\right) - \left({x_e^1}^2 + {x_n^1}^2\right)\right] \end{aligned} \tag{4.60}$$

This forms the basis for Chan's estimation approach. If $I \geq 4$, the system is overdetermined, and the LS estimate is obtained by the same approach as (4.49). This provides an initial estimate of the state $\mathbf{x}_a$ for the source position.

- Step 2: WLS Refinement

The first step does not account for the error covariance of TDOA measurements, which are correlated due to the common reference station. The second step refines the estimate using a WLS solution, which incorporates the covariance matrix $\mathbf{\Sigma}$ of measurement errors. The WLS estimate is given by

$$\mathbf{x}_a = \left(\mathbf{H}_a^{\mathrm{T}} \mathbf{\Sigma}^{-1} \mathbf{H}_a\right)^{-1} \mathbf{H}_a^{\mathrm{T}} \mathbf{\Sigma}^{-1} \mathbf{y}_a \tag{4.61}$$

where $\mathbf{\Sigma}$ is the error covariance matrix, given by

$$\mathbf{\Sigma}_a = \begin{bmatrix} \sigma_{2,1}^2 & \sigma_1^2 & \cdots & \sigma_1^2 \\ \sigma_1^2 & \sigma_{3,1}^2 & \cdots & \sigma_1^2 \\ \vdots & \vdots & \ddots & \vdots \\ \sigma_1^2 & \sigma_1^2 & \cdots & \sigma_{I,1}^2 \end{bmatrix} \tag{4.62}$$

where $\sigma_{i,1}^2 = \sigma_i^2 + \sigma_1^2$ represents the variance of each range difference measurement and σ_1^2 represents the error contribution from the reference station. By applying the WLS estimate, the final position estimate is obtained, which is more statistically efficient and approaches the maximum likelihood solution for Gaussian noise.

It is worth noting that the quadratic constraint enforcement can yield two symmetric intersection solutions in some cases (geometrically, two possible mirror positions relative to the anchor configuration). Additional information or constraints (such as an initial guess, known bounds, or a third measurement in the 2-D case) can help to identify the correct one. In Chan's original formulation with the minimum three base stations, the ambiguity is resolved by selecting the positive root for d_a^1 (distance must be positive, and typically the smaller root corresponds to the feasible solution within the deployment area) [52]. When more than three stations are used, the LS step tends to bias the solution toward the correct position given all the measurements, and the quadratic correction step then fine-tunes the range consistency.

Chan's algorithm, originally developed for one-way TOA measurements, can be effectively applied to two-way TOA-based localization, such as RTT and Wi-Fi FTM. In these methods, the mobile agent measures the RTT between itself and multiple anchors, deriving the one-way TOA by dividing by two. Mathematically, two-way TOA still provides range estimates to known anchors, preserving the core structure of the TOA equations used in Chan's algorithm. The algorithm's key steps, linearizing the range equations, applying LS estimation, and refining the solution using WLS, remain unchanged. This makes it a practical, closed-form approach that avoids the convergence issues of nonlinear LS methods. Chan's algorithm is a robust choice for two-way TOA positioning, offering a balance between computational efficiency and accuracy. When combined with noise mitigation techniques or filtering methods, it can serve as a reliable foundation for RTT-based indoor localization in Wi-Fi, UWB, and other RF systems.

4.4 Angle-Based Indoor Positioning

Comparing with TOA and TDOA, the angle-based methods are much less relying on the time synchronization between the beacons and the agent. The required accuracy of time synchronization is similar to the required output rate of the positioning solutions. In addition, the angle measurement can directly be used to determine the location of the agent. The angle-based measurement focused is the AOA from the beacons to the agent while the equations derived below can also be applied to the methods for the AOD.

AOA-based positioning methods leverage bearing measurements to estimate an agent's location. Among the various techniques used, LS estimation (LSE), the orthogonal vector estimator (OVE), and the pseudolinear estimator (PLE) are three prominent approaches, each with distinct mathematical formulations, computational requirements, and performance characteristics.

LSE is an iterative method that refines an initial position estimate by minimizing the squared residuals between measured and expected AOA values [53]. It provides the highest accuracy when measurement noise is Gaussian and can incorporate WLS for improved estimation when noise variances are known. However, LSE requires an initial guess and has higher computational complexity due to its iterative nature. It is particularly suitable for high-precision localization applications where accuracy is prioritized over computation speed.

OVE introduces a vector-based formulation, solving for the position in a closed-form manner without requiring iterations [54]. By minimizing errors perpendicular to bearing lines, it offers higher accuracy than PLE while maintaining reasonable computational efficiency. However, it remains sensitive to poor geometric configurations and noise. OVE is well-suited for applications that require a balance between computational efficiency and accuracy, such as sensor fusion and autonomous navigation.

PLE, in contrast, is a linearized LS estimator that approximates the AOA equations for direct computation [55]. It is computationally the fastest among the three methods, making it ideal for real-time positioning and low-power embedded systems. However, PLE introduces bias, especially under high noise conditions or poor sensor geometries. It is often used for initial position estimation or scenarios where computational efficiency is critical.

The following sections provide the detailed mathematical formulations of these methods, starting from LSE, followed by the OVE, and concluding with the PLE. Their comparison is given in Table 4.6.

4.4.1 Model and Estimation: IWLS Estimation

In a 3-D positioning scenario, an AOA measurement provides two angular components, the elevation angle θ^i_{ϖ} which measures the vertical angle (i.e., inclination) between the LOS vector (to the agent) and the horizontal plane (North-East plane), and the azimuth angle ψ^i_{ϖ} which measures the horizontal bearing relative to a reference direction (e.g., North in the later explanations), as shown in Figure 4.5.

The AOA measurement models establish a trigonometric relationship between the agent's position and the beacon's position. These can be expressed as:

$$\sin\theta^i_a = \frac{x^i_u - x_{u,a}}{\sqrt{\left(x^i_e - x_{e,a}\right)^2 + \left(x^i_n - x_{n,a}\right)^2 + \left(x^i_u - x_{u,a}\right)^2}} = f^i\left(\mathbf{x}_a\right) \quad (4.63)$$

Table 4.6
Comparison for the Models and Estimation Algorithms Used for AOA-Based Positioning

Criterion	LSE	OVE	PLE
Type of solution	Iterative solution (requires refinement)	Closed-form (direct algebraic)	Closed-form (linearized LS)
Computational complexity	High (iterative matrix inversion, convergence required)	Moderate (matrix inversion required, but no iteration)	Low (simplified LS, minimal matrix operations)
Handling of nonlinearity	Solves nonlinear equations iteratively via Taylor series expansion	Uses vector-based formulation to preserve geometric relationships	Linearizes nonlinear AOA equations, introducing approximation errors
Robustness to noise	High (converges to best estimate under Gaussian noise)	Moderate (perpendicular fitting mitigates some noise effects)	Low (bias increases with noise, sensitive to outliers)
Accuracy	High (optimal under Gaussian noise)	Moderate to high (better than PLE, but still affected by bias)	Moderate (biased under high noise and poor geometry)
Initial guess required?	Yes (convergence depends on a good initial position)	No (direct computation)	No (direct computation)
Sensitivity to geometric configuration	High (convergence may fail if initial guess is poor or beacons are aligned)	Moderate (less sensitive than LSE but still geometry-dependent)	High (poor geometry causes large errors due to linearization effects)
Solution stability	Moderate (depends on iterative convergence, may be unstable for poor initial guesses)	High (closed-form and stable, but error grows with noise)	High (fast and repeatable, but biased)
Application scenarios	High-accuracy applications requiring precise positioning (e.g., indoor navigation, UWB localization)	Applications needing balance between computational efficiency and accuracy (e.g., sensor fusion, autonomous vehicles)	Real-time, low-computation environments where a fast but approximate solution is acceptable (e.g., initial positioning, quick AOA fixes)

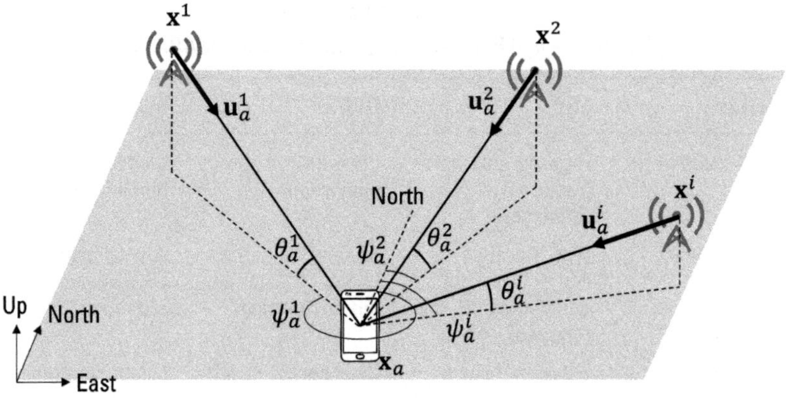

Figure 4.5 The measurements of AOA across the beacons.

$$\tan\psi_a^i = \frac{x_e^i - x_{e,a}}{x_n^i - x_{n,a}} = g^i\left(\mathbf{x}_a\right) \tag{4.64}$$

The measurement vector of each beacon $i = 1, \cdots, I$ can be formulated as

$$\mathbf{z}_a^i = \begin{bmatrix} \sin\theta_a^i \\ \tan\psi_a^i \end{bmatrix} \tag{4.65}$$

Incorporating the measurements from I beacons, the AOA measurement vector is defined as

$$\mathbf{z}_a = \begin{bmatrix} \mathbf{z}_a^1 \\ \vdots \\ \mathbf{z}_a^I \end{bmatrix} \tag{4.66}$$

Same as the direct range-based method, the state vector is defined in the form of (4.35). With the AOA measurements from multiple beacons, the agent position can be estimated by solving the geometry equations (4.63) and (4.64) for all I beacons. By applying the first-order Taylor series on (4.63), the trigonometric relationship can be rewritten as

$$\begin{aligned} f^i\left(\mathbf{x}_a\right) \approx\ & f^i\left(\mathbf{x}_{a(0)}\right) + \frac{\partial f^i\left(\mathbf{x}_{a(0)}\right)}{\partial x_{e,a}}\delta x_{e,a} + \frac{\partial f^i\left(\mathbf{x}_{a(0)}\right)}{\partial x_{n,a}}\delta x_{n,a} \\ & + \frac{\partial f^i\left(\mathbf{x}_{a(0)}\right)}{\partial x_{u,a}}\delta x_{u,a} \end{aligned} \tag{4.67}$$

where $\mathbf{X}_{a(0)}$ with the coordinate $(x_{e,a(0)}, x_{n,a(0)}, x_{u,a(0)})$ is the initial estimate of the agent's position, $\delta x_{e,a}, \delta x_{n,a}, \delta x_{u,a}$ are increments from the initial position to the next estimated agent position, the first-order partial derivative terms can be calculated as:

$$\frac{\partial f^i\left(\mathbf{x}_{a(0)}\right)}{\partial x_{e,a}} = \frac{\left(x_u^i - x_{u,a(0)}\right)\left(x_e^i - x_{e,a(0)}\right)}{\left[\left(x_e^i - x_{e,a(0)}\right)^2 + \left(x_n^i - x_{n,a(0)}\right)^2 + \left(x_u^i - x_{u,a(0)}\right)^2\right]^{3/2}} \tag{4.68}$$

$$\frac{\partial f^i\left(\mathbf{x}_{a(0)}\right)}{\partial x_{n,a}} = \frac{\left(x_u^i - x_{u,a(0)}\right)\left(x_n^i - x_{n,a(0)}\right)}{\left[\left(x_e^i - x_{e,a(0)}\right)^2 + \left(x_n^i - x_{n,a(0)}\right)^2 + \left(x_u^i - x_{u,a(0)}\right)^2\right]^{3/2}} \tag{4.69}$$

$$\frac{\partial f^i\left(\mathbf{x}_{a(0)}\right)}{\partial x_{u,a}} = \frac{-\left(x_e^i - x_{e,a(0)}\right)^2 - \left(x_n^i - x_{n,a(0)}\right)^2}{\left[\left(x_e^i - x_{e,a(0)}\right)^2 + \left(x_n^i - x_{n,a(0)}\right)^2 + \left(x_u^i - x_{u,a(0)}\right)^2\right]^{3/2}} \tag{4.70}$$

Similarly, the trigonometric relationship of the azimuth angle and the agent position can be applied with the first-order Taylor series expansion, in which (4.64) can be approximated by

$$g^i\left(\mathbf{x}_a\right) \approx g^i\left(\mathbf{x}_{a(0)}\right) + \frac{\partial g^i\left(\mathbf{x}_{a(0)}\right)}{\partial x_{e,a}}\partial x_{e,a} + \frac{\partial g^i\left(\mathbf{x}_{a(0)}\right)}{\partial x_{n,a}}\partial x_{n,a} + \frac{\partial g^i\left(\mathbf{x}_{a(0)}\right)}{\partial x_{u,a}}\partial x_{u,a} \tag{4.71}$$

$$\frac{\partial g^i\left(\mathbf{x}_{a(0)}\right)}{\partial x_{e,a}} = \frac{-1}{x_e^i - x_{e,a(0)}} \tag{4.72}$$

$$\frac{\partial g^i\left(\mathbf{x}_{a(0)}\right)}{\partial x_{n,a}} = \frac{x_e^i - x_{e,a(0)}}{\left(x_n^i - x_{n,a(0)}\right)^2} \tag{4.73}$$

$$\frac{\partial g^i\left(\mathbf{x}_{a(0)}\right)}{\partial x_{u,a}} = 0 \tag{4.74}$$

Based on the measurement vector and the state vector, the above relationships with I beacons can be represented as:

$$\mathbf{z}_a - \mathbf{z}_{a(0)} = \mathbf{H}_{a(0)} = \delta\mathbf{x}_a \tag{4.75}$$

$$\mathbf{H}_{a(0)} = \begin{bmatrix} \frac{\partial f^1(\mathbf{x}_{a(0)})}{\partial x_{e,a}} & \frac{\partial f^1(\mathbf{x}_{a(0)})}{\partial x_{n,a}} & \frac{\partial f^1(\mathbf{x}_{a(0)})}{\partial x_{u,a}} \\ \frac{\partial g^1(\mathbf{x}_{a(0)})}{\partial x_{e,a}} & \frac{\partial g^1(\mathbf{x}_{a(0)})}{\partial x_{n,a}} & 0 \\ \vdots & \vdots & \vdots \\ \frac{\partial f^1(\mathbf{x}_{a(0)})}{\partial x_{e,a}} & \frac{\partial f^1(\mathbf{x}_{a(0)})}{\partial x_{n,a}} & \frac{\partial f^1(\mathbf{x}_{a(0)})}{\partial x_{u,a}} \\ \frac{\partial g^1(\mathbf{x}_{a(0)})}{\partial x_{e,a}} & \frac{\partial g^1(\mathbf{x}_{a(0)})}{\partial x_{n,a}} & 0 \end{bmatrix} \tag{4.76}$$

Note that $\mathbf{z}_{a(0)}$ consisting of $\mathbf{z}^i_{a(0)} = [f^i(\mathbf{x}_{a(0)}) \;\; g^i(\mathbf{x}_{a(0)})]^{\mathrm{T}}$ is the approximate observation vector calculated based on the initial agent position $\mathbf{x}_{a(0)}$ with (4.63) and (4.64). To this step, we formed the linearized model between the AOA measurements and agent's state based on an initial guess. If the number of beacons is more than 3, then the estimation becomes an overdetermined problem. As mentioned earlier in this section, the noises of the measurements are assumed Gaussian random noise. Thus, the LSE can be achieved by

$$\delta\mathbf{x}_a = \left(\mathbf{H}_{a(0)}{}^{\mathrm{T}}\mathbf{H}_{a(0)}\right)^{-1}\mathbf{H}_{a(0)}{}^{\mathrm{T}}\delta\mathbf{z}_a \tag{4.77}$$

If the noise standard deviations are available, the WLS estimation can be conducted to achieve a MLE.

$$\delta\mathbf{x}_a = \left(\mathbf{H}_{a(0)}{}^{\mathrm{T}}\mathbf{W}_a\mathbf{H}_{a(0)}\right)^{-1}\mathbf{H}_{a(0)}{}^{\mathrm{T}}\mathbf{W}_a\delta\mathbf{z}_a$$

$$\mathbf{W}_a = \begin{bmatrix} \left(\sigma^2_{f^1(\mathbf{x}_{a(0)})}\right)^{-1} & 0 & \cdots & 0 & 0 \\ 0 & \left(\sigma^2_{g^1(\mathbf{x}_{a(0)})}\right)^{-1} & \cdots & 0 & 0 \\ \vdots & \vdots & \ddots & \vdots & \vdots \\ 0 & 0 & \cdots & \left(\sigma^2_{f^1(\mathbf{x}_{a(0)})}\right)^{-1} & 0 \\ 0 & 0 & \cdots & 0 & \left(\sigma^2_{g^1(\mathbf{x}_{a(0)})}\right)^{-1} \end{bmatrix} \tag{4.78}$$

where $\sigma^2_{f^1(\mathbf{x}_{a(0)})}$ and $\sigma^2_{g^1(\mathbf{x}_{a(0)})}$ representing the variance of the AOA measurements corresponding to the elevation angle and the azimuth angle, respectively.

However, the derived equation (4.64) becomes unstable when the azimuth angle ψ is 90° or 270°, resulting in singular points in the measurement model. This issue arises because the denominator in the azimuth angle equation approaches 0, leading to numerical instability. In practice, 90° AOA measurements are common. For example, if the agent is positioned directly East of the beacon, the azimuth angle is ψ = 90°, creating a singularity in the Jacobian matrix. To overcome this issue, spatial rotations in 3-D space can be parametrized using unit quaternions. Quaternions provide a 4-D representation of 3-D rotations, avoiding the singularities that occur in Euler angle-based formulations. The derivation of this approach can be found in [60], and the corresponding Jacobian matrix remains the same as in (4.76). It is important to note that the derivation of quaternion-based transformations can be mathematically complex. However, it effectively eliminates singularities and ensures stability in AOA-based positioning models.

4.4.2 Model and Estimation: OVE

The OVE is a closed-form solution for estimating an agent's position using AOA measurements from multiple beacons. Unlike LSE, which requires iterative refinement, OVE formulates the localization problem as a linear system and provides a direct algebraic solution.

OVE is called "orthogonal" because it minimizes perpendicular distance errors between the estimated agent location and the bearing vectors. The method assumes that the true agent position lies along the intersection of bearing lines, but due to noise, the actual intersection is imperfect. OVE computes the best-fit location by solving for the agent's position in an LS sense.

Let's start with the measurement model in OVE, it uses unit direction vector representation. Similar to LSE, each beacon i measures the elevation angles θ_a^i and the azimuth angles ψ_a^i of the agent's position. These angles define a unit direction vector from beacon i to the agent, as

$$\eta_a^i = \begin{bmatrix} \cos\theta_a^i \sin\psi_a^i \\ \cos\theta_a^i \cos\psi_a^i \\ \sin\theta_a^i \end{bmatrix} \quad (4.79)$$

The true agent position $\mathbf{x}_a$, as defined by (4.15) in the ENU coordinate, must lie along the bearing line defined by each beacon. This relationship can be expressed as:

$$\mathbf{x}_a = \mathbf{x}^i + \lambda_i \eta_a^i \tag{4.80}$$

where $\mathbf{x}^i$ is the known coordinate of the beacon at (x_e^i, x_n^i, x_u^i), and λ_i is an unknown scaling factor, representing the agent's distance from the beacon. It defines a ray in 3-D space. As two skew lines in 3-D do not intersect (only if in the same plane), measurements (i.e., line of ray) from at $I \geq 3$ beacons are needed to obtain an exact intersection indicating the agent's location, or forms an overdetermined system to be solved by LS.

To derive a linear system for positioning, a popular approach is multiplying a unit vector $\mathbf{s}_a^i$ orthogonal to η_a^i on both sides to eliminate the distance factor, giving

$$\mathbf{s}_a^{i\,\mathrm{T}} \mathbf{x}_a = \mathbf{s}_a^{i\,\mathrm{T}} \mathbf{x}^i \tag{4.81}$$

where $\mathbf{s}_a^i \cdot \eta_a^i = 0$ (orthogonal vectors), $\|\mathbf{s}_a^i\| = 1$ (unit vector), and the superscript T denotes the transpose of a vector. A straightforward candidate of η_a^i is to rotate $\mathbf{s}_a^i$ with 90° in terms of its related angular parameter, that is, increasing θ_a^i by $\pi/2$ gives

$$\mathbf{s}_a^i = \begin{bmatrix} \cos\left(\theta_a^i + \dfrac{\pi}{2}\right)\sin\psi_a^i \\ \cos\left(\theta_a^i + \dfrac{\pi}{2}\right)\cos\psi_a^i \\ \sin\left(\theta_a^i + \dfrac{\pi}{2}\right) \end{bmatrix} = \begin{bmatrix} -\sin\theta_a^i \sin\psi_a^i \\ -\sin\theta_a^i \cos\psi_a^i \\ \cos\theta_a^i \end{bmatrix} \tag{4.82}$$

Thus, for each beacon, the geometrical relationship from (4.81) can be rewritten as

$$\begin{aligned} &-\sin\theta_a^i \sin\psi_a^i x_{e,a} - \sin\theta_a^i \cos\psi_a^i x_{n,a} + \cos\theta_a^i x_{u,a} \\ &= -\sin\theta_a^i \sin\psi_a^i x_e^i - \sin\theta_a^i \cos\psi_a^i x_n^i + \cos\theta_a^i x_u^i \end{aligned} \tag{4.83}$$

$$\underbrace{\begin{bmatrix} -\sin\theta_a^1 \sin\psi_a^1 & -\sin\theta_a^1 \cos\psi_a^1 & \cos\theta_a^1 \\ -\sin\theta_a^2 \sin\psi_a^2 & -\sin\theta_a^2 \cos\psi_a^2 & \cos\theta_a^2 \\ \vdots & \vdots & \vdots \\ -\sin\theta_a^I \sin\psi_a^I & -\sin\theta_a^I \cos\psi_a^I & \cos\theta_a^I \end{bmatrix}}_{\mathbf{S}_a} \begin{bmatrix} x_{e,a} \\ x_{n,a} \\ x_{u,a} \end{bmatrix}$$

$$= \underbrace{\begin{bmatrix} -\sin\theta_a^1 \sin\psi_a^1 x_e^1 - \sin\theta_a^1 \cos\psi_a^1 x_n^1 + \cos\theta_a^1 x_u^1 \\ -\sin\theta_a^2 \sin\psi_a^2 x_e^2 - \sin\theta_a^2 \cos\psi_a^2 x_n^2 + \cos\theta_a^2 x_u^2 \\ \vdots \\ -\sin\theta_a^I \sin\psi_a^I x_e^I - \sin\theta_a^I \cos\psi_a^I x_n^I + \cos\theta_a^I x_u^I \end{bmatrix}}_{\mathbf{z}_a} \tag{4.84}$$

which forms a linear equation associating the AOA measurements and the location of the agent. By collecting the linear equations from all beacons, the geometrical relationship can be established in a matrix form, as follows.

Similar to the preceding overdetermined problem, the agent's position can be resolved by an LSE, using

$$\mathbf{x}_a = \left(\mathbf{S}_a^{\mathrm{T}} \mathbf{S}_a\right)^{-1} \mathbf{S}_a^{\mathrm{T}} \mathbf{z}_a \tag{4.85}$$

which provides the closed-form 3-D positioning solution with OVE. Note that OVE is derived by transforming the nonlinear relationship in a linear manner, which inherits a bias during the estimation. Specifically, because $\mathbf{S}_a$ depends on the AOA measurements (θ and ψ) with noises, it is also correlated with the noise in $\mathbf{z}_a$, making OVE a biased estimator [56]. This bias is generally related to the measurement noise covariance as well as the scale of the ranging measurement. For example, if the agent is very far from the beacons, small angular errors would lead to large position errors, and the linear approximation mislocates the intersection systematically.

4.4.3 Model and Estimation: 3-D PLE

Considering the 2-D (in the horizontal East-North plane) position estimation based on AOA measurements, the objective is to search for an agent position that minimizes the error compared to each beacon's AOA measurements according to the geometrical relationship in (4.80). A straightforward approach is the PLE, which establishes linear equations based on the geometrical relationship between angular measurements and the agent's location and solves the equations from all available beacons. More specifically, the geometrical relationship can be described in terms of East and North directions, respectively, as follows:

$$x_{e,a} - x_e^i = \lambda_i \sin\psi_a^i \tag{4.86}$$

$$x_{n,a} - x_n^i = \lambda_i \cos\psi_a^i \tag{4.87}$$

If $\cos\psi_a^i \neq 0$, the unknown distance λ_i in (4.87) can be substituted by the relationship in (4.86), yielding

$$x_{n,a} - x_n^i = \frac{x_{e,a} - x_e^i}{\sin\psi_a^i}\cos\psi_a^i \tag{4.88}$$

By multiplying $\sin\psi_a^i$ on both sides, the angular relationship can be rewritten as

$$\cos\psi_a^i x_{e,a} - \sin\psi_a^i x_{n,a} = \cos\psi_a^i x_e^i - \sin\psi_a^i x_n^i \tag{4.89}$$

which represents the line of bearing in Cartesian form. It is essentially the equation of a line passing through (x_e^i, x_n^i) with a normal vector $(-\cos\psi_a^i, \sin\psi_a^i)$. We can verify that this equation is satisfied by any point $(x_{e,a}, x_{e,a})$ that lies on the line of bearing. Thus, each AOA measurement provides one linear equation of the form (4.89) relating $x_{e,a}$ and $x_{n,a}$, which can be combined to form a linear system:

$$\underbrace{\begin{bmatrix} \cos\psi_a^1 & -\sin\psi_a^1 \\ \cos\psi_a^2 & -\sin\psi_a^2 \\ \vdots & \vdots \\ \cos\psi_a^I & -\sin\psi_a^I \end{bmatrix}}_{\mathbf{S}_a} \begin{bmatrix} x_{e,a} \\ x_{n,a} \end{bmatrix} = \underbrace{\begin{bmatrix} \cos\psi_a^1 x_e^1 - \sin\psi_a^1 x_n^1 \\ \cos\psi_a^2 x_e^2 - \sin\psi_a^2 x_n^2 \\ \vdots \\ \cos\psi_a^I x_e^I - \sin\psi_a^I x_n^I \end{bmatrix}}_{\mathbf{z}_a} \tag{4.90}$$

If $I = 2$ and the two lines are not parallel, this system has an exact solution (the intersection point). With $I > 2$, the system is overdetermined (more equations than unknowns), and an exact solution generally does not exist due to the noise causing inconsistency. In that case, we seek an LS solution that minimizes the sum of squared errors, which yields the positioning solution as

$$\begin{bmatrix} x_{e,a} \\ x_{n,a} \end{bmatrix} = \left(\mathbf{S}_a^{\mathrm{T}}\mathbf{S}_a\right)^{-1}\mathbf{S}_a^{\mathrm{T}}\mathbf{z}_a \tag{4.91}$$

When the localization is in 3-D, one straightforward extension of the PLE is to decouple the problem into horizontal and vertical components [57]. The idea is to first estimate the East-North coordinates using only the azimuth angles (ignoring elevation) and then estimate the Up coordinate using the elevation angles and the previously estimated horizontal position, namely the 3-D PLE.

For the horizontal position, by ignoring the elevation angle θ^i_{a}, each beacon provides an azimuth angle ψ^i_{a}, while the agent's position is projected at $(x_{e,a}, x_{n,a})$ on the horizontal plane. Thus, the 3-D estimation problem is reduced into 2-D, which can be solved by the preceding PLE using (4.90).

The second step is to estimate the Up coordinate based on the horizontal positioning solution and the elevation angle measurement θ^i_a from each beacon. Considering the geometry for a given beacon i, the elevation relates the difference in height to the horizontal distance from the beacon, which can be described by a tangent function:

$$\tan\theta^i_a = \frac{x^i_u - x_{u,a}}{\sqrt{\left(x^i_e - x_{e,a}\right)^2 + \left(x^i_n - x_{n,a}\right)^2}} \tag{4.92}$$

where the Up coordinate can be estimated by a specific beacon's measurement, using

$$x^i_{u,a} = x^i_u - \tan\theta^i_a\sqrt{\left(x^i_e - x_{e,a}\right)^2 + \left(x^i_n - x_{n,a}\right)^2} \tag{4.93}$$

or generally

$$x^i_{u,a} = x^i_u - \left\|\mathbf{x}_a(1:2) - \mathbf{x}^i(1:2)\right\|\tan\theta^i_a \tag{4.94}$$

where $x^i_{u,a}$ is the Up coordinate estimation from the measurement of beacon i, the bracket (1:2) denotes the vector only incorporating the first and second entries (East and North). Due to the measurement noise, the estimation results from different beacons may not be consistent. Thus, a sensible approach is to average these to reduce the effect of noise, using

$$x_{u,a} = \frac{1}{I}\sum_{i=1}^{I}\left(x^i_u - \left\|\mathbf{x}_a(1:2) - \mathbf{x}^i(1:2)\right\|\tan\theta^i_a\right) \tag{4.95}$$

By combining $x_{u,a}$ with the preceding estimated $x_{e,a}$ and $x_{n,a}$ from the first step, the 3-D positioning solution for the agent can be obtained. In many cases, if the noise is not too large, $x^i_{u,a}$ will not vary drastically with i, so the average is reasonable. One could also incorporate a weighted average approach (e.g., giving more weight to measurements from closer receivers, which tend to have more accurate elevation information), if the measurement quality is known.

Note that the 2-D PLE solution could be significantly biased, especially in the case with large bearing noise or unfavorable geometry. The reason is that the linear equation (4.89) was derived by effectively treating ψ_a^i as exact when forming $\mathbf{S}_a$ and $\mathbf{z}_a$, thereby "pushing" the nonlinearity into the error term. In other words, we replaced the true nonlinear measurement equation related to ψ_a^i with an arctangent function, with a linear equation plus an approximation error. A certain bias is introduced depending on the bearing noise variance and the target-to-sensor geometry. Thus, the 3-D PLE solution also inherits the same bias issue as the 2-D PLE for the horizontal part, and the subsequent averaging for $x_{u,a}$ can introduce an additional error if the estimation of ($x_{e,a}$, $x_{n,a}$) is biased or if the geometry varies significantly among the measurements. Nonetheless, the PLE is attractive for its simplicity and often serves as an initial guess for more advanced iterative algorithms (such as Gauss-Newton or EKFs), but sacrifices accuracy for computational efficiency. While it is widely used in real-time applications, it suffers from bias and sensitivity to noise.

4.5 Challenges for Radio Signal-Based Indoor Positioning

Looking at (4.20), the factors affecting the accuracy of estimation can be separated into two groups, the qualities of the measurement vector $\mathbf{z}_a$ and the linearized measurement model matrix $\mathbf{H}_a$. The next two subsections discuss the quality of the measurement and the two subsections after that discuss the quality of the model.

4.5.1 Thermal Noise on the Oscillator

Accurate positioning using TDOA and AOA methods relies on precise time and phase measurements. Because oscillators serve as the time-keeping elements in wireless devices, their stability directly affects the accuracy of these positioning techniques. However, real-world oscillators are subject to thermal noise, leading to random fluctuations in their frequency and phase over time. These variations introduce timing uncertainty, which degrades the precision of TDOA measurements and, in some cases, impacts multidevice AOA estimations.

Before considering the effects of noise, oscillators undergo calibration to correct their initial frequency offset. This process aligns the oscillator's frequency to a reference standard, reducing systematic errors. However, calibration alone cannot eliminate random frequency fluctuations, which persist due to the fundamental physical limitations of the oscillator. Even a well-calibrated oscillator exhibits phase noise and frequency noise, which introduce small, unpredictable errors in the timing of received signals. These residual noise components are particularly problematic in TDOA systems, where even nanosecond-level timing errors translate to substantial localization inaccuracies.

In the context of RF-based positioning, phase noise refers to short-term random variations in the phase of an oscillator's output signal. It is commonly represented in the frequency domain as a power spectral density (PSD), denoted as $S_{c\phi}(f)$, which quantifies how the oscillator's phase stability degrades at different frequency offsets. These phase fluctuations manifest as timing jitter in the time domain, causing slight variations in when signal peaks and zero-crossings occur. Because TDOA positioning depends on precise timestamps, any timing jitter can lead to errors in distance estimation.

Similarly, frequency noise describes small, random deviations in the oscillator's output frequency over time. These deviations accumulate and cause the oscillator to drift slightly away from its nominal frequency. Frequency noise and phase noise are closely related—phase noise is essentially the integral of frequency noise over time, meaning that frequency instability leads to increased phase uncertainty as time progresses.

The random nature of these noise sources can be mathematically modeled using a state-space representation of oscillator drift. The time-offset error and its rate of change (drift) evolve over time according to the following process noise covariance matrix:

$$\mathbf{Q}_{\text{clock}} = \begin{bmatrix} S_{c\phi}\tau + \frac{1}{3}S_{cf}\tau^3 & \frac{1}{2}S_{cf}\tau^2 \\ \frac{1}{2}S_{cf}\tau^2 & S_{cf}\tau \end{bmatrix}, \quad \begin{cases} S_{c\phi} = c^2 \cdot \frac{h_0}{2} \\ S_{cf} = c^2 \cdot 2\pi^2 \cdot h_{-2} \end{cases} \tag{4.96}$$

where τ is the update interval, $S_{c\phi}$ is the oscillator phase noise PSD, and S_{cf} is the oscillator frequency noise PSD. The first and second rows are corresponding to the time offset and its drift, respectively. According to [58], the coefficients of the TCXO are $h_0 = 2 \times 10^{-19}$ and $h_{-2} = 2 \times 10^{-20}$. If manufacturer-provided oscillator noise specifications are available, these values should be substituted to ensure an accurate system model.

The presence of oscillator noise affects positioning accuracy in several ways:

1. *TDOA sensitivity to timing errors:* In TDOA systems, each receiver's clock must be synchronized to the nanosecond level. Any oscillator-induced timing jitter directly translates to range estimation errors. For

instance, a 1-ns timing uncertainty causes a 30-cm error in range calculations, as radio waves travel approximately 0.3m per nanosecond. If the oscillator exhibits significant phase noise, these errors accumulate over multiple measurements, reducing overall localization precision.

2. *AOA impact in multidevice systems:* While single-device AOA measurements (e.g., an antenna array on a single receiver) are not directly impacted by absolute oscillator drift, multireceiver AOA systems or TDOA-AOA hybrid systems rely on synchronized phase references across multiple nodes. In such cases, oscillator phase noise introduces drift between receivers, causing angle estimation errors that propagate into position calculations.
3. *Long-term drift and synchronization requirements:* Even if a system starts with well-calibrated oscillators, long-term drift can cause increasing timing discrepancies. High-precision systems must periodically resynchronize oscillators using external references, such as GPSDOs or common-view time transfer techniques. If synchronization updates are infrequent, the accumulated phase noise may exceed acceptable error thresholds.

To minimize the impact of oscillator noise on TDOA/AOA positioning, several mitigation techniques can be employed:

- *High-stability oscillators:* Using low-phase-noise oscillators such as oven-controlled crystal oscillators (OCXOs) or atomic clocks significantly reduces timing uncertainty. However, these components increase cost and power consumption.
- *Frequent synchronization:* Periodic clock synchronization (e.g., through over-the-air timing beacons) helps to correct drift before it degrades positioning accuracy.
- *Adaptive filtering:* Advanced filtering techniques, such as EKF, incorporate oscillator noise models to compensate for timing errors dynamically. These methods are particularly useful when oscillator noise characteristics are well understood and can be modeled accurately.
- *Dual-frequency compensation:* Some positioning systems use dual-frequency signals to estimate and correct oscillator drift in real time.

4.5.2 Multipath and NLOS Effects

TDOA and AOA rely on the assumption that signals travel in a straight-line, LOS path between the transmitter and the receiver. However, in realistic environments, obstacles such as walls, buildings, and furniture cause the transmitted signal to reflect, diffract, or scatter, resulting in multipath propagation. In some cases, the direct LOS path is completely obstructed, leading to NLOS reception, where only reflected or diffracted signals reach the receiver.

These propagation effects introduce significant errors in positioning calculations. As illustrated in Figure 4.6, multipath propagation causes multiple signal copies to arrive at the receiver at different times, distorting ranging and angle measurements. In NLOS scenarios, the absence of the direct path results in systematic biases in estimated distances and angles. These errors increase the uncertainty of position estimation and degrade the overall accuracy of TDOA and AOA-based localization systems.

This section discusses the challenges posed by multipath and NLOS propagation, followed by two common mitigation approaches: channel estimation and ray tracing-based modeling.

4.5.2.1 Impact on TDOA and AOA Positioning

TDOA-based positioning estimates a signal source's location by measuring the TDOA between multiple receivers. However, multipath and NLOS conditions disrupt these measurements in different ways:

- *Multipath effects:* When a signal travels via multiple paths (direct and reflected), the receiver might detect multiple arrival peaks in the received waveform. The TDOA algorithm may lock onto the wrong peak, leading to incorrect time differences and a shifted position estimate.

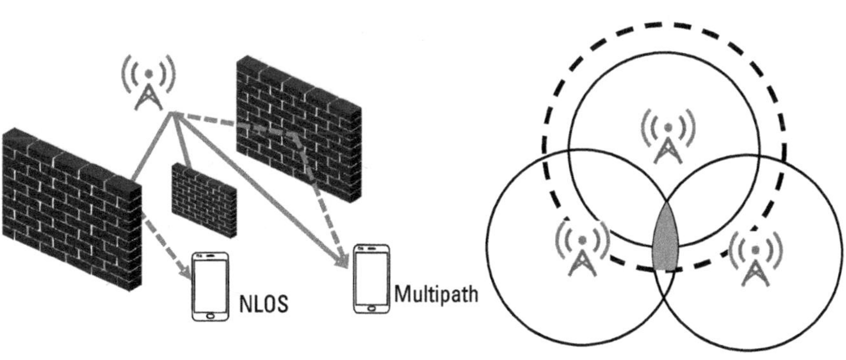

(a) (b)

Figure 4.6 (a) Multipath and NLOS. (b) How the reflection affects the positioning trilateration.

- *NLOS effects:* If the direct path is blocked, the receiver only detects reflected signals. Because these reflections travel a longer path than the true LOS signal, the measured arrival time is delayed, introducing a positive bias in the estimated distance. This leads to an overestimation of range differences, which distorts the computed hyperbolic position solution. For example, in an indoor environment, a TDOA-based system may estimate a transmitter's position to be tens or even hundreds of meters away from its actual location due to NLOS propagation errors.

AOA-based positioning systems estimate the transmitter's location by measuring the angle at which a signal arrives at multiple receivers equipped with antenna arrays. However, multipath and NLOS introduce errors in direction estimation:

- *Multipath effects:* When a receiver detects both direct and reflected signals, its AOA algorithm might be confused by multiple arriving directions. If the strongest received signal is a reflection, the system may incorrectly infer that the transmitter is located in the wrong direction.
- *NLOS effects:* In pure NLOS conditions, the direct path is completely blocked, and the receiver only detects reflected signals. Consequently, the estimated arrival angle corresponds to the direction of the reflecting surface rather than the true transmitter location.

In urban and indoor environments, AOA errors due to NLOS can result in severe mislocalization, as signals might appear to originate from walls or objects rather than the actual source.

4.5.2.2 Channel Estimation for Multipath Mitigation

One approach to mitigating multipath effects is channel estimation, which involves analyzing the CIR to separate direct and reflected signal components. This method is embedded within the receiver's signal processing algorithms. The idea is to estimate the "channels," which means the signal paths from a beacon to an agent's receiver. As a standard form of the received multipath signal, consider that one channel of the LOS path and the $M-1$ channel of multipaths can be written as

$$s(t) = \sum_{m=0}^{M-1} a_m e^{-j\phi_m} r(t - \tau_m) \tag{4.97}$$

where a_m and ϕ_m are the amplitude and phase of the mth signal propagation path, τ_m is the corresponding propagation delay, and $r(t)$ is the transmitted signal. By correlating the received signal s with the local signal replica of r, the time delay τ and the phase angle ϕ of the $M-1$ multipath channels can be estimated using

$$\left(\hat{\tau},\hat{\phi}\right) = \underset{\tau,\phi}{\arg\max}\ \text{correlation}\left(s,r\left(\tau,\phi\right)\right) \tag{4.98}$$

where the hat denotes the estimated variable. The first detected peak in the estimated channel impulse response typically corresponds to the LOS path (if present), while later peaks represent multipath components. Advanced methods such as the MLE and multipath estimating delay lock loop (MEDLL) refine these estimates to enhance accuracy. A review of various channel estimation techniques for multipath mitigation is available in [59]. Although channel estimation can help to identify early-arriving signal components (which are likely LOS), it does not always eliminate NLOS bias. If the direct path is completely blocked, the earliest detected signal will still be an indirect reflection, leading to unavoidable positioning errors.

4.5.3 Distribution of Beacons

A couple of the most frequently asked questions of setting up a beacon-based indoor positioning system are: Where should I place the beacons? Does the location of beacons affect the performance of my indoor positioning system? Intuitively, the answer is yes, but the real question is how to quantify this relationship. This section aims to answer this question. Considering the TOA and TDOA-based point positioning methods, both methods can form a state and measurement equation as follows:

$$\mathbf{x}_a = \left(\mathbf{H}_a^{\mathrm{T}}\mathbf{H}_a\right)^{-1}\mathbf{H}_a^{\mathrm{T}}\mathbf{z}_a \tag{4.99}$$

As shown in (4.99), the state $\mathbf{x}_a$ is estimated by two factors: (1) the quality of the measurements, $\mathbf{z}_a$ and (2) the LOS vectors from beacons relative to the agent a, $\mathbf{H}_a$ which assumes a linear relationship. To theoretically derive their quantitative impacts, the following analysis is used. If there is an random noise $\omega_{\mathbf{z}_a}$ in the measurement vector where each component follows a zero mean Gaussian distribution $\mathcal{N}(0, \sigma_{\mathbf{z}_a}^2)$, it will also result in a zero mean Gaussian random noise $\omega_{\mathbf{x}_a}$ in the state estimation, in which $\omega_{\mathbf{x}_a} = (\mathbf{H}_a^{\mathrm{T}}\mathbf{H}_a)^{-1}\mathbf{H}_a^{\mathrm{T}}\omega_{\mathbf{z}_a}$. Then the noise covariance of the state $\mathbf{C}(\mathbf{x}_a)$ is calculated as

$$\mathbf{C}\left(\mathbf{x}_a\right) = \mathbb{E}\left[\omega_{\mathbf{x}_a}\omega_{\mathbf{x}_a}^{\mathrm{T}}\right] \tag{4.100}$$

where $\mathbb{E}[\cdot]$ denotes the expectation calculation. To substitute the state estimation noise $\omega_{\mathbf{x}_a}$ with the linear relationship, we obtain,

$$\begin{aligned}\mathbf{C}(\mathbf{x}_a) &= \mathbb{E}\left[\mathbf{K}\omega_{\mathbf{z}_a}\left(\mathbf{K}\omega_{\mathbf{z}_a}\right)^{\mathrm{T}}\right]\\ &= \mathbb{E}\left[\left(\mathbf{H}_a{}^{\mathrm{T}}\mathbf{H}_a\right)^{-1}\mathbf{H}_a{}^{\mathrm{T}}\omega_{\mathbf{z}_a}\omega_{\mathbf{z}_a}{}^{\mathrm{T}}\mathbf{H}_a\left(\mathbf{H}_a{}^{\mathrm{T}}\mathbf{H}_a\right)^{-1}\right]\\ &= \left(\mathbf{H}_a{}^{\mathrm{T}}\mathbf{H}_a\right)^{-1}\mathbf{H}_a{}^{\mathrm{T}}\mathbf{C}(\mathbf{z}_a)\mathbf{H}_a\left(\mathbf{H}_a{}^{\mathrm{T}}\mathbf{H}_a\right)^{-1}\\ \text{where } \mathbf{K} &= \left(\mathbf{H}_a{}^{\mathrm{T}}\mathbf{H}_a\right)^{-1}\mathbf{H}_a{}^{\mathrm{T}}\end{aligned} \tag{4.101}$$

Assuming that $\mathbf{z}_a$ are i.i.d. (independent and identically distributed) random variables, the noise covariance of measurement vector $\mathbf{z}_a$ becomes

$$\mathbf{C}(\mathbf{z}_a) = \mathbf{I}\cdot\sigma_{\mathbf{z}_a}^2 \tag{4.102}$$

where $\mathbf{I}$ denotes identity matrix. Then (4.102) becomes

$$\mathbf{C}(\mathbf{x}_a) = \left(\mathbf{H}_a{}^{\mathrm{T}}\mathbf{H}_a\right)^{-1}\sigma_{\mathbf{z}_a}^2 \tag{4.103}$$

As can be seen, the noise covariance becomes only related to a standardized measurement noise $\sigma_{\mathbf{z}_a}^2$ and $(\mathbf{H}_a{}^{\mathrm{T}}\mathbf{H}_a)^{-1}$. For ranging-based measurements, general speaking, the state vector is $\mathbf{x}_a = [x_{e,a} \quad x_{n,a} \quad x_{u,a}]^T$. Recall that the $\mathbf{C}(\mathbf{x}_a)$ indicates the covariance between each pair of elements in the East, North, and up directions. Then the overall variance of the state estimation in the NED coordinate can be approximated by the sum of its variances in each directions, using

$$\begin{aligned}\sigma_{x_{\mathrm{enu},a}}^2 &= \sigma_{x_{e,a}}^2 + \sigma_{x_{n,a}}^2 + \sigma_{x_{u,a}}^2\\ \sigma_{x_{\mathrm{enu},a}} &= \sqrt{\sigma_{x_{e,a}}^2 + \sigma_{x_{n,a}}^2 + \sigma_{x_{u,a}}^2} = \sqrt{\mathrm{tr}\left(\mathbf{C}(\mathbf{x}_a)\right)}\end{aligned} \tag{4.104}$$

where $\mathrm{tr}(\cdot)$ denotes the trace of a square matrix that sums the elements on the main diagonal. Substituting (4.102) to (4.103),

$$\sigma_{x_{\mathrm{enu},a}} = \sqrt{\mathrm{tr}\left(\left(\mathbf{H}_a{}^{\mathrm{T}}\mathbf{H}_a\right)^{-1}\sigma_{\mathbf{z}_a}^2\right)} = \sqrt{\mathrm{tr}\left(\left(\mathbf{H}_a{}^{\mathrm{T}}\mathbf{H}_a\right)^{-1}\right)}\cdot\sigma_{\mathbf{z}_a} \tag{4.105}$$

For the simplicity, we describe the components in $(\mathbf{H}_a{}^{\mathrm{T}}\mathbf{H}_a)^{-1}$ as follows:

$$\left(\mathbf{H}_a^{\mathrm{T}}\mathbf{H}_a\right)^{-1} = \begin{bmatrix} \kappa_{x_e x_e} & \kappa_{x_e x_n} & \kappa_{x_e x_u} \\ \kappa_{x_n x_e} & \kappa_{x_n x_n} & \kappa_{x_n x_u} \\ \kappa_{x_u x_e} & \kappa_{x_u x_n} & \kappa_{x_u x_u} \end{bmatrix} \tag{4.106}$$

Then

$$\sigma_{x_{\mathrm{enu},a}} = \sqrt{\kappa_{x_e x_e} + \kappa_{x_n x_n} + \kappa_{x_u x_u}} \cdot \sigma_{\mathbf{z}_a} \tag{4.107}$$

Considering (4.107), if $\sqrt{\kappa_{x_e x_e} + \kappa_{x_n x_n} + \kappa_{x_u x_u}}$ equals κ, then 1m of standardized measurement noise will produce $\kappa \times$ 1m of the positioning noise. In the GNSS field, $\sqrt{\kappa_{x_e x_e} + \kappa_{x_n x_n} + \kappa_{x_u x_u}}$ is well-known as dilution of precision (DOP) as the DOP factor will dilute (transform the domain of) the measurement uncertainty into the positioning uncertainty This book also adopts the same logic and calls it DOP. Analogous to (4.104), the DOP can be divided based on the direction of the state concerned.

$$\begin{aligned} \sigma_{x_{\mathrm{horizontal},a}} &= \sqrt{\sigma^2_{x_{e,a}} + \sigma^2_{x_{n,a}}} = \sqrt{\kappa_{x_e x_e} + \kappa_{x_n x_n}} \cdot \sigma_{\mathbf{z}_a} = HDOP \cdot \sigma_{\mathbf{z}_a} \\ \sigma_{x_{\mathrm{vertical,a}}} &= \sqrt{\sigma^2_{x_{u,a}}} = \sqrt{\kappa_{x_u x_u}} \cdot \sigma_{\mathbf{z}_a} = VDOP \cdot \sigma_{\mathbf{z}_a} \end{aligned} \tag{4.108}$$

HDOP and *VDOP* are therefore used to indicate that 1m of measurement error will be diluted (mapped) to the positioning error in the horizontal and vertical directions, respectively. Thus, the smaller the *HDOP* and *VDOP* are, the less affected the positioning result will be by the measurement errors. The immediate next issue is how to make them smaller.

The DOP is calculated based on the measurement matrix $\mathbf{H}_a$. Taking TOA as an example, $\mathbf{H}_a = [\mathbf{h}_a^1 \ \cdots \ \mathbf{h}_a^i \ \cdots \ \mathbf{h}_a^I]^T$ and $\mathbf{h}_a^i$ denotes the unit LOS vector between the beacon i and the agent a. This indicates that the distribution of the beacons related to the agents is very related to the DOP. This idea can be easily understood by referring to the illustration given in Figure 4.7.

To echo back to the question mentioned in the beginning of this section, the simulation of placing the beacons at different locations can be achieved by using (4.108). Zwirello et al. [60] analyzed the change of HDOP and VDOP using different distributions of beacons as shown in Figure 4.8. As can be seen, the DOP values changes as the change of the agent receiver's location. In the other words, based on a predefined distribution of beacons, some areas could have larger DOP than others. Readers should aware this in designing a beacon-based indoor positioning system.

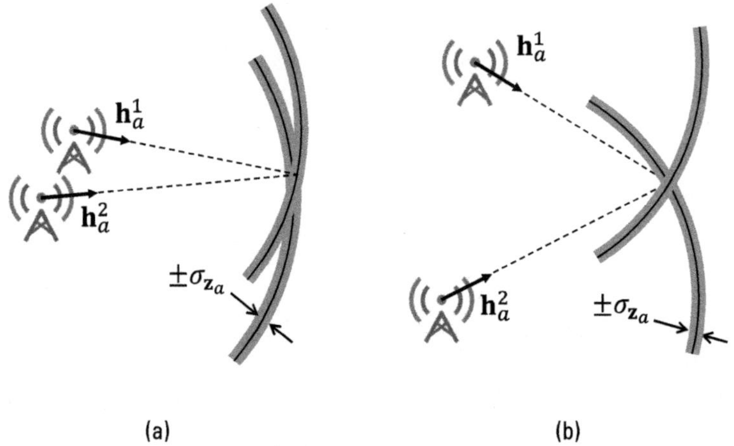

Figure 4.7 Beacon distributions with: (a) a higher DOP, and (b) a lower DOP.

The discussion in this section can apply to not only the TOA ranging and RSS ranging-based point positioning method but also to the TDOA-ranging and AOA-based methods by simply replacing the measurement matrix $\mathbf{H}_a$.

4.5.4 Nonlinearity Caused by an Initial Guess

In this chapter, we have assumed the iterative least squares (ILS) estimation can successfully solve the agent receiver's position based on a major assumption; the initial guess of the state, $\mathbf{x}_{a(0)}$, is sufficiently close to the actual position of the agent, $\mathbf{x}_{gt,a}$. This assumption is critical because ILS relies on a linearized measurement model, which only holds when the estimated position is near the true location. Figure 4.9 shows the importance of this assumption. As can be seen, if the initial guess is not accurate enough, the assumption $\mathbf{h}^i_{a(0)} \approx \mathbf{h}^i_{gt,a}$ cannot be made. If the distance between the initial guess and the ground truth location of the agent is relatively small compared to the distance between the beacons to the agent, the assumption can still hold. In the other words, the required accuracy of the initial guess to solve the point-positioning methods is related to the size of the indoor venue of a designed indoor positioning system.

However, the radio-signal based point positioning also plays a role to provide initial guess for some multisensor integrated indoor positioning systems. Thus, the stringent requirement on the initial guess may not be appropriate to those systems. An alternative solution is to consider higher-order terms when

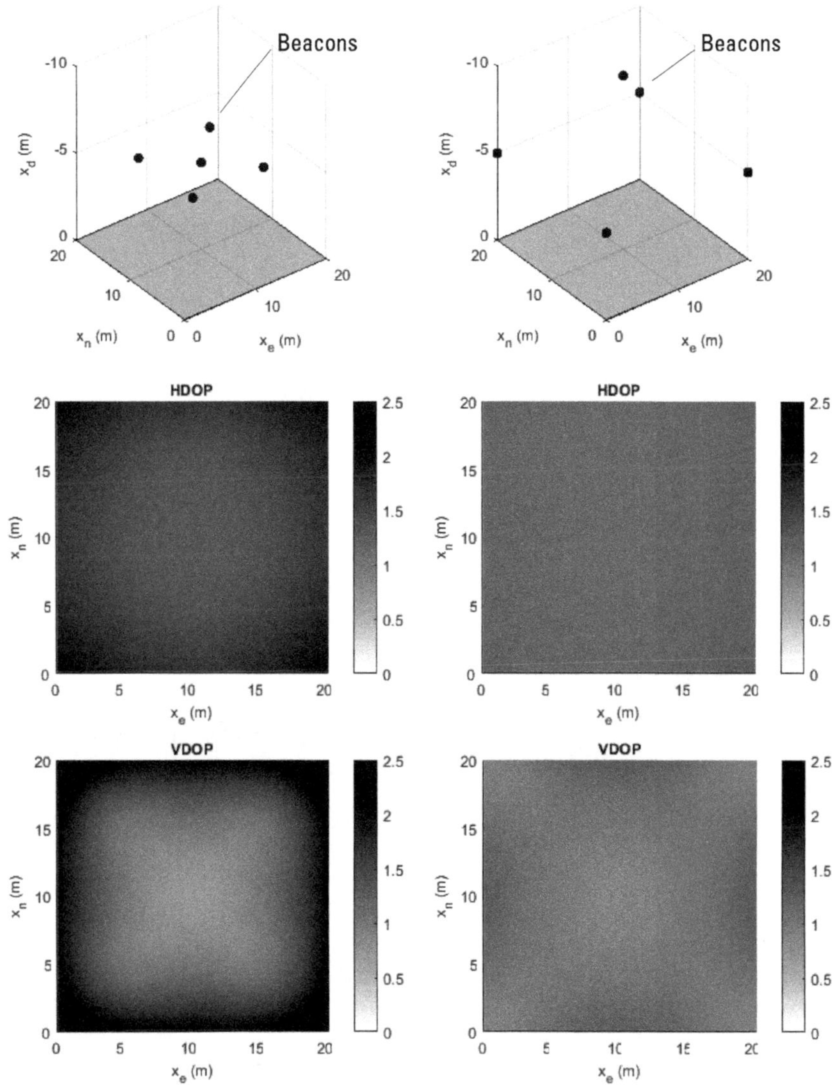

Figure 4.8 HDOPs and VDOPs under different beacon distributions [60].

linearizing the measurement models. The higher-order terms are usually used to describe the model with higher nonlinearity. As shown in (4.16) and (4.36), the TOA and TDOA measurement models are nonlinear. Taking the TOA one as an example, if we consider a second-order term in the linearization via Taylor series expansion, the linearized measurement model becomes:

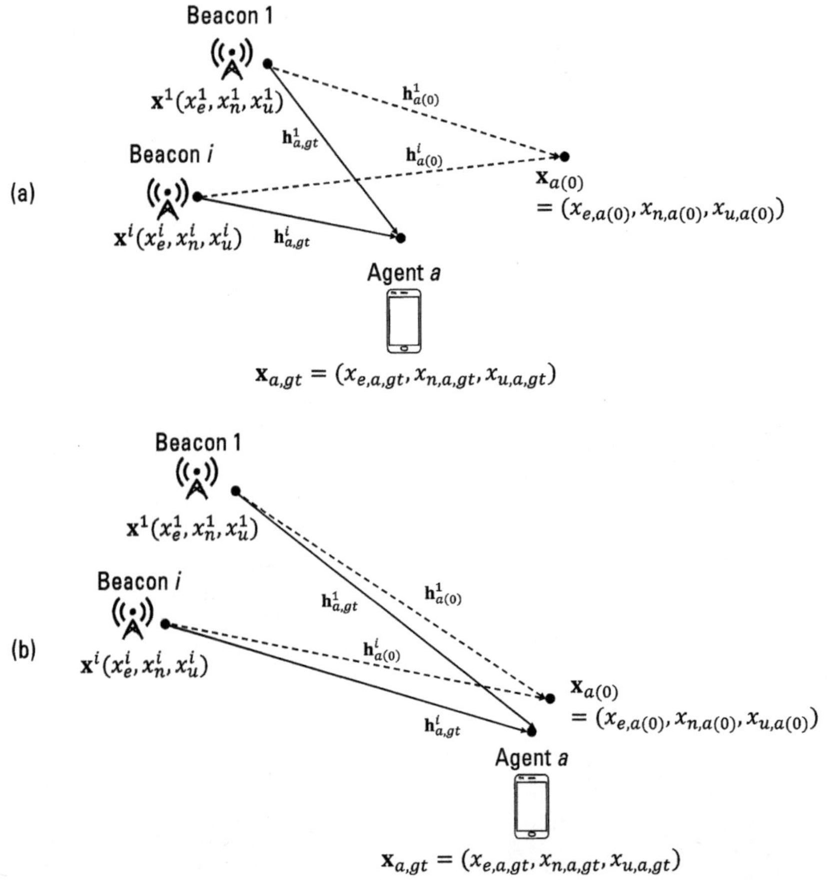

Figure 4.9 Illustrations of the accuracy of the initial guess used for the ILS method: (a) poor initial guess, and (b) accurate initial guess.

$$
\begin{aligned}
d_a^i(\mathbf{x}_a) \approx\ & d_a^i(\mathbf{x}_{a(0)}) + \frac{\partial d_a^i}{\partial x_{e,a}}\delta x_{e,a} + \frac{\partial d_a^i}{\partial x_{n,a}}\delta x_{n,a} + \frac{\partial d_a^i}{\partial x_{u,a}}\delta x_{u,a} \\
& + \frac{\partial^2 d_a^i}{\partial x_{e,a}^2}\delta x_{e,a}^2 + \frac{\partial^2 d_a^i}{\partial x_{n,a}^2}\delta x_{n,a}^2 + \frac{\partial^2 d_a^i}{\partial x_{u,a}^2}\delta x_{u,a}^2 \\
=\ & d_a^i(\mathbf{x}_{a(0)}) + \mathbf{h}_{a(0)}^i(\mathbf{x}_a - \mathbf{x}_{a(0)}) + \mathbf{g}_{a(0)}^i(\mathbf{x}_a - \mathbf{x}_{a(0)})^2
\end{aligned}
\tag{4.109}
$$

where $\mathbf{g}_{a(0)}^i$ denotes the second-order term of the linearized measurement model. Then the ILS can be conducted by:

$$d_a^i\left(\mathbf{x}_a\right)-d_a^i\left(\mathbf{x}_{a(0)}\right)=\mathbf{h}_{a(0)}^i\left(\mathbf{x}_a-\mathbf{x}_{a(0)}\right)+\mathbf{g}_{a(0)}^i\left(\mathbf{x}_a-\mathbf{x}_{a(0)}\right)^2$$
$$\delta\mathbf{z}_a=\mathbf{H}_{a(0)}\delta\mathbf{x}_a+\mathbf{G}_{a(0)}\delta\mathbf{x}_a^2 \quad (4.110)$$

Another direction to handle the issue of a poor initial guess is to apply the robust estimation such as an M-estimator and coarse-to-fine approach introduced in Chapter 2. By relaxing the nonconvexity (in the cost of a higher computational load in general) of the LS problem, the impact caused by the poor initial guess can be mitigated.

References

[1] Giovanelli, D., et al., "Bluetooth-Based Indoor Positioning Through ToF and RSSI Data Fusion," *2018 International Conference on Indoor Positioning and Indoor Navigation (IPIN)*, September 24–27, 2018, pp. 1–8.

[2] Hengyotmark, S., et al., "Pseudo-Ranging Based on Round-Trip Time of Bluetooth Low Energy Beacons," in *Recent Advances in Information and Communication Technology*, P. Meesad, S. Sodsee, and H. Unger, (eds.), New York: Springer International Publishing, 2018, pp. 202–211.

[3] Luo, H., et al., "Integration of GNSS and BLE Technology with Inertial Sensors for Real-Time Positioning in Urban Environments," *IEEE Access*, Vol. 9, 2021, pp. 15744–15763.

[4] Yu, N., et al., "A Precise Dead Reckoning Algorithm Based on Bluetooth and Multiple Sensors," *IEEE Internet of Things Journal*, Vol. 5, No. 1, 2018, pp. 336–351.

[5] Sheng, Z., and J. K. Pollard, "Position Measurement Using Bluetooth," *IEEE Transactions on Consumer Electronics*, Vol. 52, No. 2, 2006, pp. 555–558.

[6] Monfared, S., et al., "AoA-Based Iterative Positioning of IoT Sensors with Anchor Selection in NLOS Environments," *IEEE Transactions on Vehicular Technology*, Vol. 70, No. 6, 2021, pp. 6211–6216.

[7] Ye, F., et al., "A Low-Cost Single-Anchor Solution for Indoor Positioning Using BLE and Inertial Sensor Data," *IEEE Access*, Vol. 7, 2019, pp. 162439–162453.

[8] Gezici, S., et al., "Localization Via Ultra-Wideband Radios: A Look at Positioning Aspects for Future Sensor Networks," *IEEE Signal Processing Magazine*, Vol. 22, No. 4, 2005, pp. 70–84.

[9] Dardari, D., et al., "Ranging with Ultrawide Bandwidth Signals in Multipath Environments," *Proceedings of the IEEE*, Vol. 97, No. 2, 2009, pp. 404–426.

[10] Alavi, B., and K. Pahlavan, "Modeling of the TOA-Based Distance Measurement Error Using UWB Indoor Radio Measurements," *IEEE Communications Letters*, Vol. 10, No. 4, 2006, pp. 275–277.

[11] Alsindi, N. A., B. Alavi, and K. Pahlavan, "Measurement and Modeling of Ultrawideband TOA-Based Ranging in Indoor Multipath Environments," *IEEE Transactions on Vehicular Technology*, Vol. 58, No. 3, 2009, pp. 1046–1058.

[12] Taponecco, L., A. A. D. Amico, and U. Mengali, "Joint TOA and AOA Estimation for UWB Localization Applications," *IEEE Transactions on Wireless Communications*, Vol. 10, No. 7, 2011, pp. 2207–2217.

[13] Luo, Y., and C. L. Law, "Indoor Positioning Using UWB-IR Signals in the Presence of Dense Multipath with Path Overlapping," *IEEE Transactions on Wireless Communications*, Vol. 11, No. 10, 2012, pp. 3734–3743.

[14] Joon-Yong, L., and R. A. Scholtz, "Ranging in a Dense Multipath Environment Using an UWB Radio Link," *IEEE Journal on Selected Areas in Communications*, Vol. 20, No. 9, 2002, pp. 1677–1683.

[15] Adams, J. C., et al., "Ultra-Wideband for Navigation and Communications," *2001 IEEE Aerospace Conference Proceedings (Cat. No. 01TH8542)*, Vol. 2, March 10–17, 2001, pp. 2/785–2/792.

[16] Angelis, G. D., A. Moschitta, and P. Carbone, "Positioning Techniques in Indoor Environments Based on Stochastic Modeling of UWB Round-Trip-Time Measurements," *IEEE Transactions on Intelligent Transportation Systems*, Vol. 17, No. 8, 2016, pp. 2272–2281.

[17] Zhu, D., and K. Yi, "EKF Localization Based on TDOA/RSS in Underground Mines Using UWB Ranging," *2011 IEEE International Conference on Signal Processing, Communications and Computing (ICSPCC)*, September 14–16, 2011, pp. 1–4.

[18] Bocquet, M., C. Loyez, and A. Benlarbi-Delai, "Using Enhanced-TDOA Measurement for Indoor Positioning," *IEEE Microwave and Wireless Components Letters*, Vol. 15, No. 10, 2005, pp. 612–614.

[19] Bottigliero, S., et al., "A Low-Cost Indoor Real-Time Locating System Based on TDOA Estimation of UWB Pulse Sequences," *IEEE Transactions on Instrumentation and Measurement*, Vol. 70, 2021, pp. 1–11.

[20] Mahfouz, M. R., et al., "Investigation of High-Accuracy Indoor 3-D Positioning Using UWB Technology," *IEEE Transactions on Microwave Theory and Techniques*, Vol. 56, No. 6, 2008, pp. 1316–1330.

[21] Pierucci, L., and P. J. Roig, "UWB Localization on Indoor MIMO Channels," *2005 International Conference on Wireless Networks, Communications and Mobile Computing*, Vol. 2, June 13–16, 2005, pp. 890–894.

[22] Zhang, Y., et al., "High Resolution 3-D Angle of Arrival Determination for Indoor UWB Multipath Propagation," *IEEE Transactions on Wireless Communications*, Vol. 7, No. 8, 2008, pp. 3047–3055.

[23] Wen, F., and C. Liang, "Fine-Grained Indoor Localization Using Single Access Point with Multiple Antennas," *IEEE Sensors Journal*, Vol. 15, No. 3, 2015, pp. 1538–1544.

[24] Zhao, B., et al., "A Tensor-Based Joint AoA and ToF Estimation Method for Wi-Fi Systems," *IEEE Wireless Communications Letters*, Vol. 10, No. 11, 2021, pp. 2543–2546.

[25] Golden, S. A., and S. S. Bateman, "Sensor Measurements for Wi-Fi Location with Emphasis on Time-of-Arrival Ranging," *IEEE Transactions on Mobile Computing*, Vol. 6, No. 10, 2007, pp. 1185–1198.

[26] Ma, C., et al., "Wi-Fi RTT Ranging Performance Characterization and Positioning System Design," *IEEE Transactions on Mobile Computing*, Vol. 21, No. 2, 2022, pp. 740–756.

[27] Retscher, G., "Fundamental Concepts and Evolution of Wi-Fi User Localization: An Overview Based on Different Case Studies," *Sensors*, Vol. 20, No. 18, 2020, p. 5121, https://www.mdpi.com/1424-8220/20/18/5121.

[28] Diggelen, F. V., R. Want, and W. Wang, "How to Achieve 1-Meter Accuracy in Android," *GPS World*, July 3, 2018, https://www.gpsworld.com/how-to-achieve-1-meter-accuracy-in-android/.

[29] Kumarasiri, R., et al., "An Improved Hybrid RSS/TDOA Wireless Sensors Localization Technique Utilizing Wi-Fi Networks," *Mobile Networks and Applications*, Vol. 21, No. 2, 2016, pp. 286–295.

[30] Xie, T., et al., "A Wi-Fi-Based Wireless Indoor Position Sensing System with Multipath Interference Mitigation," *Sensors*, Vol. 19, No. 18, 2019, p. 3983, https://www.mdpi.com/1424-8220/19/18/3983.

[31] Dwivedi, S., et al., "Positioning in 5G Networks," *IEEE Communications Magazine*, Vol. 59, No. 11, 2021, pp. 38–44.

[32] Huang, J., J. Liang, and S. Luo, "Method and Analysis of TOA-Based Localization in 5G Ultra-Dense Networks with Randomly Distributed Nodes," *IEEE Access*, Vol. 7, 2019, pp. 174986–175002.

[33] Koivisto, M., et al., "Joint Device Positioning and Clock Synchronization in 5G Ultra-Dense Networks," *IEEE Transactions on Wireless Communications*, Vol. 16, No. 5, 2017, pp. 2866–2881.

[34] Peral-Rosado, J. A. D., et al., "Exploitation of 3D City Maps for Hybrid 5G RTT and GNSS Positioning Simulations," *ICASSP 2020—2020 IEEE International Conference on Acoustics, Speech and Signal Processing (ICASSP)*, May 4–8, 2020, pp. 9205–9209.

[35] Deng, Z., et al., "A TDOA and PDR Fusion Method for 5G Indoor Localization Based on Virtual Base Stations in Unknown Areas," *IEEE Access*, Vol. 8, 2020, pp. 225123–225133.

[36] Menta, E. Y., et al., "On the Performance of AoA–Based Localization in 5G Ultra–Dense Networks," *IEEE Access*, Vol. 7, 2019, pp. 33870–33880.

[37] "IEEE Standard for a Precision Clock Synchronization Protocol for Networked Measurement and Control Systems," *IEEE Std. 1588-2019 (Revision of IEEE Std 1588-2008)*, 2020, pp. 1–499.

[38] Lipinski, M., "The White Rabbit Project," February 19, 2024, https://white-rabbit.web.cern.ch/.

[39] Mills, D., et al., "Network Time Protocol Version 4: Protocol and Algorithms Specification," RFC5905, 2010, https://www.rfc-editor.org/rfc/rfc5905.html.

[40] GPS.gov, "GPS Standard Positioning Service (SPS) Performance Standard," 2020, https://www.gps.gov/technical/ps/.

[41] International Telecommunication Union, "G.8262.1: Timing Characteristics of Enhanced Synchronous Equipment Slave Clock," 2022, https://www.itu.int/rec/T-REC-G.8262.1-202211-I/en.

[42] Zhang, F., et al., "Ultrawideband-Based Real-Time Positioning with Cascaded Wireless Clock Synchronization Method," *IEEE Internet of Things Journal*, Vol. 11, No. 9, 2024, pp. 16731–16745.

[43] Blankenship, Y., and Z. Zou, "Paving the Way for a Wireless Time Sensitive Networking Future," Ericsson, 2023, https://www.ericsson.com/en/blog/2023/9/paving-the-way-for-wireless-time-sensitive-networking-future.

[44] Lindh, J., *Application Report: Bluetooth Lowenergy Beacons*, Texas Instruments, 2016.

[45] Petovello, M., "Clock Offsets in GNSS Receiver " *InsideGNSS*, 2011.

[46] Goldsmith, A., *Wireless Communications*, Cambridge, UK: Cambridge University Press, 2005.

[47] Zanella, A., "Best Practice in RSS Measurements and Ranging," *IEEE Communications Surveys & Tutorials*, Vol. 18, No. 4, 2016, pp. 2662–2686.

[48] Patwari, N., et al., "Relative Location Estimation in Wireless Sensor Networks," *IEEE Transactions on Signal Processing*, Vol. 51, No. 8, 2003, pp. 2137–2148.

[49] Cheung, K. W., et al., "Least Squares Algorithms for Time-of-Arrival-Based Mobile Location," *IEEE Transactions on Signal Processing*, Vol. 52, No. 4, 2004, pp. 1121–1130.

[50] Fang, B. T., "Simple Solutions for Hyperbolic and Related Position Fixes," *IEEE Transactions on Aerospace and Electronic Systems*, Vol. 26, No. 5, 1990, pp. 748–753.

[51] Chan, Y. T., and K. C. Ho, "A Simple and Efficient Estimator for Hyperbolic Location," *IEEE Transactions on Signal Processing*, Vol. 42, No. 8, 1994, pp. 1905–1915.

[52] Reza, R. I., *Data Fusion for Improved TOA/TDOA Position Determination in Wireless Systems*, Virginia Tech, 2000.

[53] Foy, W., "Position-Location Solutions by Taylor-Series Estimation," *IEEE Transactions on Aerospace and Electronic Systems*, Vol. AES-12, No. 2, 1976, pp. 187–194.

[54] Doğançay, K., "Bearings-Only Target Localization Using Total Least Squares," *Signal Processing*, Vol. 85, No. 9, 2005, pp. 1695–1710.

[55] Lingren, A. G., and K. F. Gong, "Position and Velocity Estimation Via Bearing Observations," *IEEE Transactions on Aerospace and Electronic Systems*, Vol. AES-14, No. 4, 1978, pp. 564–577.

[56] Doğançay, K., "On the Bias of Linear Least Squares Algorithms for Passive Target Localization," *Signal Processing*, Vol. 84, No. 3, 2004, pp. 475–486.

[57] Dogancay, K., and G. Ibal, "Instrumental Variable Estimator for 3D Bearings-Only Emitter Localization," *2005 International Conference on Intelligent Sensors, Sensor Networks and Information Processing*, December 5–8, 2005, pp. 63–68.

[58] Brown, R. G., and P. Y. C. Hwang, *Introduction to Random Signals and Applied Kalman Filtering*, Third Edition, New York: John Wiley & Sons, 1997.

[59] Feng, R., et al., "Wireless Channel Parameter Estimation Algorithms: Recent Advances and Future Challenges," *China Communications*, Vol. 15, No. 5, 2018, pp. 211–228.

[60] Zwirello, L., et al., "UWB Localization System for Indoor Applications: Concept, Realization and Analysis," *Journal of Electrical and Computer Engineering*, Vol. 2012, 2012, p. 849638.

5

Indoor Positioning Using Feature-Matching Methods

Previous chapters discussed range-based and geometric positioning techniques such as TOA, TDOA, and AOA methods, as well as the algorithms using estimators, filters, and optimizers. In this chapter, we focus on feature-matching methods for indoor positioning, which do not require explicit time or angle measurements. Instead, these methods leverage distinctive features or patterns observed in the environment and match them to a pre-collected database or learned model to infer location. Feature-matching approaches, sometimes categorized as "scene analysis" methods in contrast to geometric triangulation and proximity methods [1], encompass a broad category of techniques, including location fingerprinting, pattern recognition algorithms, and machine learning-based models. The common idea is to use the unique signatures of an environment, such as RF signal strengths, magnetic field anomalies, visual landmarks, or other sensor patterns as fingerprints of specific locations. During an offline phase, these fingerprints are collected and stored with their associated ground-truth locations. Then, in the online phase, a user measures the features at an unknown location and a matching algorithm attempts to find the best match in the database or model, thereby estimating the user's location.

This chapter provides a comprehensive introduction of indoor positioning using feature matching. We begin by covering the theoretical foundations and mathematical formulations of fingerprinting methods, including deterministic approaches (e.g., nearest-neighbor matching) and probabilistic approaches (e.g., Bayesian estimation). We then describe pattern recognition techniques, highlighting how classical classification and regression algorithms can be applied to indoor localization. Following that, we explore modern machine learning-based approaches, especially deep learning models that have emerged as powerful tools for building location estimators from complex and heterogeneous sensor data. We also identify key challenges unique to feature-matching methods, such as environmental dynamics, scalability of data collection, and domain adaptation and discuss how current research addresses these issues.

It is important to note that feature-matching methods are fundamentally different from the geometric approaches covered in earlier chapters. Rather than computing positions from distance or angle measurements and solving geometric equations, fingerprinting and learning-based methods treat localization as a pattern-matching or inference problem. This enables the use of readily available signals (such as Wi-Fi received signal strength) without requiring additional infrastructure, but it also introduces new challenges, which we will discuss in Section 5.4. By the end of this chapter, the reader should have a clear understanding of how indoor localization can be achieved through feature matching, the theoretical basis of these techniques, and their practical considerations.

5.1 Fundamentals of Fingerprinting for Indoor Localization

Fingerprinting is one of the most prevalent feature-matching approaches for indoor positioning. In a fingerprinting system, the environment is mapped in advance to create a fingerprint database. Each entry in this database consists of a location coordinate (often represented as a point on a floor plan) and a feature vector that captures the observable signals or patterns at that location. During the online phase, the system measures the same type of feature vector at an unknown location and compares it against the stored fingerprints to find the best match. The estimated position is then inferred to be the location associated with the most similar fingerprint in the database [2].

5.1.1 Principles of Location Fingerprinting

The basic principle of location fingerprinting is that different locations in an indoor environment often have distinguishable signatures:

- *RF signal features:* The most classic example is Wi-Fi RSS fingerprinting [3]. Wireless access points (APs) are distributed around a building, and, at any given location, the set of RSS values from these APs can serve as a fingerprint. Because of attenuation by walls and distance, each location yields a different vector of RSS measurements. Similar to Wi-Fi APs, Bluetooth Low Energy (BLE) beacons broadcast signals that can be measured (for example, as RSS values or proximity estimates). A fingerprint can thus include BLE signal strengths or the set of nearby beacon identifiers, and these BLE features are often combined with Wi-Fi fingerprints to improve coverage and accuracy [4]. CIR or channel frequency responses (CFR) (especially from UWB or the channel state information in Wi-Fi) is an emerging area [5]. These signals could potentially improve accuracy to submeter levels by distinguishing locations based on subtle differences in channel patterns that narrowband RSS alone could not capture.
- *Magnetic field anomalies:* Indoor environments contain structural steel and electrical equipment that disturb the Earth's magnetic field. The magnetic field intensity and direction at a point can act as a fingerprint [6]. This concept has been utilized by systems that map indoor magnetic signatures and later match a user's magnetometer readings to that map.
- *Light and sound:* Characteristics of ambient light or acoustic signals (e.g., ultrasound beacons or background noise profiles) can be used in certain scenarios as location signatures.
- *Visual landmarks:* If a camera is available, visual features (such as the presence of certain markers, posters, or the scene captured by the camera) can be treated as a fingerprint of location. For example, an image taken in one hallway will have different visual features than an image taken in another hallway.
- *Heterogeneous data fusion:* Often multiple modalities are combined, such as using Wi-Fi signals together with BLE, inertial sensor patterns (such as step and turn patterns), or barometric pressure (for floor detection). These multisensor fingerprints offer higher robustness.

The fingerprinting approach generally consists of two phases, as shown in Figure 5.1. In the offline survey (also known as the training phase), a site survey is conducted where an operator (or an automated robot or crowdsourcing participants) collects data at various reference locations. At each reference point (RP), the feature vector $\mathbf{f}_{\mathrm{RP}}$ is recorded along with the known coordinates

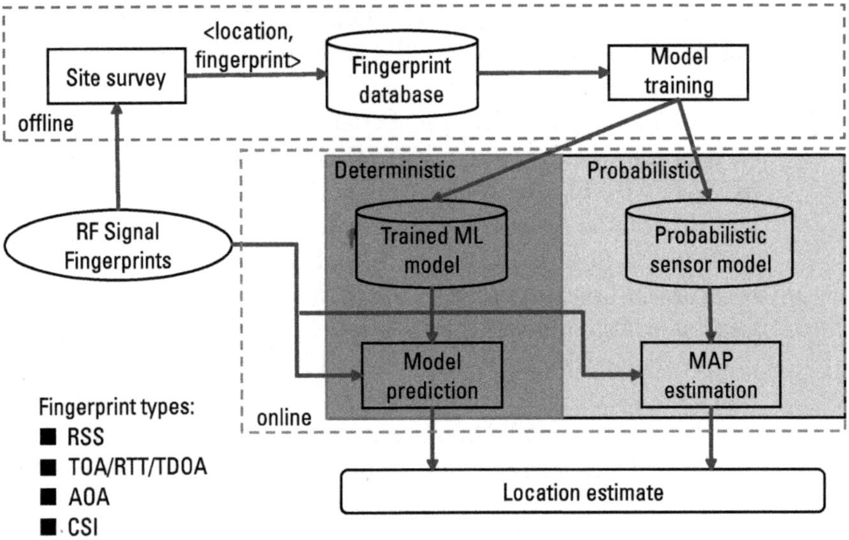

Figure 5.1 The general framework of fingerprinting-based indoor positioning with RF signals.

$\mathbf{x}_{RP}$. The set of all such pairs $\{(\boldsymbol{x}_i, \boldsymbol{f}_i)\}_{i=1}^{M}$ forms the fingerprint database, where M is the number of reference points surveyed. The offline survey also forms the correspondence between $\boldsymbol{x}_i$ and $\boldsymbol{f}_i$, which can be established either deterministically or probabilistically. Then, in the online positioning (also known as the query phase), a user at an unknown location obtains a measurement $\mathbf{z}$ (the same type of feature vector as in the fingerprints, e.g., a vector of RSS values). The system then compares $\mathbf{z}$ to all (or a subset of) stored fingerprints $\{\mathbf{f}_i\}$ and computes a similarity or distance metric. The location estimate $\hat{\mathbf{x}}$ is determined based on the best match according to this metric. This basic procedure was first demonstrated in early Wi-Fi localization systems such as RADAR by Bahl and Padmanabhan [3], which used empirical signal strength maps to achieve meter-level accuracy. A key assumption is that the environment remains relatively static between the offline and online phases so that the fingerprints remain valid over time.

5.1.2 Mathematical Formulation of the Fingerprint-Matching Problem

We can formalize the fingerprint-based positioning problem as follows. Let $\mathbf{x} \in \mathbb{R}^d$ denote the unknown true location of the user (for example, $d = 2$ for 2-D floor plan coordinates, or $d = 3$ if elevation is included). Let $\mathbf{z} \in \mathbb{R}^N$

denote the observed feature vector at that location (e.g., the vector of N signal strength readings). We assume an offline database of M reference points, with known locations $\{\mathbf{x}_i: i = 1, \ldots, M\}$ and their corresponding fingerprint feature vectors $\{\mathbf{f}_i: i = 1, \ldots, M\}$.

One straightforward approach to estimate $\mathbf{x}$ is to find the reference point whose fingerprint is closest to the observed $\mathbf{z}$ in terms of some distance metric. That is,

$$i^* = \arg\min_{1 \le i \le M} D\left(\mathbf{z}, \mathbf{f}_i\right) \tag{5.1}$$

where $D(\ ,\ )$ is a distance measure such as Euclidean distance $D(\mathbf{a}, \mathbf{b}) = \sqrt{\sum_{n=1}^{N}(a_n - b_n)^2}$ or Manhattan distance $D(\mathbf{a}, \mathbf{b}) = \sum_{n=1}^{N}|a_n - b_n|$. The estimated position is then $\hat{\mathbf{x}} = \mathbf{x}_{i^*}$ (i.e., the coordinates of that nearest-neighbor reference point). This is the simplest form of deterministic fingerprint matching, often referred to as the Nearest Neighbor (NN) algorithm in pattern recognition.

In practice, using just one nearest neighbor can be susceptible to measurement noise or an outlier fingerprint. A common extension is the k-Nearest Neighbors (k-NN) approach, which finds the k closest fingerprints to $\mathbf{z}$ and then either chooses the most frequent location among them (if reference points are grouped by region) or, more commonly, computes a weighted average of their coordinates:

$$\hat{\mathbf{x}} = \frac{\sum_{i \in \mathcal{K}(\mathbf{z})} w_i \mathbf{x}_i}{\sum_{i \in \mathcal{K}(\mathbf{z})} w_i} \tag{5.2}$$

where $\mathcal{K}(\mathbf{z})$ is the index set of the k nearest reference points to $\mathbf{z}$, and w_i is a weight assigned to neighbor i. The weights w_i can be chosen, for example, as $w_i = 1/(\mathbf{z}, \mathbf{f}_i)$ (inverse distance weighting) so that closer matches have a higher influence on the final estimate. Equation (5.2) then yields an interpolated position, which can mitigate the quantization effect of returning only previously surveyed points.

Figure 5.2 illustrates the three steps to obtain a position fix using the fingerprinting method. The fingerprint-matching problem can also be cast in a probabilistic framework. Let the location $\mathbf{x}$ be modeled as a random variable and the observed fingerprint $\mathbf{z}$ as another random variable. During the offline phase, one can build statistical models or empirical histograms of $\mathbf{Z}_i$ for each reference location. For example, at reference point i, we might model the distribution of the feature vector as $P(\mathbf{z}|\mathbf{x}_i)$ (or the likelihood of

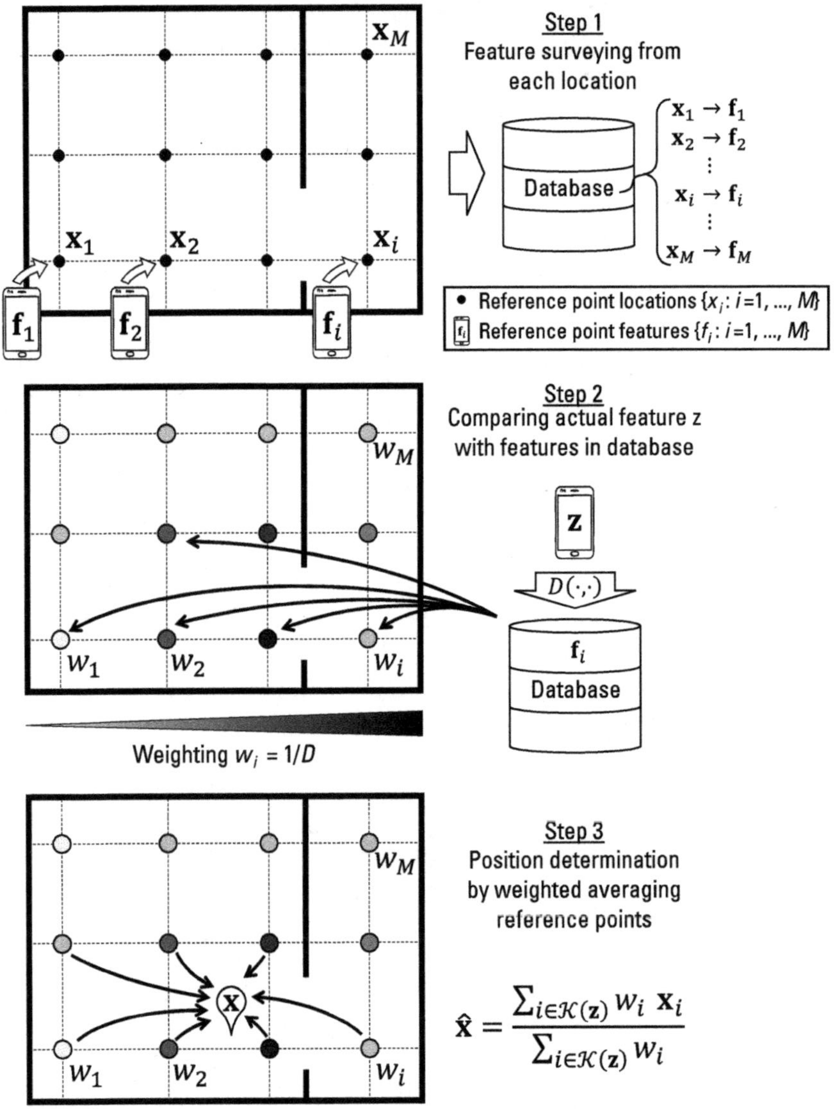

Figure 5.2 Illustration of fingerprint positioning using the deterministic method of KNN.

observing certain signal measurements given the user is at $\mathbf{x}_i$). With enough survey samples, one can empirically estimate these likelihoods. At runtime, given a new observation $\mathbf{z}$, the goal is to infer the location. By Bayes' rule, the posterior probability that the user is at location $\mathbf{x}_i$ given observation $\mathbf{z}$ is:

$$P(\mathbf{x}_i|\mathbf{Z}_i = \mathbf{z}) = \frac{P(\mathbf{Z}_i = \mathbf{z}|\mathbf{x}_i)P(\mathbf{x}_i)}{P(\mathbf{Z}_i = \mathbf{z})} \quad (5.3)$$

where $P(\mathbf{x}_i)$is the prior probability of being at $\mathbf{x}_i$ (which might be assumed uniform if no prior knowledge is available) and $P(\mathbf{Z}_i = \mathbf{z})$ is the total probability of $\mathbf{z}$ under all locations. Since $P(\mathbf{Z}_i = \mathbf{z})$ is the same for all candidate locations, the location that maximizes the posterior is also the one that maximizes the likelihood $P(\mathbf{Z}_i = \mathbf{z}|\mathbf{x}_i)$ (if priors are equal, meaning an uniform distribution is assumed). Thus, the maximum likelihood (ML) or maximum a posteriori (MAP) estimate is:

$$i^* = \arg\max_{1 \le i \le M} P(\mathbf{Z}_i = \mathbf{z}|\mathbf{x}_i) \quad (5.4)$$

and $\hat{\mathbf{x}} = \mathbf{x}_{i^*}$ as before. This probabilistic formulation underpins many radio fingerprinting algorithms that use statistical models of signal strength. For instance, the Horus system by Youssef and Agrawala utilized a probabilistic approach to account for Wi-Fi RSS variability [7], modeling the distribution of RSS at each reference point with histograms or parametric distributions and computing $P(\mathbf{x}_i|\mathbf{z})$ to make a decision.

The advantage of the probabilistic approach is that it provides a measure of confidence or likelihood for each candidate location, which can be useful to derive uncertainty estimates or to combine with other information (e.g., combining with inertial sensors via a Bayesian filter, although we do not cover filtering in this chapter). One can also derive a probabilistic location estimate as a weighted average of reference point coordinates:

$$\hat{\mathbf{x}} = \sum_{i=1}^{M} P(\mathbf{x}_i|\mathbf{Z}_i = \mathbf{z}) \cdot \mathbf{x}_i \quad (5.5)$$

which effectively gives an expectation of the user's location under the posterior distribution. However, in practice, this sum often includes many negligible probabilities and a few dominant ones (especially if the fingerprints are well-separated), so a calculation based on the top k candidates is typically sufficient.

Compared to Figure 5.2 (the illustration based on a deterministic model), the illustration of the one using probabilistic method is shown in Figure 5.3.

5.1.3 Deterministic Versus Probabilistic Fingerprinting

To illustrate the difference between deterministic and probabilistic fingerprinting, consider a simple scenario. Suppose that the feature vector is

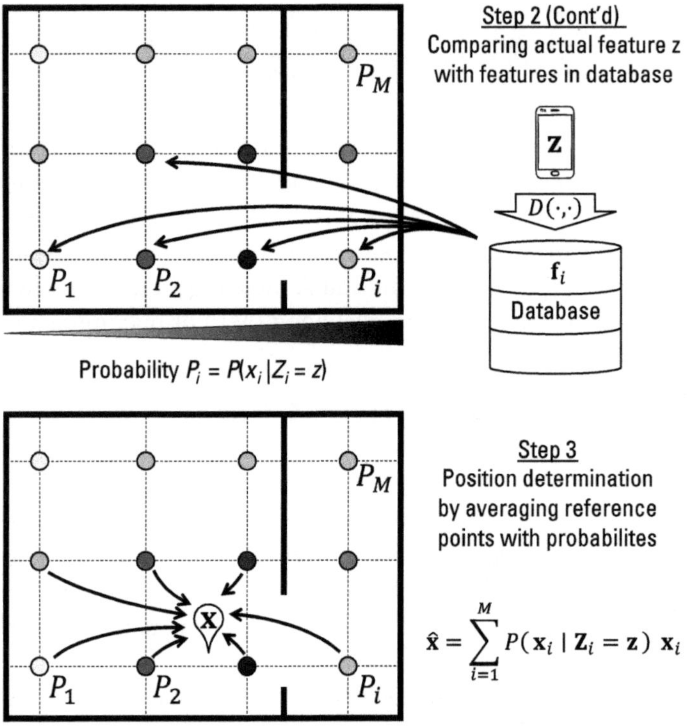

Figure 5.3 Illustration of fingerprint positioning using the probabilistic method of MAP.

one-dimensional ($N = 1$) measuring, say, the RSS from a single Wi-Fi AP. Three reference points A, B, and C have mean RSS values of −40 dBm, −60 dBm, and −80 dBm, respectively, at locations a_A, x_B, x_C. In a deterministic approach, if a user measures $z = -58$ dBm, one might compute distances: $|-58 - (-40)| = 18$, $|-58 - (-60)| = 2$, and $|-58 - (-80)| = 22$ for locations A, B, and C, respectively. The nearest neighbor is B (distance 2), so we pick location x_B. A probabilistic approach would model that each reference point's RSS is a random variable (due to fading and noise). If at B the distribution is centered at −60 with some variance, we could compute the likelihood of −58 at each point: $L_A = p(Z_A = -58|x_A)$, $_{LB} = p(Z_B = -58|x_B)$, and $L_C = p(Z_C = -58|x_C)$. If we assume Gaussian distributions (which is used in many practical applications) for simplicity:

$$L_i = \frac{1}{\sqrt{2\pi}\sigma_i} \exp\left(-\frac{\left(-58 - \mu_i\right)^2}{2\sigma_i^2} \right) \tag{5.6}$$

where μ_i is the mean RSS at reference i and σ_i is the standard deviation. Even if μ_B is slightly different from the observation, a larger σ_B (wider distribution) could make L_B comparable to L_A if A had a smaller variance but the value is far from μ_A. In any case, the outcome is typically still that x_B has the highest probability (unless −58 was an unlikely reading for B and more likely for A due to overlapping distributions).

Deterministic methods such as *k*-NN are simpler to implement and often perform well if the feature space distances correlate with physical distance. Probabilistic methods can yield better results when properly calibrated because they account for the variability of measurements; for example, far-from-router areas might have very fluctuating signals that a deterministic method could mishandle, whereas a probabilistic method might down-weight those readings due to broad distributions. Some systems in the literature compare these approaches and sometimes hybridize them (e.g., using *k*-NN to get a set of candidates and then applying Bayesian weighting) [8].

5.1.4 Advanced Enhancements in Fingerprinting

Over the years, many enhancements to basic fingerprinting have been proposed. We briefly mention a few notable ones:

- *Signal processing and filtering:* Before matching, raw measurements can be filtered or averaged. For instance, taking multiple RSS scans and using a moving average can reduce random noise. Some approaches also filter out unstable AP signals (heuristically) or use techniques like Kalman filtering on the measurement sequence to exclude those with higher residuals.
- *Dimensionality reduction:* When N (number of features) is large (e.g., dozens of APs), algorithms such as principal component analysis (PCA) can be applied to focus on the most informative linear combinations of signals [9]. This can speed up matching and sometimes improve accuracy by ignoring redundant dimensions. Nonlinear reduction such as autoencoders (in the ML context) have also been used, as discussed in Section 5.2.
- *Clustering and indexing of fingerprints:* To scale the fingerprints to large environments, the database can be clustered or indexed to avoid comparing against all M fingerprints every time. One method is to partition the area into regions (using, e.g., *k*-means clustering on the fingerprint feature space) and only search within the region whose centroid is closest to $\mathbf{z}$. Another is to use tree structures or hashing

for high-dimensional search. These techniques improve computational efficiency, which is important for real-time systems.

- *Continuous space estimation:* Instead of assuming that the user's location must coincide with a surveyed point, various interpolation schemes create a continuous radio map. For example, one can use Gaussian process regression to model the signal strength as a continuous function of location [10]. This yields a predictive mean and variance of the signal at any coordinate, which can then be used in a likelihood formula to find the most likely location via continuous optimization. Such methods blend fingerprinting with radio propagation modeling, offering a trade-off between purely empirical and theoretical approaches. Ideally, these techniques can reduce the effort in building the offline database.
- *Crowdsourcing and automated updates:* A major challenge for fingerprinting is keeping the database fresh (i.e., up-to-date) as environments change. Modern approaches allow users' devices to contribute new fingerprints as they move around (with either opportunistic or explicit reporting of their locations), thereby continually updating the database. The system might use the incoming data to detect changes (e.g., an AP moved or a new AP added) and adapt. Although this enters the realm of localization infrastructure management rather than the core algorithm, it is facilitated by robust fingerprint matching that can handle a degree of variation.

Fingerprinting has proven to be effective in many real-world deployments, achieving typical accuracies on the order of 5 to 10m for Wi-Fi-based systems in office buildings [3]. However, its performance can degrade if the environment or device conditions differ significantly from the training data. This motivates more adaptive and learning-based approaches, which we cover in later sections.

5.2 Pattern Recognition Approaches for Indoor Positioning

The problem of indoor localization via feature matching can be viewed through the lens of pattern recognition. In this view, the goal is to recognize the user's location based on observed patterns (features) and how one would classify an object based on its attributes. Many classical pattern recognition algorithms can be applied to fingerprint data to improve accuracy or reduce computational load, beyond the basic nearest-neighbor matching discussed earlier.

In this section, we discuss how indoor localization can benefit from established pattern recognition techniques. We cover classification approaches that treat location estimation as a discrete classification problem, regression approaches that predict continuous coordinates, and feature engineering techniques that extract more robust patterns from raw sensor data. We also highlight how some methods explicitly leverage domain-specific patterns, such as building layout constraints or motion patterns.

5.2.1 Location Estimation as a Classification Problem

One way to frame indoor positioning is as a multiclass classification problem. Suppose that the area of interest is divided into a finite set of C distinct locations or zones (these could be cells in a grid, specific rooms, specific floors, or simply the set of reference points themselves). Each class $class \in (1, 2, \ldots, Classes\}$ corresponds to a location or region in the building. During the offline phase, we gather feature vectors for each class. In the online phase, given an observed feature vector $\mathbf{z}$, the task is to predict to which class (location) it belongs.

There are standard classification algorithms that can be trained for this task:

- *Decision trees and random forests:* A decision tree classifier can be trained on the fingerprint data with location classes as labels. The tree will learn thresholds on feature values to split the space into regions corresponding to different locations. Decision trees are human-interpretable to some extent (e.g., they might learn rules such as "if AP1 signal > −50 dBm and AP2 signal > −60 dBm, then Location = Lobby"). An ensemble of trees (i.e., random forests) often improves accuracy by reducing overfitting, especially given the high-dimensional and possibly noisy fingerprint data.
- *Support vector machines (SVM):* SVMs seek to find hyperplanes in the feature space that separates classes with maximum margin. For indoor localization, one can train an SVM to distinguish between every pair of location classes. However, SVMs are more naturally binary classifiers; multiclass SVM approaches (one-versus-one or one-versus-rest) can be applied if the number of distinct locations is not too large. In a large building with hundreds of reference points, using SVM directly can be computationally intensive and may require kernel methods to handle class complexity. Nevertheless, for smaller-scale problems or for distinguishing between broader areas, SVMs have been applied.

- *Bayesian classification:* As introduced in the fingerprinting section (Section 5.1.2), one can take a Bayesian pattern recognition approach. Naive Bayes classifiers, which assume feature independence, have sometimes been used with quantized signal features. More complex Bayesian networks or even discriminative models such as logistic regression (which essentially fits a logistic function to approximate $P(\text{location} = \textit{class}|\mathbf{z})$) can also be used. These approaches are closely related to the probabilistic fingerprinting discussion; the difference is often in whether the model is generative (model $P(\mathbf{Z}|\mathbf{x})$ as in Naive Bayes) or discriminative (directly model decision boundaries as in logistic regression or SVM).

Each of these classifiers has advantages and disadvantages. Simpler classifiers such as decision trees are easy to implement and can handle large numbers of classes (locations) but might not capture subtle joint patterns in the features. More complex ones such as SVM or Bayesian methods can potentially be more accurate if tuned well, but they might require significant training data and careful parameter selection (e.g., kernel choice for SVM, or distribution assumptions for Bayes). Empirically, studies have shown that these advanced classifiers can improve localization accuracy in challenging scenarios. For example, Brunato and Battiti demonstrated that SVM-based matching yielded better performance than a weighted k-NN baseline in a WLAN localization testbed [11]. However, such gains come with increased computational costs and complexity of model training.

One practical consideration when applying classification algorithms is that the number of classes (distinct location labels) can be very high if one treats every reference point as a separate class. If M reference points exist, then $\textit{Classes} = M$ in that formulation. Some locations might be very similar in their fingerprints, effectively making the classification problem extremely difficult for those classes (they are not linearly separable or even easily nonlinearly separable). To address this, one strategy is to reduce the granularity of classes by grouping nearby reference points into one class or region, meaning clustering. This sacrifices some precision (the system will output a region or cell rather than an exact point), but it can make the classification task tractable. Subsequently, a finer positioning within that region can be done by other means (perhaps a local fingerprint interpolation or an additional sensor cue). This two-step hierarchical approach, as demonstrated in Figure 5.4, is sometimes used in large buildings: first classify the area (which floor, building, or wing or which sector of a floor), and then pinpoint within that area.

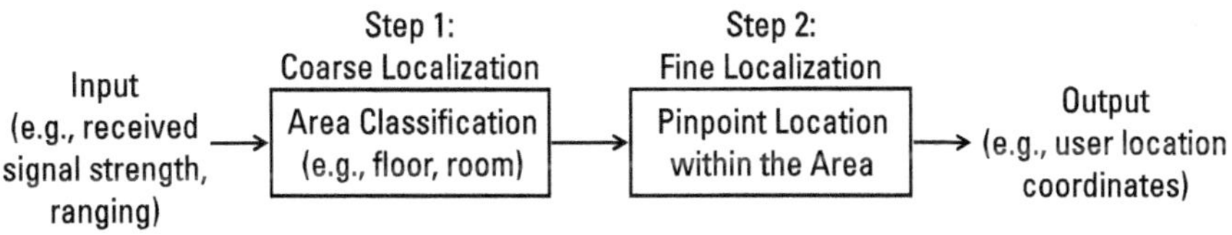

Figure 5.4 The two-step hierarchical approach for indoor positioning.

5.2.2 Regression Approaches for Coordinate Estimation

Instead of classifying the location into a discrete set, one can attempt to learn a direct mapping from features to continuous coordinates (x, y) (and z if height is included). This is a regression problem: given a feature vector $\mathbf{f}$, predict $\mathbf{x} = [x, y]$. Classical regression algorithms can be employed. Here we introduce popular methods:

- *Multivariate linear regression:* In theory, one could fit a linear model $\hat{\mathbf{x}} = W\mathbf{x} + \mathbf{b}$ where W is a $2 \times N$ matrix (for 2-D output) and $\mathbf{b}$ is a bias vector. N is the input feature numbers (e.g., number of APs). Each row of W and the corresponding bias would form a linear combination of the input features to predict x and y. However, indoor propagation is highly nonlinear with respect to position, so a linear model will generally not fit well except perhaps in very small, simple areas. That said, linear regression might be used in a piecewise manner if the environment is partitioned.
- *Gaussian process regression:* As mentioned in the fingerprinting section, Gaussian processes (GP) can model the mapping from location to signal, but, conversely, they can also be used to infer location from signal by inverting the relationship in a probabilistic sense. If we have a priori that the location coordinates are a function of the features, a GP could be trained in the reversed way (although typically it is easier to model signal as function of location, not vice versa, due to dimensionality differences). Instead, one might train two GPs, one for x and one for y, each as a function of the feature vector. However, directly using a high-dimensional feature vector as input to a GP might be impractical for large N.
- *Support vector regression (SVR):* The regression counterpart of SVM can be used to predict continuous coordinates. SVR attempts to find a function (nonlinear if using a kernel) that fits within an ϵ-tube (which defines the tolerance for errors in the prediction of numerical

values) of the training data and is as flat as possible, meaning minimizing complexity. For localization, one could train two SVR models, one for x and one for y (or use a multi-output SVR if available). This method has been explored in some studies with moderate success [12]; it often requires careful tuning of kernel parameters. SVR tends to be slower than a simple neural net or tree once trained, because it involves computing distances to support vectors. However, one advantage is that SVR inherently performs a form of regularization, which can prevent overfitting.

- *Ensemble methods:* Aside from random forests (which we covered as a pattern recognition classifier), other ensemble approaches such as Gradient Boosting Machines (GBM) or Adaptive Boosting (AdaBoost) have been tried for indoor location classification. AdaBoost can combine many weak classifiers into a stronger one, and some experiments used decision stumps (one-level decision trees) boosted to create a robust location predictor [8]. The ensemble approach can handle the multiclass nature and often yields good accuracy, but, similar to SVM, the computational load might be higher, and the model can become a bit of a black box.

The regression perspective is appealing because it directly aims to minimize location error during training. However, pure regression on high-dimensional, sparse data (the surveyed fingerprints are only known at reference points) can be tricky. Often, a classification step (coarse localization) is followed by regression (fine localization), or regression models are used to refine a classification or nearest-neighbor result. When using these regression methods, the model training is typically done offline with the collected fingerprint dataset. One must be careful to avoid overfitting between the input features and the reference points. A common practice is to evaluate the model on validation data collected at distinct locations or at times different from the training phase to ensure that the model generalizes.

Another interesting direction in pattern recognition is semi-supervised learning or manifold learning. In some cases, one may collect a lot of unlabeled sensor data (no ground truth position) while only some fraction is labeled (with ground truth). Algorithms such as self-organizing maps or manifold alignment might be used to cluster the unlabeled data, potentially revealing the structure of the space [13]. Then a smaller number of labeled points can anchor those clusters to physical locations. This is beyond the scope of our chapter to discuss, but it is worth noting as a pattern recognition approach, especially when labeled data (ground-truth locations) are expensive to get.

5.2.3 Feature Engineering and Dimensionality Reduction

In pattern recognition, how you represent the data (feature engineering) is often as important as the choice of algorithm. For indoor positioning, the raw features might be, for example, a list of RSS values from various APs. However, these raw features can be transformed or expanded to improve performance. Next are popular methods. M denotes the total fingerprint data epoch, and N denotes the number of features in each fingerprint data epoch.

- *Feature selection:* Not all AP signals may be informative. Some may be too weak or sporadic to be useful, or some may be highly correlated with others. A common step is to select a subset of the strongest or most stable AP signals to use in the fingerprint. This reduces M and often improves robustness. Algorithms such a greedy forward selection [14] or backward elimination [15] can be employed, using cross-validation error as a criterion to pick the best subset of features for location prediction.
- *PCA:* As mentioned, PCA can reduce noise. One might take the collected $M \times N$ matrix of fingerprints and compute its principal components and then represent each fingerprint by the first k principal component coefficients. If most of the variance (information) is retained in those k components, then matching or classification can be done in this reduced k-dimensional space. Importantly, PCA is an unsupervised method (it does not use location information in finding components, just the feature correlations), so it aims to capture the structure of the feature data, which might align with spatial differences but not always. Still, it can help when N is very large or if signals are very noisy.
- *Discriminant analysis:* Methods such as linear discriminant analysis (LDA) explicitly try to project data into a lower-dimensional space that maximizes separability between classes (locations). If treating each reference point as a class, LDA could, in principle, yield a projection that emphasizes differences between reference points. However, LDA typically requires more samples than dimensions for stability, which can be an issue if $N > M$ (only few epochs of data are collected) or if there are many classes. In practice, one might use LDA on a higher-level grouping (such as room-level classes) to get features that separate rooms well.

Physical constraints of the environment can also be incorporated into pattern-based localization. For example, after a fingerprint-based estimate is obtained, it can be adjusted to the nearest valid location on the floor plan (e.g., shifting an estimate that falls inside a wall to the closest open area). In this way, pattern matching is augmented with geographic context, improving the realism and feasibility of the position estimate.

Lastly, we note that any pattern recognition approach should ultimately be evaluated on how well it estimates location, not just the classification accuracy. For instance, if a classification approach predicts the wrong room but that room is adjacent to the correct one, the location error might be small if the user was near the boundary. At this time, an error uncertainty metric would reflect a small error, whereas classification accuracy would mark it as a failure. Therefore, the design of the system often involves a combination of classification and regression thinking. Many competitions and benchmarks in indoor localization (such as the annual IPIN competition with different tracks of data corresponding to different user scenarios [16]) use accuracy metrics such as mean error distance, median error, or percentage of estimates within 1m, 2m, and 5m, which guide how algorithms are tuned.

5.3 Deep Learning-Based Approaches

Deep learning techniques have increasingly been applied to indoor positioning, building upon the pattern recognition approaches with more advanced models and training methodologies. In many cases, the line between pattern recognition and machine learning is blurred, as techniques such as decision trees or SVM are indeed part of machine learning. However, in this chapter, we use the term "machine learning-based approaches" to emphasize methods that involve an explicit training phase to learn complex models (often with many parameters) from data, including neural networks and deep learning, which have shown promising results in recent years [17]. Unlike the simpler methods that might use direct matching or relatively straightforward classification boundaries, advanced deep learning models can capture intricate nonlinear relationships in fingerprint data, potentially improving accuracy and robustness. They can also naturally handle multimodal inputs and incorporate contextual information if available.

In this section, we first discuss deep learning approaches, such as recurrent neural network and convolutional neural networks, that have been adapted for fingerprint-based positioning. We also address practical considerations such as training data requirements and model generalization.

5.3.1 Algorithms for Localization

Deep learning refers to neural networks with multiple layers that can automatically learn hierarchical feature representations. A typical neural network architecture might have N input neurons (for the N-dimensional feature vector), one or more hidden layers with nonlinear activation functions (e.g., sigmoid or ReLU), and output neurons corresponding to either coordinates or class labels. In a typical coordinate-regression network, there would be 2 (or 3) output nodes for x, y (or z), and the network is trained to minimize mean squared error between the outputs and the true coordinates. In a classification network, there would be M output nodes (for M location classes or reference points), with a softmax activation and cross-entropy loss, so the network is trained to output a probability distribution over locations given the input. In 2002, Battiti et al. showed that neural networks could achieve comparable or slightly better accuracy than k-NN when trained well [18], although, in the early 2000s, the size of networks and amount of data were limited.

A deep learning approach can combine the advantages of both classification and regression. If a classification approach yields an estimated class (such as a specific reference point identification), one can post-process that into coordinates (by mapping that identification to known coordinates, or using a centroid if the class was a region). If the model yields a continuous coordinate, that is straightforward, but sometimes a hybrid output is used (e.g., an NN that outputs both a classification distribution and a coordinate). Figure 5.5 shows a classic flowchart on how to apply deep learning algorithms to indoor feature-matching localization.

The popular deep learning networks [19] are briefly introduced here and their structures are illustrated in Figure 5.6.

- *Fully connected neural networks (FCNN):* Also known as multilayer perceptron (MLP) with many hidden layers. These can be seen as an extension of the simple NN approach mentioned above, but with more layers and neurons. In practice, simply stacking more layers on an FCNN for fingerprinting can yield some gains, but there is a risk of overfitting without enormous data. Researchers have used techniques like dropout (randomly dropping units during training) and L_2 regularization to improve generalization. Some deep networks use one-hot encoding of location identification as output with a softmax and then convert to coordinates if needed (for example, by outputting the weighted centroid of the highest-probability classes). This network

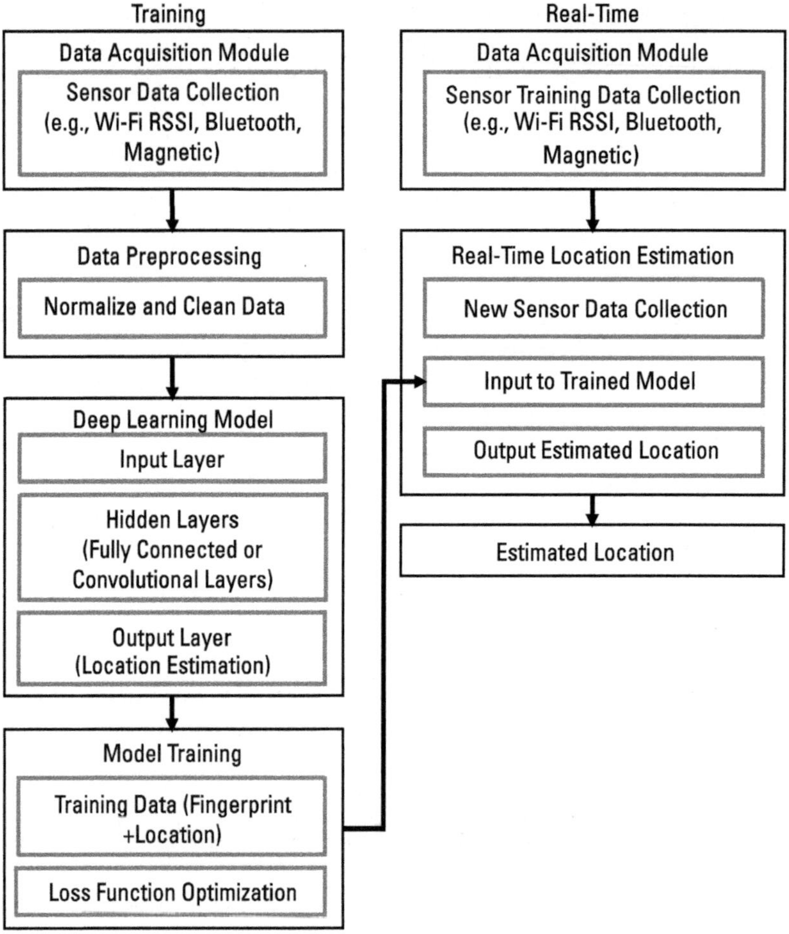

Figure 5.5 An illustration of a deep learning model for fingerprint-based indoor positioning.

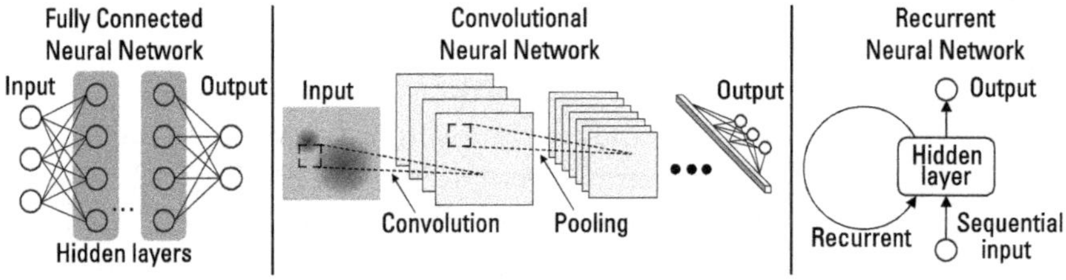

Figure 5.6 Popular deep learning network architectures [19].

is classic and usually a good starting point for researchers who just stepped into the field.

- *Convolutional neural networks (CNN):* CNNs are typically used for grid or image-like data. One innovative approach in Wi-Fi fingerprinting is to treat the list of RSS readings as a sort of 1-D image or as a 2-D pseudoimage. For example, one can arrange the signals from different APs into a 2-D matrix (perhaps sorting APs by their physical location or frequency channel) and treat that as an image in which pixel intensity corresponds to signal strength. Then a CNN can be applied to learn spatial filters on this matrix [20]. Some studies have shown that CNNs slightly outperform FCNN for fingerprint data, possibly because they can capture the spatial correlation of the local patterns (such as a group of APs that often vary together). It is also worth mentioning that, in some cases, framing the fingerprint data as an image can incorporate knowledge of physical layout. One approach is to create a 2-D grid overlay on the floor plan and "paint" the signal strengths onto this grid at the reference point locations, effectively making a sparse image for each AP or each fingerprint. Then one might use a CNN to regress from this image to the user's position. This sort of spatially aware deep learning incorporates knowledge of the building's geometry. However, it requires preprocessing to align data with coordinates.
- *Recurrent neural networks (RNN):* If a user is walking, one can consider the sequence of recent fingerprints. RNNs or long short-term memory (LSTM) networks can take a time series of observations and output an updated location estimate, effectively learning to smooth or track the user. This borders on tracking (which might involve filtering), but purely data-driven sequence models can be trained to incorporate the user's motion patterns or the temporal stability of signals. For example, a user walking down a hallway will have gradually changing Wi-Fi signals; an RNN could learn to interpret a sequence such as [–60, –58, –55,...] from a particular AP as indicative of moving toward that AP, thus adjusting position estimates accordingly. This is an interesting direction where machine learning is used for what traditionally might be handled by a Kalman filter or particle filter.
- *Transformer networks (transformers):* Transformers are a newer deep learning architecture that utilizes self-attention mechanisms to model complex dependencies in data. Unlike RNNs, which process sequences step by step, transformers process the entire input simultaneously, allowing them to capture long-range dependencies more effectively. In the context of indoor localization, transformers have been explored

for both Wi-Fi fingerprinting and multisensor fusion. One approach is to treat each signal source (e.g., an AP or beacon) as a "token" and let the transformer dynamically learn which signals are most relevant to estimating location [21]. This makes it inherently adaptive to varying signal conditions. Another application is using transformers for sequential fingerprint data, where the model processes a time-series of RSS readings and learns temporal patterns for improved localization accuracy. Compared to LSTMs, transformers excel at handling long-term dependencies while being more parallelizable, making them efficient for batch processing. However, standard transformers require significant resources, so lightweight adaptations (e.g., pruning attention heads, quantized models) are necessary for real-time mobile deployment. Recent studies have indicated that transformer-based models can outperform LSTMs and CNNs in complex environments, particularly when exploring the correlation between continuous inputs [21]. Nonetheless, their data hunger and computational overhead remain challenges, making them most suitable for large-scale, high-accuracy applications rather than lightweight, on-device inference.

- *Autoencoders and unsupervised feature learning:* Autoencoders are neural networks trained to compress and reconstruct data. In fingerprinting, an autoencoder can be used to learn a lower-dimensional representation of the high-dimensional feature vector (similar to PCA but potentially nonlinear). One could train an autoencoder on all the fingerprint vectors $\{\mathbf{f}_i\}$ to get, say, a 10-dimensional code that captures most of the information. Then use those 10-dimensional codes as features for a simpler localization algorithm (such as k-NN or a small classifier). The benefit is noise reduction and focusing on key variations. Furthermore, if one has a large amount of unlabeled data (fingerprints without known locations), an autoencoder can be trained on that to improve the representation, and then a smaller labeled set is used for supervised learning on top of the learned features (this is a form of unsupervised pre-training).

Deep learning models often achieve high accuracy in the training environment but can suffer if the environment changes (e.g., furniture moved, AP transmit power changed). One solution is to periodically retrain or fine-tune the model with new data. Another approach is to design the model outputs in a way that provides uncertainty information. For example, a Bayesian neural network or an ensemble of networks could give a distribution over position rather than a point estimate, which can be useful in practice to know when the model is confident versus when it is unsure.

Because deep models can be data-hungry, there have been efforts to create large public databases of indoor localization fingerprints. For example, the UJIIndoorLoc-Mag dataset [22] contains over 20,000 Wi-Fi fingerprints collected in a multibuilding campus and has been used to benchmark many algorithms including deep networks. Results from such evaluations show deep learning to be among the top performers, often achieving median errors around 1 to 2m in those scenarios, which is a notable improvement over traditional methods on the same data [17].

Furthermore, recent research has explored few shot and transfer learning techniques to enable models to generalize to new buildings or floors with very little new training data, which is crucial for making deep learning solutions more practical in dynamic real-world deployments. For example, in practice, one may have a trained model in one building and want to use it in another with some adaptation. Transfer learning in deep networks might involve taking a model trained on one large dataset and fine-tuning it on another smaller dataset, rather than starting from scratch. This has been popular in image-based localization (taking networks pre-trained on large image datasets and fine-tuning on specific building images). For radio fingerprinting, transfer learning might involve transferring knowledge of general signal patterns. Some researchers have also looked at simulation-to-real transfer: train a deep network on simulated radio maps (using ray tracing or other propagation models for many virtual environments) and then adapt it to real measurements with minimal real data [23]. This is still an emerging area.

5.4 Key Challenges and Future Directions

While integrating features matching with machine learning methods have shown great promise for indoor positioning, they also face a number of challenges. In this section, we outline some key challenges and discuss ongoing research directions aimed at addressing them.

5.4.1 Data Scalability and Deployment Effort

A recurring theme with machine learning approaches is the reliance on data and one major challenge with fingerprinting methods is the effort needed to deploy them in a new environment. Collecting a fine-grained fingerprint database is labor-intensive. If a building has many floors and rooms, surveying every area with sufficient density of reference points can be time-consuming. In the early stages of the first author's career, collecting fingerprints for a classroom already took him a half-day as shown in Figure 5.7 [24].

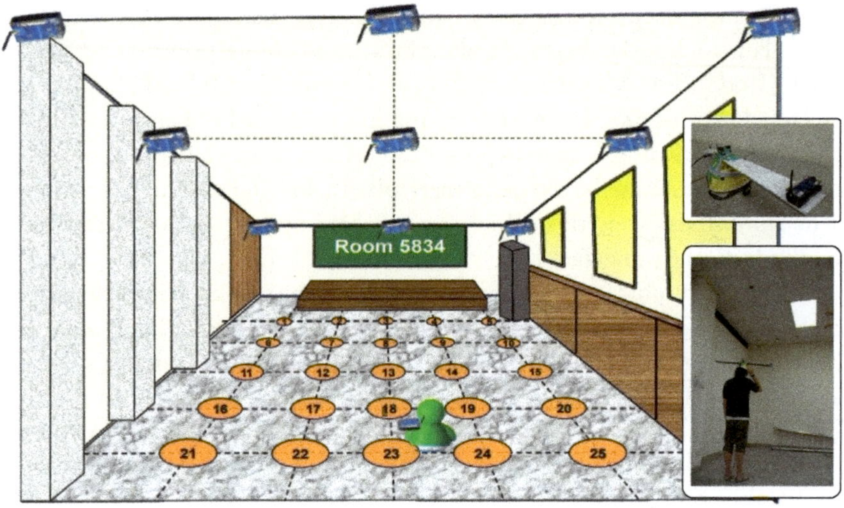

Figure 5.7 An illustration of collecting fingerprint data from 9 ZigBee transceivers at 25 grid points. The data collector stands at each of the grid points for 5 minutes.

The larger the area, the more data storage and computation are required for matching. Techniques such as data augmentation are used to synthetically increase the diversity of training data without additional surveys. For example, one could add random noise to existing fingerprints to simulate measurement noise or perturb a known location's fingerprint by interpolating with a neighbor's to simulate an in-between location. In vision, data augmentation (rotating, scaling images) is common; for RSS fingerprints, adding small Gaussian noise or dropping out a random AP reading (to simulate a temporary signal loss) could help the model to be robust to those events [25].

To make things harder, a variety of devices might measure signals differently (due to calibration offsets, different antenna gains). If the training data is collected with one device and a user uses another, the performance can degrade. Approaches to handle this include calibrating the devices (if possible) or including data from multiple device types in training, but this is not too cost-effective. Another approach is to use relative features such as the rank of signals or to normalize the feature vector (for example, subtract the mean and divide by the standard deviation of RSS values for that scan) so that differences in overall sensitivity are mitigated. However, this normalization method does not always work if the two devices generate measurements with very different probabilistic behaviors.

In addition, indoor environments are not static. Furniture can be moved, doors opened or closed, people come and go (causing signal shadowing and changing multipath), and electronic devices might be added or removed (introducing new signals or interference). A fingerprint collected at one time may gradually become outdated. This concept is known as temporal degradation of fingerprints. Studies have shown that Wi-Fi fingerprints can remain usable over months, but accuracy might slowly worsen as the environment changes [26]. For magnetic fingerprints, large changes are less frequent (the building structure would have to change), but small changes such as new electronic equipment can perturb readings. Periodic recalibration could be conducted to update the database. For instance, update readings in areas that users frequent the most. More recently, approaches such as crowdsourcing (users contribute data) have been proposed to reduce the burden on a dedicated surveyor [27], but then ensuring the quality and proper labeling of crowd-collected data becomes an issue. These are discussed in Chapter 9.

Even if we frequently update the database to refine the model, users and surrounding dynamics can affect the real-time reading of the user device. The robust function mentioned in Chapter 2 can also alleviate this. Another good idea is to increase the data heterogeneity (e.g., using multiple technologies including Wi-Fi, cellular, magnetic) to increase the chance that some signals remain stable even if others fluctuate.

Other than the problems of data, the computational cost could also be an issue. Training a deep neural network can be computationally intensive (see Figure 5.8). Evaluating a neural network (especially if small to medium-sized) on a mobile device can be feasible due to optimized libraries and the relatively low-dimensional input. Some systems offload computation to a server (cloud localization services) to run heavy algorithms and send results back to the device. The choice may depend on privacy (sending raw sensor data out) versus device capability and latency requirements. In summary, machine learning has brought powerful tools to indoor positioning, enabling high accuracy in

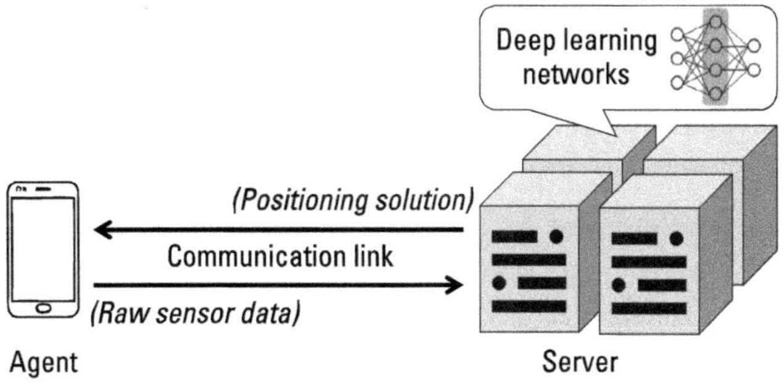

Figure 5.8 An illustration of conducting computationally intensive positioning algorithms by a centralized server with a communication link.

fingerprint-based systems. However, the success of these approaches is tightly coupled with data quality.

5.4.2 Context Awareness and Semantic Localization

Another direction is to incorporate context or semantic information. If the system knows, for example, that the user is on the second floor (maybe from an elevator sensor or user input in an app), it can eliminate matches to first-floor fingerprints entirely, improving accuracy. Context can come from user behavior (if the user is known to be in a meeting scheduled in a certain room, the system might give higher probability to that room). These ideas go beyond pure signal matching and into the realm of context-aware computing. They highlight that feature-matching algorithms can be part of a larger intelligent system that uses multiple sources of information.

For example, a sudden change in barometric pressure reading combined with a distinctive Wi-Fi signature might indicate that a user moved from one floor to another. Similarly, detecting a pattern of several consecutive door openings could hint that the user is moving through a corridor or checking rooms. By fusing such contextual cues with traditional fingerprints, the system can refine its estimates and even provide meaningful place labels (e.g., "in the elevator" or "in conference room 2") rather than just coordinates. This means that the semantic can be used for localization and that is actually how we human being navigate ourselves intrinsically. Pattern recognition can be applied to sensor data to recognize such semantic places, and machine learning can fuse different sensors to improve place recognition. For example, an algorithm might learn that a certain pattern of walking (detected by accelerometer) followed by a sharp drop in light level (entering a dark room) plus certain Wi-Fi signals corresponds to entering a specific lab room. This is quite complex to engineer manually, but deep learning could potentially learn these correlations if given enough training data with semantic labels.

5.4.3 Unavailability of Fingerprints: Physical Model-Based

Due to the cost and labor, the surveying of fingerprints may not be feasible for the database development. An alternative approach to build such a database is via physical model predictions, for example, simulating signal propagation paths for fingerprints on each location based on a 3-D environmental model. This method relies on prior knowledge of surrounding buildings, walls, and obstacles, as well as an appropriate physical model of the measurement features for realistic fingerprints that capable of determining user's location.

This approach will be demonstrated via an example of indoor positioning by radio signals.

In an indoor environment, radio signal propagation can be modeled (simulated) by ray-tracing algorithms. Ray-tracing algorithms compute the possible reflection and diffraction paths that a signal could take before reaching the receiver. This approach follows the law of reflection, where the angle of incidence equals the angle of reflection. Using a known 3-D map of the environment, ray-tracing can: (1) identify NLOS conditions by determining whether a direct path exists, (2) predict reflection paths that could lead to multipath errors, and (3) correct TDOA and AOA errors by incorporating known reflection-induced delays and angles. With the modeled ray traces, the multipath and NLOS can be detected. Furthermore, they can be regarded as additional features for indoor positioning. This is because the reflections and deflections are usually distinctive spatially, meaning that the multipath effects are very different if the location of the agent is different.

To utilize the modeled ray traces, the position-hypothesized method, such as a particle filter, is popularly developed. This can be illustrated by Figure 5.9. First, several hypothesized states $\mathbf{x}_{p,a}$ of the agent a are distributed based

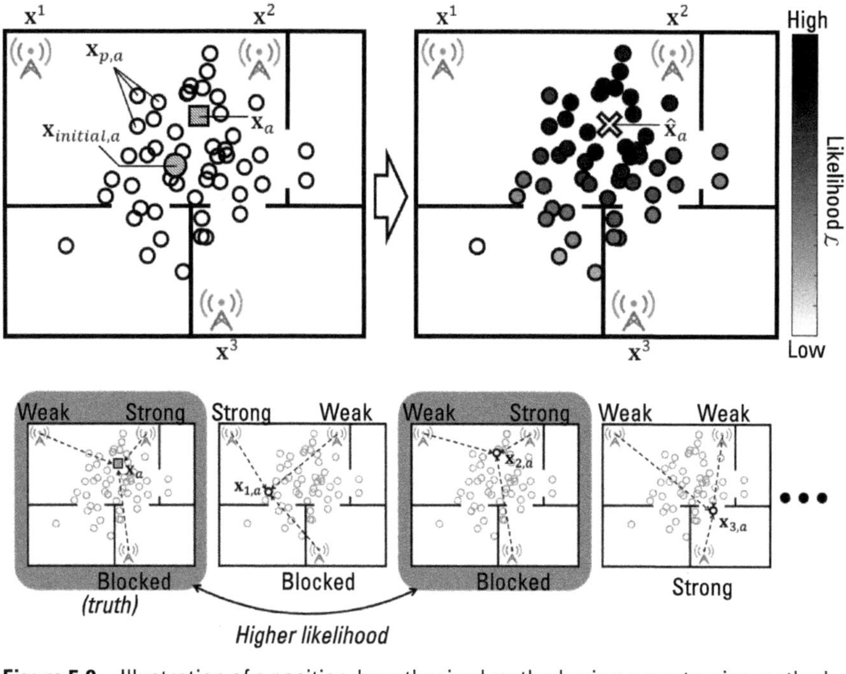

Figure 5.9 Illustration of a position-hypothesized method using a ray-tracing method.

on an initial guess ($\mathbf{x}_{\text{initial}}$, which can be provided by the LS methods mentioned earlier in Chapter 4 or commercial solutions). They can be distributed as grid points or as Gaussian random variables. The key is the uncertainty of the initial guess that has to be considered for distributing the hypothesized states, which is as follows:

$$\mathbf{x}_{p,a} \sim \mathcal{N}\left(\mathbf{x}_{\text{initial},a}, \sigma_{\text{initial},a}{}^{2}\right) \tag{5.7}$$

where $\sigma_{\text{initial},a}$ denotes the standard deviation of the error in the initial guess $\mathbf{x}_{\text{initial},a}$. Then, if one assumes that each measurement z_a^i in $\mathbf{z}_a$ follows independent Gaussian distribution, $z_a^i \sim \mathcal{N}(h_{\text{raytrace}}(\mathbf{x}_a, \mathbf{x}^i), \sigma_{\text{meas\_noise}}{}^2$, where $h_{\text{raytrace}}(\mathbf{x}_a, \mathbf{x}^i)$ denotes the simulated measurement from $\mathbf{x}^i$ to $\mathbf{x}_a$ by the ray-tracing model. The probability density function of each measurement can be represented by

$$f\left(z_a^i \middle| \mathbf{x}_a, \mathbf{x}^i, \sigma_{\text{meas\_noise}}\right) = \frac{1}{\sqrt{2\pi\sigma_{\text{meas\_noise}}{}^2}} e^{-\frac{1}{2}\left(\frac{z_a^i - h_{\text{raytrace}}\left(\mathbf{x}_a, \mathbf{x}^i\right)}{\sigma_{\text{meas\_noise}}}\right)^2} \tag{5.8}$$

For I measurements, the joint probability density function can be written as

$$\begin{aligned} f\left(z_a^1, \ldots, z_a^I \middle| \mathbf{x}_a, \mathbf{x}^1 \ldots, \mathbf{x}^I, \sigma_{\text{meas\_noise}}\right) &= \prod_{i=1}^{I} f\left(z_a^i \middle| \mathbf{x}_a, \mathbf{x}^i, \sigma_{\text{meas\_noise}}\right) \\ &= \left(\frac{1}{2\pi\sigma_{\text{meas\_noise}}{}^2}\right)^{I/2} e^{\left(-\frac{\sum_{i=1}^{I}\left(z_a^i - h_{\text{raytrace}}\left(\mathbf{x}_a, \mathbf{x}^i\right)\right)^2}{2\sigma_{\text{meas\_noise}}{}^2}\right)} \end{aligned} \tag{5.9}$$

Therefore, the likelihood of each hypothesized state can be represented by

$$\begin{aligned} &\mathcal{L}\left(\mathbf{x}_{p,a} \middle| \mathbf{z}_a, \mathbf{x}^1 \ldots, \mathbf{x}^I, \sigma_{\text{meas\_noise}}\right) \\ &= \left(\frac{1}{2\pi\sigma_{\text{meas\_noise}}{}^2}\right)^{I/2} e^{\left(-\frac{\sum_{i=1}^{I}\left(z_a^i - h_{\text{raytrace}}\left(\mathbf{x}_{p,a}, \mathbf{x}^i\right)\right)^2}{2\sigma_{\text{meas\_noise}}{}^2}\right)} \end{aligned} \tag{5.10}$$

The ray-tracing model can be used modeling the TOA, TDOA, RSS, and AOA. There are many commercial solutions on ray-tracing algorithms that can directly be used in (5.8). Finally, by considering all the likelihoods, the agent's state can be estimated using a weighted average calculation, where the weighting of each hypothesized state $\mathbf{x}_{p,a}$ is described by its likelihood:

$$\hat{\mathbf{x}}_a = \frac{\sum_p \mathbf{x}_{p,a} \cdot \mathcal{L}\left(\mathbf{x}_{p,a} | \mathbf{z}_a, \mathbf{x}^1 \ldots, \mathbf{x}^I, \sigma_{\text{meas\_noise}}\right)}{\sum_p \mathcal{L}\left(\mathbf{x}_{p,a} | \mathbf{z}_a, \mathbf{x}^1 \ldots, \mathbf{x}^I, \sigma_{\text{meas\_noise}}\right)} \tag{5.11}$$

To implement this method, there are several challenges. The better handling of the following challenges we do, the better positioning performance we can obtain.

1. *The comprehensiveness of the 3-D model:* 3-D models can be described in several levels of detail [28]. For example, if the ray-tracing method cannot have the material of different plane (or even pixel) of the model, the attenuated signal strength of the traced signal cannot be too practical. This is only one attribute. If the shape of the obstacles cannot be well-described by the 3-D models, the accuracy of the ray-tracing model will be affected. Last but not least, the dynamic objects surrounding the agent may also cause reflection effects to the signal received by the agent, resulting in the degraded performance of the likelihood estimation in (5.10).
2. *The complexity of the ray-tracing model:* The ray-tracing method has to consider all the potential blockages, reflections, and diffraction, meaning that it has to consider all the edges and planes within the 3-D models. If multiple reflections during the transmission from a beacon to an agent are considered, the computational load will increase exponentially. As a result, the balance between the computational load to the accuracy of the developed ray-tracing model is also a challenge.
3. *The multimodel issue:* This position-hypothesized method is relying on the assumption that the reflection effects are distinctive between different spatially distributed hypothesized states. However, if the reflections received at two different locations are similar in a way, it may cause a multimodel issue, meaning that there are two very high likelihoods at two different locations. It will cause a problem when calculating the weighted average of the hypothesized states, that is, (5.1). This multimodel will frequently occur if the received measurements are very noisy due to the likelihood of different hypothesized states becoming similar. An approach to alleviate the issue is to analyze the likelihoods obtained. In general, the likelihoods should descend with a gradient instead of being uniform-distributed.

5.5 Conclusions

In this chapter, we explored indoor positioning techniques based on feature matching, encompassing fingerprinting, pattern recognition, and machine learning approaches. These methods provide an alternative to geometric and signal propagation-based techniques (such as TOA, TDOA, and AOA methods discussed in earlier chapters) by leveraging the unique signatures of the environment. In comparison to traditional methods, fingerprinting and learning-based approaches often excel in environments where obtaining geometric measurements (distances or angles) is impractical or where multipath effects are too complex for simple models. The feature-matching method effectively treats the environment as a black box and learns its behavior empirically. This empirical approach can surpass geometric model-based methods in complex settings, but it requires care to ensure that the learned models remain valid and adaptive to changes. However, these black boxes approach can be an issue when troubleshooting why the system gives a grossly incorrect estimate; it might be hard to reason whether it was due to an overfitted model or a rare pattern that confused it. This actually hinders the continuous advancement of deep learning-based indoor feature-matching system.

While feature matching-based indoor positioning algorithms leverage environmental signatures such as Wi-Fi, BLE, and magnetic field anomalies, these methods inherently depend on the stability of external signals. However, in many scenarios, such as underground facilities, complex indoor areas, or environments with limited infrastructure, external signals may be weak, unreliable, or completely unavailable. Another drawback of feature matching-based methods is that they are environment-specific and inefficient to adapt to new settings; a new environment still requires collecting fresh data and fine-tuning the offline-trained model. To address these limitations, proprioceptive sensors, including inertial measurement units (IMUs) and odometers, offer an alternative approach for positioning based on a user's own motion. Chapter 6 will continue to introduce the use of proprioceptive sensors for indoor positioning.

References

[1] Ou, C. W., et al., "A ZigBee Position Technique for Indoor Localization Based on Proximity Learning," *2017 IEEE International Conference on Mechatronics and Automation (ICMA)*, August 6–9, 2017, pp. 875–880.

[2] Liu, H., et al., "Survey of Wireless Indoor Positioning Techniques and Systems," *IEEE Transactions on Systems, Man, and Cybernetics, Part C (Applications and Reviews)*, Vol. 37, No. 6, 2007, pp. 1067–1080.

[3] Bahl, P., and V. N. Padmanabhan, "RADAR: An In-Building RF-Based User Location and Tracking System," *Proceedings IEEE INFOCOM 2000. Conference on Computer Communications. Nineteenth Annual Joint Conference of the IEEE Computer and Communications Societies (Cat. No.00CH37064)*, Vol. 2, March 26–30, 2000, pp. 775–784.

[4] Haverinen, J., and A. Kemppainen, "Global Indoor Self-Localization Based on the Ambient Magnetic Field," *Robotics and Autonomous Systems*, Vol. 57, No. 10, 2009, pp. 1028–1035.

[5] Wu, K., et al., "CSI-Based Indoor Localization," *IEEE Transactions on Parallel and Distributed Systems*, Vol. 24, No. 7, 2013, pp. 1300–1309.

[6] Li, B., et al., "How Feasible Is the Use of Magnetic Field Alone for Indoor Positioning?" *2012 International Conference on Indoor Positioning and Indoor Navigation (IPIN)*, November 13–15, 2012, pp. 1–9.

[7] Youssef, M., and A. Agrawala, "The Horus Location Determination System," *Wireless Networks*, Vol. 14, No. 3, 2008, pp. 357–374.

[8] Dawes, B., and K. -W. Chin, "A Comparison of Deterministic and Probabilistic Methods for Indoor Localization," *Journal of Systems and Software*, Vol. 84, No. 3, 2011, pp. 442–451.

[9] Xia, S., et al., "Indoor Fingerprint Positioning Based on Wi-Fi: An Overview," *ISPRS International Journal of Geo-Information*, Vol. 6, No. 5, 2017, p. 135, https://www.mdpi.com/2220-9964/6/5/135.

[10] Hähnel, B. F. D., and D. Fox, "Gaussian Processes for Signal Strength-Based Location Estimation," *Proceeding of Robotics: Science and Systems*, 2006.

[11] Brunato, M., and R. Battiti, "Statistical Learning Theory for Location Fingerprinting in Wireless LANs," *Computer Networks*, Vol. 47, No. 6, 2005, pp. 825–845.

[12] Jondhale, S. R., et al., "Support Vector Regression for Mobile Target Localization in Indoor Environments," *Sensors*, Vol. 22, No. 1, 2022, p. 358, https://www.mdpi.com/1424-8220/22/1/358.

[13] Braga, P. H. M., and H. F. Bassani, "A Semi-Supervised Self-Organizing Map for Clustering and Classification," *2018 International Joint Conference on Neural Networks (IJCNN)*, July 8–13, 2018, pp. 1–8.

[14] Deng, K., *Omega: On-Line Memory-Based General Purpose System Classifier*, Carnegie Mellon University, 1999.

[15] Wikipedia, "Stepwise Regression," 2023, https://en.wikipedia.org/wiki/Stepwise_regression.

[16] Potortì, F., "The IPIN Competition 2024," *IEEE Dataport*, 2024.

[17] Alhomayani, F., and M. H. Mahoor, "Deep Learning Methods for Fingerprint-Based Indoor Positioning: A Review," *Journal of Location Based Services*, Vol. 14, No. 3, 2020, pp. 129–200.

[18] Battiti, R., N. T. Le, and A. Villani, "Location-Aware Computing: A Neural Network Model for Determining Location in Wireless LANs," 2002.

[19] LeCun, Y., Y. Bengio, and G. Hinton, "Deep Learning," *Nature*, Vol. 521, No. 7553, 2015, pp. 436–444.

[20] Chen, H., et al., "ConFi: Convolutional Neural Networks Based Indoor Wi-Fi Localization Using Channel State Information," *IEEE Access*, Vol. 5, 2017, pp. 18066–18074.

[21] Zhang, Z., et al., "TIPS: Transformer Based Indoor Positioning System Using Both CSI and DoA of WiFi Signal," *IEEE Access*, Vol. 10, 2022, pp. 111363–111376.

[22] Torres-Sospedra, J., et al., "UJIIndoorLoc-Mag: A New Database for Magnetic Field-Based Localization Problems," *2015 International Conference on Indoor Positioning and Indoor Navigation (IPIN)*, October 13–16, 2015, pp. 1–10.

[23] Zhang, C., et al., "Federated Radio Frequency Fingerprinting with Model Transfer and Adaptation," *IEEE INFOCOM 2023—IEEE Conference on Computer Communications Workshops (INFOCOM WKSHPS)*, May 2023, pp. 1–6.

[24] Jan, S. -S., L. -T. Hsu, and W. -M. Tsai, "Development of an Indoor Location Based Service Test Bed and Geographic Information System with a Wireless Sensor Network," *Sensors*, Vol. 10, No. 4, 2010, pp. 2957–2974, https://www.mdpi.com/1424-8220/10/4/2957.

[25] Amores, E. M. L., "Data Augmentation Models for Improved Indoor Positioning Accuracy Using RSS Fingerprinting," Master's Degree in Data Science, Geographic Information Systems, Satellite Images and Geopositioning, Universitat Oberta de Catalunya, 2024, http://hdl.handle.net/10609/152082.

[26] Mendoza-Silva, G. M., et al., "Long-Term WiFi Fingerprinting Dataset for Research on Robust Indoor Positioning," *Data*, Vol. 3, No. 1, 2018, p. 3, https://www.mdpi.com/2306-5729/3/1/3.

[27] Rai, A., et al., "Zee: Zero-Effort Crowdsourcing for Indoor Localization," *Proceedings of the 18th Annual International Conference on Mobile Computing and Networking*, Istanbul, Turkey, 2012.

[28] Gröger, G., and L. Plümer, "CityGML—Interoperable Semantic 3D City Models," *ISPRS Journal of Photogrammetry and Remote Sensing*, Vol. 71, 2012, pp. 12–33.

6

Positioning by Proprioceptive Sensors and Environmental Sensors

This chapter focuses on positioning techniques based on proprioceptive sensors, such as inertial sensors and wheel speed sensors, as well as environmental sensors, including magnetometers and barometers. These sensors enable self-contained navigation using dead reckoning, an approach that estimates position based on recorded motion data from a known initial state (as illustrated in Figure 6.1, which provides an example of dead reckoning at sea). Proprioceptive sensors measure internal motion characteristics of a platform [1]. For instance, an inertial measurement unit (IMU), which consists of accelerometers and gyroscopes, can be used to compute a moving object's position by integrating acceleration and angular rate data over time. Similarly, wheel speed sensors (or encoders) provide velocity information that allows dead reckoning to track a mobile robot's displacement. Dead reckoning is particularly useful in indoor environments, where external localization signals (e.g., Wi-Fi, or UWB) may be unavailable or unreliable.

While inertial navigation and wheel odometry are commonly used, they have cumulative errors due to sensor drift, noise, and bias. Over time, small errors in acceleration, angular velocity, or wheel speed measurements accumulate, leading to significant position drift. To address this issue, sensor fusion strategies integrating data from multiple sources or motion constraints

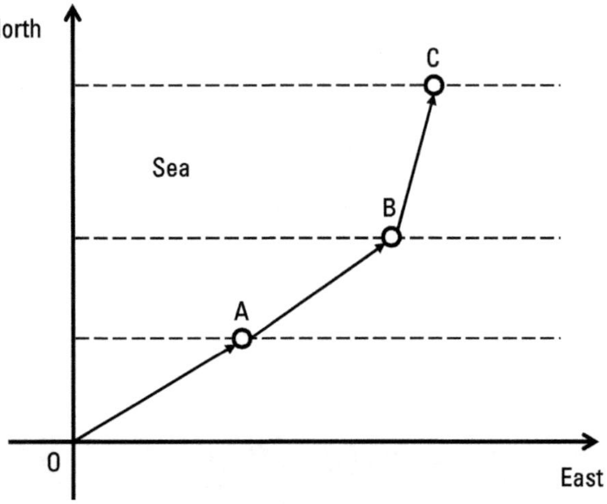

Figure 6.1 Dead reckoning at sea.

are integrated to improve positioning accuracy. For example, IMU and wheel speed sensor fusion can correct odometry errors caused by wheel slip, while zero-velocity updates (ZUPT) can help to mitigate drift in pedestrian navigation. Additionally, environmental sensors, such as magnetometers for heading estimation and barometers for altitude sensing, provide complementary information that enhances the accuracy of dead reckoning.

This chapter is organized as follows. Section 6.1 introduces inertial navigation, including the principles of inertial sensors and navigation equations that govern position estimation. Then it is followed by discussing wheel odometry and pedestrian inertial navigation. Section 6.4 introduces environmental sensors, including magnetometers for heading estimation and barometers for altitude sensing. Section 6.5 introduces error modeling and calibration. Section 6.6 explores drift mitigation techniques, covering initialization methods, and motion constraints. Finally, real-world examples and conclusions are provided.

6.1 Inertial Navigation

Inertial navigation is a self-contained navigation technique that employs measurements from inertial sensors to track the position, velocity, and attitude of an object relative to a known starting point [2]. Unlike other navigation methods, it does not rely on external references and does not emit signals to the external environment. This technology finds extensive use in indoor scenarios where continuous positioning is essential, especially when other information sources are unavailable.

6.1.1 Inertial Sensors

Inertial sensors, comprising gyroscopes and accelerometers, provide essential motion data by measuring angular velocity and acceleration, respectively. These sensors form the core components of an IMU, which enables motion tracking in 3-D space. In most navigation applications, an IMU consists of one gyroscope and one accelerometer per axis, allowing it to capture 6 degrees of freedom (6-DOF) motion.

IMUs are classified based on their performance characteristics, which determine their suitability for different applications. Table 6.1 presents a classification of IMUs by grade, detailing their typical position error, bias stability, and cost.

Strategic-grade and navigation-grade IMUs are used in weapons systems, military applications, and high-precision mapping, while commercial-grade IMUs are widely adopted for indoor positioning and consumer electronics due to their low cost and compact size [4]. These commercial IMUs rely heavily on microelectromechanical systems (MEMS) technology, which enables significant miniaturization and integration into mobile devices, robots, and even wearable systems.

Table 6.1
Classification of IMUs Based on Performance [3]

Grade	Strategic	Navigation	Tactical	Commercial
Position error	30–100 m/h	1 nmi*/h or 0.5 m/s	10–20 nmi/h	Large variation
Gyroscope bias instability	0.0001–0.001°/h	<0.01°/h	1–10°/h	0.1°/s
Accelerometer bias instability	0.1–1 μg	<100 μg	1–5 mg	100–1,000 μg
Cost	≈$1 million	≈$100,000	$2,000–$50,000	<$1 (accelerometers) <$10 (gyroscopes)

*1 nautical mile (nmi) ≈ 1,851m.

MEMS-based inertial sensors are commonly used in indoor positioning due to their small size, low power consumption, and affordability. They operate based on physical principles that enable the measurement of motion parameters:

MEMS gyroscopes use the Coriolis effect to measure angular velocity. When a mass m moves with velocity $\mathbf{v}$ within a rotating system, a Coriolis force $\mathbf{F}_c$ is exerted perpendicular to the motion direction. This force is given by:

$$\mathbf{F}_{\mathbf{c}} = -2 \cdot m \cdot (\omega \times \mathbf{v}) \tag{6.1}$$

where ω is the angular velocity vector of the rotating object. MEMS gyroscopes use vibrating elements to detect this force, which is then converted into an electrical signal proportional to angular velocity. Figure 6.2 illustrates the working principle of a MEMS gyroscope.

MEMS accelerometers measure specific force by detecting the displacement of a proof mass, which is suspended by a spring mechanism. When subjected to acceleration, the proof mass shifts, and the resulting deflection is proportional to the applied acceleration. The displacement is measured using capacitive, piezoelectric, or optical sensing techniques, as shown in Figure 6.3. MEMS accelerometers provide motion tracking, but their measurements include gravitational acceleration, requiring compensation in positioning applications.

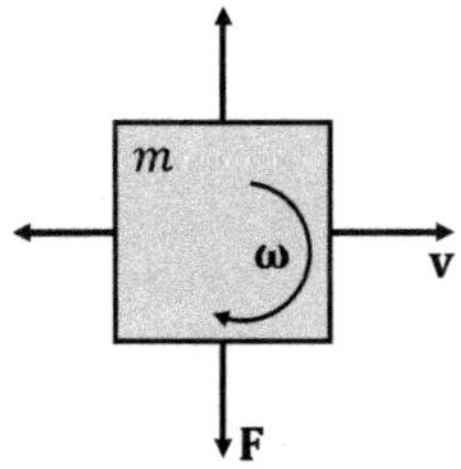

Figure 6.2 A vibrating mass gyroscope [5].

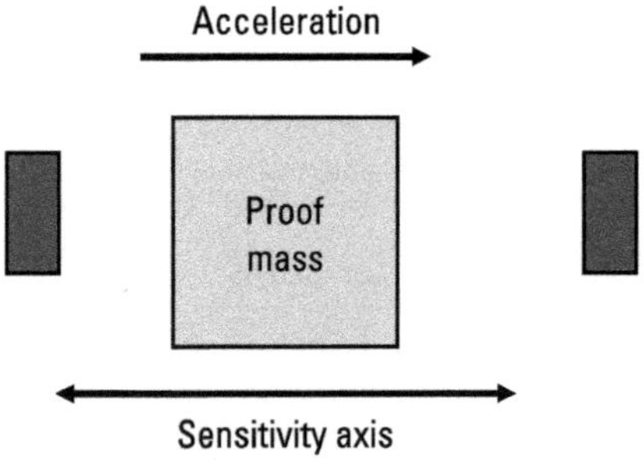

Figure 6.3 A mechanical accelerometer [6].

6.1.2 Inertial Navigation Equations

Inertial navigation is a dead reckoning technique that estimates an object's position, velocity, and attitude based on measurements from gyroscopes and accelerometers [7]. A system that integrates inertial sensors for navigation is referred to as an Inertial Navigation System (INS). INS can be categorized into two types:

1. *Stable-platform INS:* The inertial sensors are mounted on a gimballed platform that remains mechanically stabilized with respect to an inertial reference frame. While this setup reduces computational complexity, it comes with high cost, large volume, and complex maintenance requirements.
2. *Strap-down INS (SINS):* The IMU is directly attached to the moving object, and its orientation is updated numerically using onboard computing. This removes the need for mechanical gimbals, reducing cost and increasing reliability. Due to these advantages, strap-down inertial navigation is more widely adopted in modern applications [3].

MEMS-based IMUs are commonly used for indoor navigation due to their small size and affordability. However, MEMS-IMUs have higher noise levels and cannot detect the Earth's rotation rate (≈15°/h). Additionally, in low-dynamic environments such as indoor settings, traditional INS algorithms have been simplified to accommodate the limited motion characteristics of the user or robot [6]. The generalized framework of strap-down inertial navigation using MEMS-IMUs is shown in Figure 6.4. This framework consists of three core modules: attitude update (determining orientation), velocity update (determining speed and direction), and position update (determining displacement).

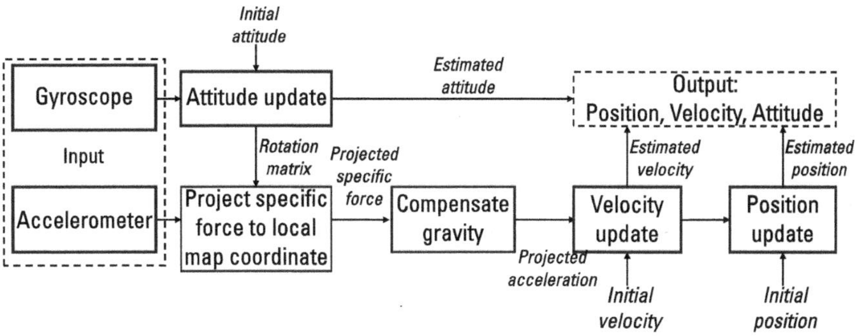

Figure 6.4 The generalized framework of strap-down inertial navigation based on MEMS-IMU.

6.1.2.1 Attitude Update

Attitude estimation is the first step in strap-down inertial navigation. The attitude of an object is commonly represented using Euler angles, rotation matrices (direction cosine matrices (DCM)), and quaternions. Among these, quaternions are preferred due to their numerical stability, computational efficiency, and lack of singularities (which occur in Euler angles) [8]. The kinematic equation for a quaternion-based attitude update is given by:

$$\dot{\mathbf{q}}_{k-1} = \frac{1}{2} \cdot \mathbf{\Omega}_{k-1} \cdot \mathbf{q}_{k-1} \tag{6.2}$$

where $\mathbf{q}_k$ is the quaternion at the instant t_k. $\mathbf{\Omega}_{k-1}$ is a skew-symmetric matrix constructed from the angular velocity vector $\boldsymbol{\omega}_{k-1}^B$, which is measured by the gyroscope. At the initial instant, the quaternion $\mathbf{q}_0$ is determined through the initial alignment. In general, for navigation-grade IMUs and above, all attitudes (roll, pitch, and heading) can be estimated using the gyroscope and accelerometer. For lower-grade IMUs, however, the heading must be estimated with the aid of additional sensors, such as a magnetometer [6].

The angular velocity matrix $\mathbf{\Omega}_{k-1}$ is given by:

$$\mathbf{\Omega}_{k-1} = \begin{bmatrix} 0 & -\omega_{x,k-1}^B & -\omega_{y,k-1}^B & -\omega_{z,k-1}^B \\ \omega_{x,k-1}^B & 0 & \omega_{z,k-1}^B & -\omega_{y,k-1}^B \\ \omega_{y,k-1}^B & -\omega_{z,k-1}^B & 0 & \omega_{x,k-1}^B \\ \omega_{z,k-1}^B & \omega_{y,k-1}^B & -\omega_{x,k-1}^B & 0 \end{bmatrix} \tag{6.3}$$

where $\omega_{k-1}^B = [\omega_{x,k-1}^B, \omega_{y,k-1}^B, \omega_{z,k-1}^B]^{\mathrm{T}}$ is the angular velocity expressed in the local body coordinate.

The quaternion update shown in (6.2) can be rewritten in discrete form:

$$\mathbf{q}_k = \mathbf{q}_{k-1} + \frac{1}{2} \cdot \mathbf{\Omega}_{k-1} \cdot \mathbf{q}_{k-1} \cdot \Delta t \tag{6.4}$$

where $\mathbf{q}_k$ is the current attitude estimate. In practical systems, the angular velocity ω is measured by gyroscopes includes bias and noise:

$$\tilde{\omega}_{k-1}^B = \boldsymbol{\omega}_{k-1}^B + \mathbf{b}_G^B + \mathbf{n}_G \tag{6.5}$$

where $\boldsymbol{\omega}_{k-1}^B$ is the true value of angular velocity. $\mathbf{b}_G^B$ is the gyroscope bias (assumed constant over short durations, meaning $\dot{\mathbf{b}}_G^B = 0$). $\mathbf{n}_G$ is the Gaussian white noise; each component in $\mathbf{n}_G$ satisfies $N(0, \sigma_G^2)$.

To improve accuracy, the gyroscope bias $\mathbf{b}_G^B$ is estimated during static periods by averaging multiple gyroscope readings. The corrected angular velocity is:

$$\hat{\boldsymbol{\omega}}_{k-1}^B = \tilde{\boldsymbol{\omega}}_{k-1}^B - \hat{\mathbf{b}}_G^B \tag{6.6}$$

which is then used in the quaternion update equations (6.2). Finally, the current estimate of the attitude $\hat{\mathbf{q}}_k$ can be obtained.

6.1.2.2 Velocity Update

Once the attitude is determined, the next step is to compute velocity. The velocity update equation is given by:

$$\mathbf{v}_k = \mathbf{v}_{k-1} + \left(\left(\mathbf{C}_M^B\right)^{\mathrm{T}} \mathbf{a}_{k-1}^B - \mathbf{g}^M\right)\Delta t \tag{6.7}$$

where $\mathbf{v}_k$ is the velocity at the instant t_k. $\mathbf{a}_{k-1}^B$ is the specific force (including motion acceleration and gravity) measured by the accelerometer, and $\mathbf{C}_M^B$ is the rotation matrix from the local map coordinate to local body coordinate. $\mathbf{g}^M$ is the gravity vector, expressed as $\mathbf{g}^M = [0, 0, g]^{\mathrm{T}}$ expressed in local map coordinate. The gravity varies with the location on Earth, and its calculation can be approximated as:

$$g = 9.7803 \cdot \left[1 + 0.0053024 \cdot \sin^2 \psi - 0.000005 \cdot \sin^2(2 \cdot \psi)\right] \tag{6.8}$$

where ψ is the latitude. In practical systems, the specific force is given by accelerometers, which is modeled as:

$$\tilde{\mathbf{a}}_{k-1}^B = \mathbf{a}_{k-1}^B + \mathbf{b}_A^B + \mathbf{n}_A \tag{6.9}$$

where $\mathbf{a}_{k-1}^B$ is the true value of the specific force. $\mathbf{b}_A^B$ is the accelerometer bias (assumed constant over short durations, meaning $\dot{\mathbf{b}}_A^B = 0$).$\mathbf{n}_A$ represents the white noise in accelerometers and assumed as Gaussian; each component in $\mathbf{n}_A$ satisfies $N(0, \sigma_A^2)$.

Unlike gyroscope bias, the accelerometer bias is unavailable even in static situations, as the gravity in the local body frame can affect the estimation. Therefore, the measured values of accelerometers $\tilde{\mathbf{a}}_{k-1}^B$ will be directly substituted into (6.7) when there is no external reference for bias estimation, and then the current estimate of the velocity $\hat{\mathbf{v}}_k$ can be obtained.

6.1.2.3 Position Update

The position update is achieved with the estimated velocity, which is shown as:

$$\hat{\mathbf{p}}_k = \hat{\mathbf{p}}_{k-1} + \hat{\mathbf{v}}_{k-1} \cdot \Delta t \tag{6.10}$$

where $\hat{\mathbf{p}}_k$ and $\hat{\mathbf{p}}_{k-1}$ are the current and previous estimates of the position.

6.2 Wheel Odometry for Mobile Robots

In inertial navigation, data from accelerometers is used to calculate speed, which is then used to estimate position. This method is versatile as it can work in any environment and with different objects. However, it has a significant drawback: the position error increases quickly over time because of the double integration. Researchers have used wheel speed sensors to reduce this error for position estimation [9, 10].

Wheel speed sensors measure the rotation of the wheels of a mobile robot to determine its speed and distance traveled. There are two types of wheel speed sensors: passive and active. Passive sensors are inductive analog devices that have a low signal-to-noise ratio (SNR) and are susceptible to vibration and interference. Active sensors are digital Hall devices that are widely used in navigation but are more expensive [11]. Wheel speed sensors can be used in two ways for positioning:

1. *Dead reckoning:* The speed or distance measured in the local body coordinate system is converted to the local map coordinate system, and then the converted values are summed up to calculate the position. If the robot has only one wheel speed sensor, it needs other sensors, such as a gyroscope or IMU, to provide the heading for the conversion.
2. *Information fusion:* The speed measured by the wheel speed sensors is combined with other sensors, such as IMU, as an observation. Kalman filter-based methods [12, 13] are often used for the integration.

Next we will introduce the two methods in detail.

6.2.1 Dead Reckoning Based on the Wheel Speed Sensor

Dead reckoning is a fundamental technique in robot navigation, estimating the robot's position over time using motion sensor data. In a system equipped with only a single wheel speed sensor, the direct measurement available is the radial speed of the wheel. However, this alone is insufficient for dead reckoning, as additional information is required to determine the robot's full velocity vector in the local map coordinate frame.

To achieve accurate dead reckoning, the system requires attitude measurements (e.g., from an IMU or other orientation sensors) to convert the wheel's radial speed into velocity components in the global reference frame. The generalized framework for dead reckoning based on speed measurements is illustrated in Figure 6.5.

In practical implementations, the attitude sensor (e.g., an IMU) is often physically mounted separately from the wheel speed sensor. As a result, the spatial transformation between these sensors must be accounted for. Figure 6.6 states the spatial relationship between the attitude measurement sensor (A-frame) and the wheel speed sensor (S-frame).

From Figure 6.6, we observe that there exists both a rotation between the speed sensor frame (S-frame) and the attitude sensor frame (A-frame) and a translation offset between these two sensors. Thus, to correctly transform the wheel speed measurement into the attitude coordinate frame, we must consider these extrinsic parameters (sensor rotation and translation). If the wheel speed sensor measures $\tilde{v}_{k-1}^{S}$, the measurement in the speed coordinate

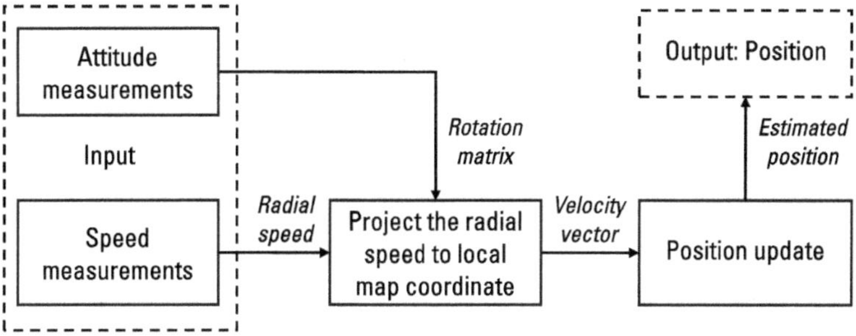

Figure 6.5 The generalized framework of dead reckoning based on speed measurements.

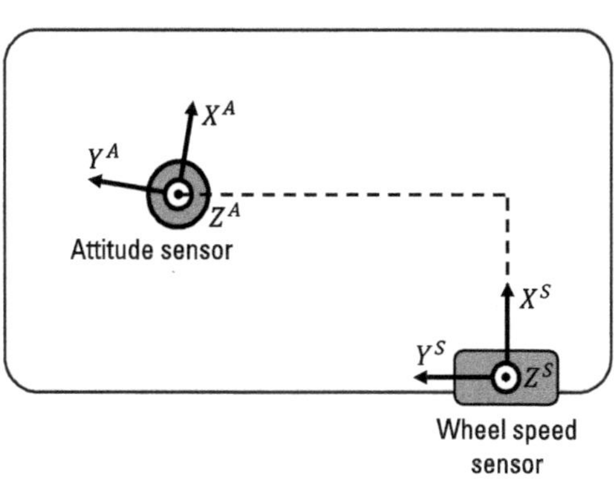

Figure 6.6 Spatial relationship between the attitude sensor and wheel speed sensor.

is $\tilde{\mathbf{v}}_{k-1}^{S} = [0, \tilde{v}_{k-1}^{S}, 0]^{\mathrm{T}}$ based on the nonholonomic constraint (NHC), which assumes that the lateral and vertical speeds of the robot are zero unless it slides or jumps. Let $\mathbf{l}^A = [l_x^A, l_y^A, l_z^A]^{\mathrm{T}}$ be the translation between the attitude and wheel speed sensors, which represents the origin of the speed coordinate in the attitude coordinate. Also, let $\mathbf{C}_S^A$ be the rotation from the speed to the attitude coordinate. Then $\tilde{\mathbf{v}}_{k-1}^{S}$ can be transformed to the attitude coordinate as follows:

$$\tilde{\mathbf{v}}_{k-1}^{A} = \hat{\mathbf{C}}_S^A \cdot \tilde{\mathbf{v}}_{k-1}^{S} - \left[\tilde{\boldsymbol{\omega}}^A\right]_\times \cdot \hat{\mathbf{l}}^A \tag{6.11}$$

where $\hat{\mathbf{C}}_S^A$ and $\hat{\mathbf{l}}^A$ are estimated extrinsic parameters. $\tilde{\boldsymbol{\omega}}^A$ is the angular velocity in the attitude coordinate system, which the attitude sensor can measure. $[\tilde{\boldsymbol{\omega}}^A]_\times$ can be expressed as:

$$\left[\tilde{\boldsymbol{\omega}}^A\right]_\times = \begin{bmatrix} 0 & -\tilde{\omega}_z^A & \tilde{\omega}_y^A \\ \tilde{\omega}_z^A & 0 & -\tilde{\omega}_x^A \\ -\tilde{\omega}_y^A & \tilde{\omega}_x^A & 0 \end{bmatrix} \tag{6.12}$$

Based on the attitude measurement, the rotation matrix $\tilde{\mathbf{C}}_A^M$ can be calculated.

$$\tilde{\mathbf{C}}_A^M = \begin{bmatrix} \left(1-2\cdot\tilde{q}_2^2-2\cdot\tilde{q}_3^2\right) & 2\cdot\left(\tilde{q}_1\cdot\tilde{q}_2-\tilde{q}_0\cdot\tilde{q}_3\right) & 2\cdot\left(\tilde{q}_1\cdot\tilde{q}_3+\tilde{q}_0\cdot\tilde{q}_2\right) \\ 2\cdot\left(\tilde{q}_1\cdot\tilde{q}_2+\tilde{q}_0\cdot\tilde{q}_3\right) & \left(1-2\cdot\tilde{q}_1^2-2\cdot\tilde{q}_3^2\right) & 2\cdot\left(\tilde{q}_2\cdot\tilde{q}_3-\tilde{q}_0\cdot\tilde{q}_1\right) \\ \left(\tilde{q}_1\cdot\tilde{q}_3-\tilde{q}_0\cdot\tilde{q}_2\right) & 2\cdot\left(\tilde{q}_2\cdot\tilde{q}_3+\tilde{q}_0\cdot\tilde{q}_1\right) & \left(1-2\cdot\tilde{q}_1^2-2\cdot\tilde{q}_2^2\right) \end{bmatrix} \tag{6.13}$$

where $\tilde{\mathbf{q}}_k = [\tilde{q}_0, \tilde{q}_1, \tilde{q}_2, \tilde{q}_3]^{\mathrm{T}}$ represents the quaternion obtained from the attitude measurement. Then $\tilde{\mathbf{v}}_{k-1}^{A}$ can be converted to the local map coordinate system as follows:

$$\hat{\mathbf{v}}_{k-1} = \tilde{\mathbf{C}}_A^M \cdot \tilde{\mathbf{v}}_{k-1}^{A} \tag{6.14}$$

Finally, the position estimate can be obtained, which is shown as:

$$\hat{\mathbf{p}}_k = \hat{\mathbf{p}}_{k-1} + \hat{\mathbf{v}}_{k-1} \cdot \Delta t \tag{6.15}$$

6.2.2 IMU and Wheel Speed Sensor Integrated Navigation

The wheel speed sensor measurement can also be used to correct the inertial navigation in a Kalman filter [14, 15] for state estimation, the details of which can be found in Chapter 3. Figure 6.7 shows the general framework of the integrated navigation with IMU and wheel speed sensor. The IMU data is used for inertial navigation and state prediction. The radial speed and NHC are used as the observation. The output consists of the position, velocity, and attitude. Next we will explain the Kalman filter formulation in detail.

The complete system state at the instant t_k consists of 16 states, which is shown as:

$$\mathbf{x}_k = \left[\mathbf{p}_k, \mathbf{v}_k, \mathbf{q}_k, \mathbf{b}_{G,k}^B, \mathbf{b}_{A,k}^B\right]^{\mathrm{T}} \tag{6.16}$$

where $\mathbf{x}_k$ includes position, velocity, attitude, gyroscope bias, and accelerometer bias. First, IMU data is utilized for inertial navigation to predict the state. The state transition equation can be expressed as:

$$\mathbf{x}_k = \mathbf{f}\left(\mathbf{x}_{k-1}, \mathbf{u}_{k-1}\right) + \mathbf{w}_{k-1} \tag{6.17}$$

where $\mathbf{u}_{k-1}$ is the input vector from IMU measurements and $\mathbf{w}_{k-1}$ is the process noise. According to (6.4), the state transition of attitude can be written as:

$$\mathbf{q}_k = \mathbf{q}_{k-1} + \frac{1}{2}\cdot\left(\tilde{\mathbf{\Omega}}_{k-1}^B - \mathbf{\Omega}_{G,k-1}^B - \mathbf{\Omega}_G\right)\cdot\mathbf{q}_{k-1}\cdot\Delta t \tag{6.18}$$

where $\tilde{\mathbf{\Omega}}_{k-1}^B$, $\mathbf{\Omega}_{G,k-1}^B$, and $\mathbf{\Omega}_G$ are formed from the components of the angular rate $\tilde{\boldsymbol{\omega}}_{k-1}^B$, angular rate bias $\mathbf{b}_{G,k-1}^B$, and noise $\mathbf{n}_G$. Then (6.18) can be rewritten as:

$$\begin{aligned} \mathbf{q}_k &= \mathbf{q}_{k-1} + \frac{1}{2}\cdot\left(\tilde{\mathbf{\Omega}}_{k-1}^B - \mathbf{\Omega}_{G,k-1}^B\right)\cdot\mathbf{q}_{k-1}\cdot\Delta t - \frac{1}{2}\cdot\mathbf{\Omega}_G\cdot\mathbf{q}_{k-1}\cdot\Delta t \\ \mathbf{q}_k &= \left[\mathbf{I}_4 + \frac{1}{2}\cdot\left(\tilde{\mathbf{\Omega}}_{k-1}^B - \mathbf{\Omega}_{G,k-1}^B\right)\cdot\Delta t\right]\cdot\mathbf{q}_{k-1} - \frac{\Delta t}{2}\cdot\Xi\cdot\mathbf{n}_G \end{aligned} \tag{6.19}$$

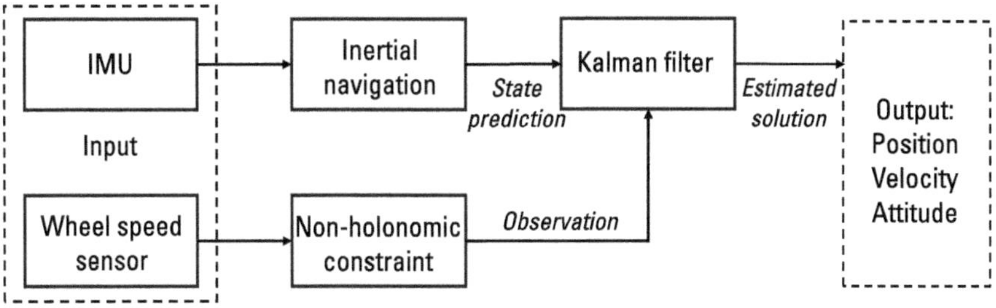

Figure 6.7 The generalized framework of IMU and wheel speed sensor integrated navigation.

where

$$\Xi = \begin{bmatrix} -q_1 & -q_2 & -q_3 \\ q_0 & -q_3 & q_2 \\ q_3 & q_0 & -q_1 \\ -q_2 & q_1 & q_0 \end{bmatrix} \tag{6.20}$$

The corresponding covariance matrix can be expressed as:

$$\begin{gathered} \boldsymbol{\Sigma}^q = \frac{\Delta t}{2} \cdot \Xi \cdot \mathbf{n}_G \cdot \left(\frac{\Delta t}{2} \cdot \Xi \cdot \mathbf{n}_G \right)^{\mathrm{T}} \\ \boldsymbol{\Sigma}^q = \left(\frac{\sigma_G \cdot \Delta t}{2} \right)^2 \cdot \begin{bmatrix} 1 - q_0^2 & -q_0 \cdot q_1 & -q_0 \cdot q_2 & -q_0 \cdot q_3 \\ -q_0 \cdot q_1 & 1 - q_1^2 & -q_1 \cdot q_2 & -q_1 \cdot q_3 \\ -q_0 \cdot q_2 & -q_1 \cdot q_2 & 1 - q_2^2 & -q_2 \cdot q_3 \\ -q_0 \cdot q_3 & -q_1 \cdot q_3 & -q_2 \cdot q_3 & 1 - q_3^2 \end{bmatrix} \end{gathered} \tag{6.21}$$

According to (6.7), the state transition of velocity can be written as:

$$\begin{gathered} \mathbf{v}_k = \mathbf{v}_{k-1} + \left[\mathbf{C}_M^{B\ \mathrm{T}} \cdot \left(\tilde{\mathbf{a}}_{k-1}^B - \mathbf{b}_{A,k-1}^B - \mathbf{n}_A \right) - \mathbf{g}^M \right] \cdot \Delta t \\ \mathbf{v}_k = \mathbf{v}_{k-1} + \left[\mathbf{C}_M^{B\ \mathrm{T}} \cdot \left(\tilde{\mathbf{a}}_{k-1}^B - \mathbf{b}_{A,k-1}^B \right) - \mathbf{g}^M \right] \cdot \Delta t - \mathbf{C}_M^{B\ \mathrm{T}} \cdot \mathbf{n}_A \cdot \Delta t \end{gathered} \tag{6.22}$$

The corresponding covariance matrix can be expressed as:

$$\begin{gathered} \boldsymbol{\Sigma}^v = \mathbf{C}_M^{B\ \mathrm{T}} \cdot \mathbf{n}_A \cdot \Delta t \cdot \left(\mathbf{C}_M^{B\ \mathrm{T}} \cdot \mathbf{n}_A \cdot \Delta t \right)^{\mathrm{T}} \\ \boldsymbol{\Sigma}^v = \left(\sigma_A \cdot \Delta t \right)^2 \cdot \mathbf{I}_3 \end{gathered} \tag{6.23}$$

According to (6.15), the state transition of position can be written as:

$$\mathbf{p}_k = \mathbf{p}_{k-1} + \mathbf{v}_{k-1} \cdot \Delta t + \mathbf{C}_M^{B\ \mathrm{T}} \cdot \mathbf{n}_A \cdot \Delta t^2 \tag{6.24}$$

The corresponding covariance matrix can be expressed as:

$$\begin{gathered} \boldsymbol{\Sigma}^p = \mathbf{C}_M^{B\ \mathrm{T}} \cdot \mathbf{n}_A \cdot \Delta t^2 \cdot \left(\mathbf{C}_M^{B\ \mathrm{T}} \cdot \mathbf{n}_A \cdot \Delta t^2 \right)^{\mathrm{T}} \\ \boldsymbol{\Sigma}^p = \left(\sigma_A \cdot \Delta t^2 \right)^2 \cdot \mathbf{I}_3 \end{gathered} \tag{6.25}$$

Therefore, $\mathbf{f}(\mathbf{x}_{k-1}, \mathbf{u}_{k-1})$ and $\mathbf{w}_{k-1}$ can be expressed as:

$$\mathbf{f}\left(\mathbf{x}_{k-1},\mathbf{u}_{k-1}\right)=\begin{bmatrix} \mathbf{p}_{k-1}+\mathbf{v}_{k-1}\cdot\Delta t \\ \mathbf{v}_{k-1}+\left[{\mathbf{C}_M^B}^{\mathrm{T}}\cdot\left(\tilde{\mathbf{a}}_{k-1}^B-\mathbf{b}_{A,k-1}^B\right)-\mathbf{g}^M\right]\cdot\Delta t \\ \left[\mathbf{I}_4+\frac{1}{2}\cdot\left(\tilde{\mathbf{\Omega}}_{k-1}^B-\mathbf{\Omega}_{G,k-1}^B\right)\cdot\Delta t\right]\cdot\mathbf{q}_{k-1} \\ \mathbf{I}_3 \\ \mathbf{I}_3 \end{bmatrix} \tag{6.26}$$

$$\mathbf{w}_{k-1}=\begin{bmatrix} {\mathbf{C}_M^B}^{\mathrm{T}}\cdot\mathbf{n}_A\cdot\Delta t^2 \\ -{\mathbf{C}_M^B}^{\mathrm{T}}\cdot\mathbf{n}_A\cdot\Delta t \\ -\frac{\Delta t}{2}\cdot\Xi\cdot\mathbf{n}_G \\ 0_{3\times1} \\ 0_{3\times1} \end{bmatrix} \tag{6.27}$$

According to (6.26), the process Jacobian can be expressed as:

$$\mathbf{F}_{k-1}=\frac{\partial\mathbf{f}\left(\mathbf{x}_{k-1},\mathbf{u}_{k-1}\right)}{\partial\mathbf{x}_{k-1}}$$

$$\mathbf{F}_{k-1}=\mathbf{I}_{16}+\begin{bmatrix} 0_3 & \mathbf{I}_3 & 0_{3\times4} & 0_3 & 0_3 \\ 0_3 & 0_3 & \partial\mathbf{v}\,\partial\mathbf{q} & 0_3 & -{\mathbf{C}_M^B}^{\mathrm{T}} \\ 0_{4\times3} & 0_{4\times3} & \frac{1}{2}\cdot\left(\tilde{\mathbf{\Omega}}_{k-1}^B-\mathbf{\Omega}_{G,k-1}^B\right) & -\frac{1}{2}\cdot\Xi & 0_{4\times3} \\ 0_3 & 0_3 & 0_{3\times4} & 0_3 & 0_3 \\ 0_3 & 0_3 & 0_{3\times4} & 0_3 & 0_3 \end{bmatrix}\cdot\Delta t \tag{6.28}$$

where

$$\partial\mathbf{v}\,\partial\mathbf{q}=2\cdot\mathbf{Q}_F\cdot\begin{bmatrix} 0 & \left(\mathbf{a}_{k-1}^B\right)^{\mathrm{T}} \\ \mathbf{a}_{k-1}^B & -\left[\mathbf{a}_{k-1}^B\right] \end{bmatrix} \tag{6.29}$$

where

$$\mathbf{Q}_F = \begin{bmatrix} q_1 & q_0 & -q_3 & q_2 \\ q_2 & q_3 & q_0 & -q_1 \\ q_3 & -q_2 & q_1 & q_0 \end{bmatrix} \tag{6.30}$$

$$\mathbf{a}_{k-1}^B = \tilde{\mathbf{a}}_{k-1}^B - \mathbf{b}_{A,\,k-1}^B$$

The process noise covariance can be expressed as:

$$\mathbf{Q}_{k-1} = \begin{bmatrix} \boldsymbol{\Sigma}^p & 0_3 & 0_{3\times4} & 0_3 & 0_3 \\ 0_3 & \boldsymbol{\Sigma}^v & 0_{3\times4} & 0_3 & 0_3 \\ 0_{4\times3} & 0_{4\times3} & \boldsymbol{\Sigma}^q & 0_{4\times3} & 0_{4\times3} \\ 0_3 & 0_3 & 0_{3\times4} & 0_3 & 0_3 \\ 0_3 & 0_3 & 0_{3\times4} & 0_3 & 0_3 \end{bmatrix} \tag{6.31}$$

Therefore, the covariance matrix prediction can be expressed as:

$$\mathbf{P}_{k|k-1} = \mathbf{F}_{k-1} \cdot \mathbf{P}_{k-1|k-1} \cdot \mathbf{F}_{k-1}^{\mathrm{T}} + \mathbf{Q}_{k-1} \tag{6.32}$$

The measurement model can be expressed as:

$$\begin{aligned} \mathbf{z}_k &= \mathbf{h}\left(\mathbf{x}_k\right) + \mathbf{r}_k \\ \tilde{\mathbf{v}}_k^A &= \mathbf{C}_M^A \cdot \mathbf{v}_k + \mathbf{r}_k \end{aligned} \tag{6.33}$$

$\mathbf{C}_M^A$ can be written as:

$$\mathbf{C}_M^A = \begin{bmatrix} \left(1 - 2 \cdot q_2^2 - 2 \cdot q_3^2\right) & 2 \cdot \left(q_1 \cdot q_2 + q_0 \cdot q_3\right) & 2 \cdot \left(q_1 \cdot q_3 - q_0 \cdot q_2\right) \\ 2 \cdot \left(q_1 \cdot q_2 - q_0 \cdot q_3\right) & \left(1 - 2 \cdot q_1^2 - 2 \cdot q_3^2\right) & 2 \cdot \left(q_2 \cdot q_3 + q_0 \cdot q_1\right) \\ 2 \cdot \left(q_1 \cdot q_3 + q_0 \cdot q_2\right) & 2 \cdot \left(q_2 \cdot q_3 - q_0 \cdot q_1\right) & \left(1 - 2 \cdot q_1^2 - 2 \cdot q_2^2\right) \end{bmatrix} \tag{6.34}$$

where $\mathbf{q}_k = [q_0, q_1, q_2, q_3]^{\mathrm{T}}$. $\mathbf{C}_M^A \cdot \mathbf{v}_k$ can be expressed as:

$$\begin{aligned} \left(\mathbf{C}_M^A \cdot \mathbf{v}_k\right)_1 &= \left(1 - 2 \cdot q_2^2 - 2 \cdot q_3^2\right) \cdot v_{x,k} + 2 \cdot \left(q_1 \cdot q_2 + q_0 \cdot q_3\right) \cdot v_{y,k} \\ &\quad + 2 \cdot \left(q_1 \cdot q_3 - q_0 \cdot q_2\right) \cdot v_{z,k} \\ \left(\mathbf{C}_M^A \cdot \mathbf{v}_k\right)_2 &= 2 \cdot \left(q_1 \cdot q_2 - q_0 \cdot q_3\right) \cdot v_{x,k} + \left(1 - 2 \cdot q_1^2 - 2 \cdot q_3^2\right) \cdot v_{y,k} \\ &\quad + 2 \cdot \left(q_2 \cdot q_3 + q_0 \cdot q_1\right) \cdot v_{z,k} \end{aligned}$$

$$\begin{aligned}\left(\mathbf{C}_M^A\cdot\mathbf{v}_k\right)_3 &= 2\cdot\left(q_1\cdot q_3+q_0\cdot q_2\right)\cdot v_{x,k}+2\cdot\left(q_2\cdot q_3-q_0\cdot q_1\right)\cdot v_{y,k}\\ &+\left(1-2\cdot q_1^2-2\cdot q_2^2\right)\cdot v_{z,k}\end{aligned} \tag{6.35}$$

where $(\mathbf{C}_M^A\cdot\mathbf{v}_k)_i$ is the ith element of $\mathbf{C}_M^A\cdot\mathbf{v}_k$, $\mathbf{v}_k=[v_{x,k}, v_{y,k}, v_{z,k}]^{\mathrm{T}}$. Taking the derivative of $\mathbf{v}_k$ and $\mathbf{q}_k$ in (6.35) yields the following form:

$$\frac{\partial\left(\mathbf{C}_M^A\cdot\mathbf{v}_k\right)}{\partial\mathbf{v}_k}=\begin{bmatrix}\left(1-2\cdot q_2^2-2\cdot q_3^2\right) & 2\cdot\left(q_1\cdot q_2+q_0\cdot q_3\right) & 2\cdot\left(q_1\cdot q_3-q_0\cdot q_2\right)\\ 2\cdot\left(q_1\cdot q_2-q_0\cdot q_3\right) & \left(1-2\cdot q_2^2-2\cdot q_3^2\right) & 2\cdot\left(q_2\cdot q_3+q_0\cdot q_1\right)\\ 2\cdot\left(q_1\cdot q_3+q_0\cdot q_2\right) & 2\cdot\left(q_2\cdot q_3-q_0\cdot q_1\right) & \left(1-2\cdot q_1^2-2\cdot q_2^2\right)\end{bmatrix} \tag{6.36}$$

$$\begin{aligned}\frac{\partial\left(\mathbf{C}_M^A\cdot\mathbf{v}_k\right)}{\partial\mathbf{q}_k} &= \begin{bmatrix}\mathbf{B}_{11} & \mathbf{B}_{12} & \mathbf{B}_{13} & \mathbf{B}_{14}\\ \mathbf{B}_{21} & \mathbf{B}_{22} & \mathbf{B}_{23} & \mathbf{B}_{24}\\ \mathbf{B}_{31} & \mathbf{B}_{32} & \mathbf{B}_{33} & \mathbf{B}_{34}\end{bmatrix}\\ \mathbf{B}_{11} &= 2\cdot q_3\cdot v_{y,k}-2\cdot q_2\cdot v_{z,k}\\ \mathbf{B}_{12} &= 2\cdot q_2\cdot v_{y,k}+2\cdot q_3\cdot v_{z,k}\\ \mathbf{B}_{13} &= -4\cdot q_2\cdot v_{x,k}+2\cdot q_1\cdot v_{y,k}-2\cdot q_0\cdot v_{z,k}\\ \mathbf{B}_{14} &= -4\cdot q_3\cdot v_{x,k}+2\cdot q_0\cdot v_{y,k}+2\cdot q_1\cdot v_{z,k}\\ \mathbf{B}_{21} &= -2\cdot q_3\cdot v_{x,k}+2\cdot q_1\cdot v_{z,k}\\ \mathbf{B}_{22} &= 2\cdot q_2\cdot v_{x,k}-4\cdot q_1\cdot v_{y,k}+2\cdot q_0\cdot v_{z,k}\\ \mathbf{B}_{23} &= 2\cdot q_1\cdot v_{x,k}+2\cdot q_3\cdot v_{z,k}\\ \mathbf{B}_{24} &= -2\cdot q_0\cdot v_{x,k}-4\cdot q_3\cdot v_{y,k}+2\cdot q_2\cdot v_{z,k}\\ \mathbf{B}_{31} &= 2\cdot q_2\cdot v_{x,k}-2\cdot q_1\cdot v_{y,k}\\ \mathbf{B}_{32} &= 2\cdot q_3\cdot v_{x,k}-2\cdot q_0\cdot v_{y,k}-4\cdot q_1\cdot v_{z,k}\\ \mathbf{B}_{33} &= 2\cdot q_0\cdot v_{x,k}+2\cdot q_3\cdot v_{y,k}-4\cdot q_2\cdot v_{z,k}\\ \mathbf{B}_{34} &= 2\cdot q_1\cdot v_{x,k}+2\cdot q_2\cdot v_{y,k}\end{aligned} \tag{6.37}$$

Then the measurement matrix can be expressed as:

$$\mathbf{H}_k=\begin{bmatrix}0_{3\times3} & \dfrac{\partial\left(\mathbf{C}_M^A\cdot\mathbf{v}_k\right)}{\partial\mathbf{v}_k} & \dfrac{\partial\left(\mathbf{C}_M^A\cdot\mathbf{v}_k\right)}{\partial\mathbf{q}_k} & 0_{3\times3} & 0_{3\times3}\end{bmatrix} \tag{6.38}$$

Let us suppose that three components of $\mathbf{z}_k$ have the same noise characteristics σ_z^2; the covariance matrix of the measurement can be expressed as:

$$\mathbf{R}_k = \sigma_z^2 \cdot \mathbf{I}_3 \tag{6.39}$$

Based on the EKF equation, the innovation covariance can be expressed as:

$$\mathbf{S}_k = \mathbf{H}_k \cdot \mathbf{P}_{k|k-1} \cdot \mathbf{H}_k^{\mathrm{T}} + \mathbf{R}_k \tag{6.40}$$

The Kalman filter gain can be obtained as:

$$\mathbf{K}_k = \mathbf{P}_{k|k-1} \cdot \mathbf{H}_k^{\mathrm{T}} \cdot \mathbf{S}_k^{-1} \tag{6.41}$$

Then the state can be updated by:

$$\mathbf{x}_{k|k} = \mathbf{x}_{k|k-1} + \mathbf{K}_k \cdot \left[\mathbf{z}_k - \mathbf{h}\left(\mathbf{x}_k\right)\right] \tag{6.42}$$

Finally, the covariance matrix can be updated by:

$$\mathbf{P}_{k|k} = \mathbf{P}_{k|k-1} - \mathbf{K}_k \cdot \mathbf{H}_k \cdot \mathbf{P}_{k|k-1} \tag{6.43}$$

6.2.3 Summary

Dead reckoning based on wheel speed sensors has a relatively simple structure, but it is affected by sensor noises and errors as sensor information is directly used for position estimation. Information fusion can filter the noise and estimate the errors online, but it has a more complex structure.

6.3 Inertial Sensor-Based Pedestrian Navigation

Unlike mobile robots, pedestrians do not have odometry-like sensors to measure their speed. However, the human body's movements usually show some regularity, which can be detected by inertial sensors attached to pedestrians. The human motion can be used to reduce the error drift of inertial sensor-based navigation, without requiring any additional sensors. Inertial sensor-based pedestrian navigation can be divided into two types:

1. *Zero velocity updates:* When pedestrians are stationary, for example, while waiting for an elevator or at a red light, their velocity can be

considered zero. This can be used as an observation to correct inertial navigation errors. However, these scenes usually occupy only a small portion of pedestrian movement, which has a limited effect. When pedestrians walk, their feet become stationary periodically, which can be detected by a foot-mounted IMU to perform the ZUPT. This method has a higher update frequency and can effectively suppress the rapid accumulation of errors.

2. *Pedestrian dead reckoning (PDR):* PDR calculates the position by detecting pedestrian steps based on accelerometer data, which are then transformed to the local map coordinate using the heading from gyroscopes or magnetometers. After that, the pedestrians' position can be obtained by accumulating the transformed steps. In this method, the step length is derived based on the step detection rather than the integration from accelerometer data, it avoids the rapid drift.

Next we will introduce the two methods in detail.

6.3.1 Zero Velocity Updates

Figure 6.8 shows the gait cycle phase of a pedestrian during the movement [16]. The gray point represents the stationary point. The stationary situation is considered as a zero-velocity observation to correct the inertial navigation velocity. Figure 6.9 shows a general application mode of the foot-mounted IMU for ZUPT.

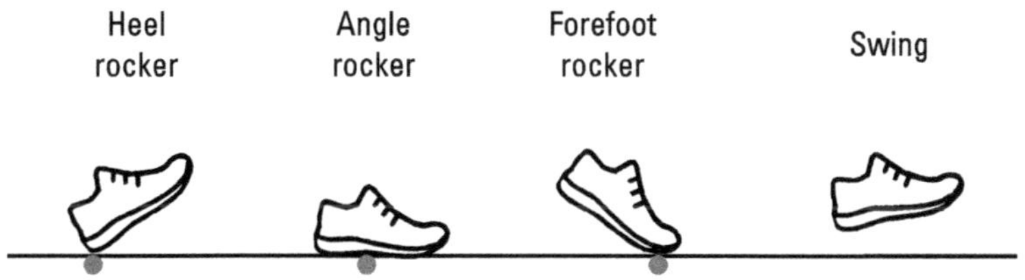

Figure 6.8 The gait cycle phase.

Figure 6.9 Foot-mounted IMU.

In ZUPT-aided inertial navigation methods, Kalman filter [17, 18] is commonly used to perform the fusion. The system state includes position, velocity, attitude, gyroscope bias, and accelerometer bias, as defined in (6.16), and its generalized framework is shown in Figure 6.10.

In stance phase detection, the system uses IMU data to recognize the stationary state by hypothesis testing [19], and the system can be considered stationary at the instant t_k based on the following equation.

$$\frac{1}{N}\cdot\sum_{l\in W_k}\left(\frac{1}{\sigma_A}\cdot\left\|\tilde{a}_l^B - g\cdot\frac{\bar{\mathbf{a}}_k^B}{\left\|\bar{\mathbf{a}}_k^B\right\|}\right\| + \frac{1}{\sigma_G}\cdot\left\|\tilde{\boldsymbol{\omega}}_l^B\right\|^2\right) < \gamma \tag{6.44}$$

where $\bar{\mathbf{a}}_k^B$ is the average of accelerometer measurements over the time window W_k of length N samples centered around instant t_k. σ_A and σ_G are the standard deviation of accelerometer and gyroscope data noise. $\tilde{\mathbf{a}}_l^B$ and $\tilde{\boldsymbol{\omega}}_l^B$ denote the accelerometer and gyroscope data at the instant t_l. γ is a zero-velocity detection threshold.

Once the system is stationary, the filter is built for state estimation. The system state and state prediction process are the same as those in IMU and wheel speed sensor integrated navigation, but the measurement model differs, which can be expressed as:

$$\begin{gathered}\mathbf{z}_k = \mathbf{h}(\mathbf{x}_k) + \mathbf{r}_k \\ \begin{bmatrix}0\\0\\0\end{bmatrix} = \begin{bmatrix}0_3 & \mathbf{I}_3 & 0_{3\times4} & 0_3 & 0_3\end{bmatrix}\cdot\mathbf{x}_k + \mathbf{r}_k \\ \mathbf{H}_k = \begin{bmatrix}0_3 & \mathbf{I}_3 & 0_{3\times4} & 0_3 & 0_3\end{bmatrix}\end{gathered} \tag{6.45}$$

Equations (6.40) to (6.43) can update the pedestrian state and its covariance matrix.

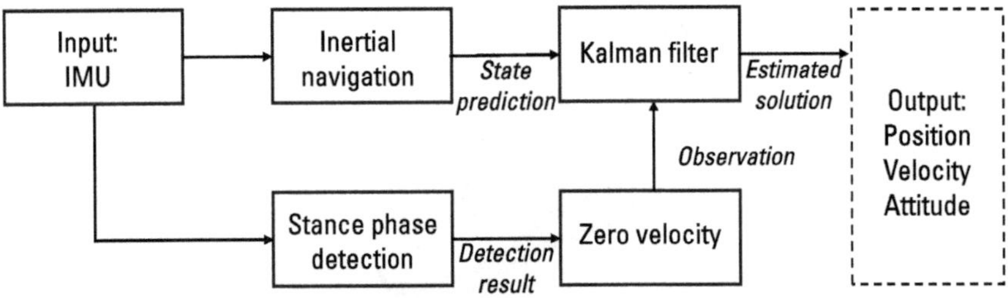

Figure 6.10 The generalized framework of ZUPT-aided inertial navigation.

6.3.2 PDR

PDR is a method that Levi and Judd first proposed in 1996 to provide autonomous navigation for foot travelers when a global positioning system (GPS) is unreliable or unavailable [20]. It resembles the dead reckoning scheme based on speed measurement, but it does not need an additional speed sensor. PDR has only one integral operation, so it can achieve higher positioning accuracy than classical inertial navigation.

PDR mainly comprises step detection, step length, and heading estimation [21], which is shown in Figure 6.11.

Step detection is the critical process of PDR, generally including the zero crossing, peak detection, autocorrelation, stance phase detection, and so on [22]. In this chapter, peak detection will be introduced to implement the detection. The total acceleration at each instance is first calculated using the following equation:

$$\tilde{a}_{k-1}^{To} = \left({\tilde{a}_{x,k-1}^{B}}^{2} + {\tilde{a}_{y,k-1}^{B}}^{2} + {\tilde{a}_{z,k-1}^{B}}^{2} \right)^{\frac{1}{2}} \tag{6.46}$$

$$a_{k-1}^{To} = a_{k-1}^{To} - g \tag{6.47}$$

where $\tilde{\mathbf{a}}_{k-1}^{B} = [\tilde{a}_{x,k-1}^{B}, \tilde{a}_{y,k-1}^{B}, \tilde{a}_{z,k-1}^{B}]^{\mathrm{T}}$ is the output of accelerometers at the instance t_{k-1}. g denotes the gravity value, calculated using (6.8). The total acceleration during the walking is shown in Figure 6.12. It can be observed that the total acceleration oscillates, allowing its peaks to be identified and marked with red points.

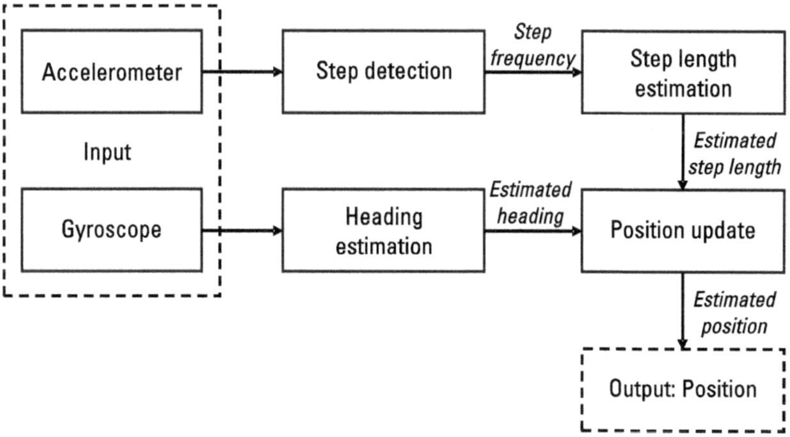

Figure 6.11 The generalized framework of pedestrian dead reckoning.

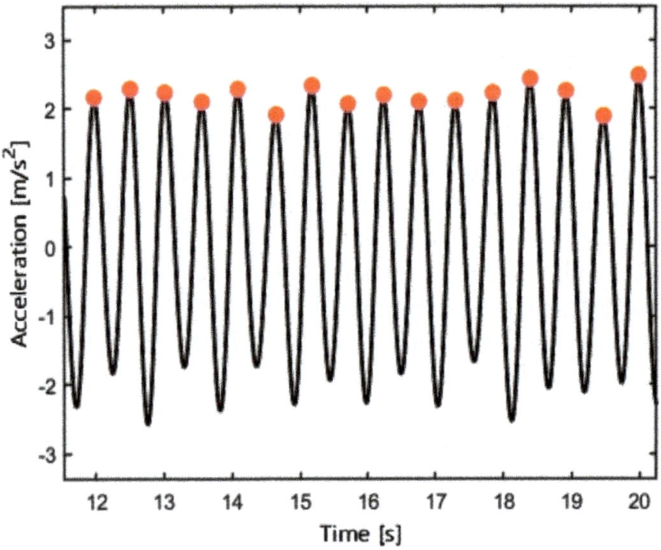

Figure 6.12 The total accelerations during the walking.

Step length estimation refers to calculating the displacement based on step detection results, mainly classified into model-based [23–25] and machine learning-based approaches [26–28]. Model-based approaches can be divided into constant, linear, and empirical models. The step length is obtained in the constant model by dividing the pedestrian walking distance by step numbers [29]. However, the estimation accuracy of this method seriously drops when pedestrian motion states greatly vary. The linear model is formed by analyzing the relationship between step length and step frequency, which will be used in this chapter to derive the step length.

The peaks are determined based on the total acceleration, and the time interval between two peaks is available. Then the step frequency can be calculated, which is shown as:

$$SF = \frac{1}{\Delta t_{pp}} \tag{6.48}$$

where SF and Δt_{pp} are step frequency and time interval between two peaks. The step length can be estimated [22], which is shown as:

$$SL = \left[0.7 + a \cdot (h - 1.75) + b \cdot \frac{(SF - 1.79) \cdot h}{1.75} \right] \cdot c \tag{6.49}$$

where SL is the step length. h is the height of the user. The coefficients a and b are two known parameters of the model, which can be set as 0.371 and 0.227. c is the personal factor that can be approximated to 1.

The step length is along the pedestrian moving direction; thus, it needs to be projected to the coordinate for navigating, which makes the heading estimation another essential issue in PDR. Integration using gyroscope data based on (6.4) is commonly utilized to obtain the heading. Then the position update model of PDR in two dimensions can be expressed as:

$$\begin{aligned} p_{x,k} &= p_{x,k-1} + SL_{k-1}\cos\psi_{k-1} \\ p_{y,k} &= p_{y,k-1} + SL_{k-1}\sin\psi_{k-1} \end{aligned} \tag{6.50}$$

where $\mathbf{P}_i = [p_{x,i}, p_{y,i}]^{\mathrm{T}}$ represents the pedestrian position at the instant t_i. SL_{k-1} and ψ_{k-1} are the step length and heading at the instance t_{k-1}.

6.3.3 Summary

ZUPT is mainly used for foot-mounted IMU at present. The advantage of this method is that it does not need to consider human kinematics, which leads to a wider range of applicability. However, the foot-mounted IMU is only suitable for special applications, as one cannot expect pedestrians in daily life to wear an IMU on their shoes.

Although PDR is an effective way to improve the accuracy with only IMU, it has two main issues: heading inconsistency and application limitation. The former means that a heading difference generally exists between the pedestrian and the device. The device measures its heading angle, but the displacement vector is along the direction of the pedestrian's movement. In most cases, the mobile device, such as the smartphone, is not fixed on the body, as it can be put in a pocket or bag or held in any posture. The performance of the PDR will degrade if the user's heading cannot be accurately estimated. The second one means that PDR can only be used in pedestrians or legged robots [30], such as robot dogs and biped robots [31], as it relies on step detection. Moreover, PDR is a person-sensitive positioning in which many factors, including heights, weights, ages, genders, leg lengths, dressings, and personal habits, affect step detection and length estimation [32].

Recently, data-driven artificial intelligence (AI) methods, such as deep learning and machine learning algorithms, have emerged as promising solutions to these challenges [33, 34]. AI-driven approaches leverage large-scale, diverse datasets to learn intricate patterns from sensor data, effectively mitigating the problems caused by heading inconsistencies and individual variations.

Furthermore, advances in sensor fusion, adaptive filtering, and personalized models powered by AI techniques offer significant potential to enhance PDR accuracy, reliability, and robustness across various usage scenarios. Looking forward, integrating AI methods with traditional PDR techniques is expected to become increasingly prevalent, paving the way for more accurate, universally adaptable, and user-friendly indoor positioning solutions.

6.4 Environmental Sensors for Enhanced Dead Reckoning

Even small errors in dead reckoning can cause significant drift, as these inaccuracies compound over time during position estimation. To mitigate this drift, dead reckoning systems incorporate environmental sensors, such as magnetometers and barometric altimeters [11]. Magnetometers measure Earth's magnetic field and serve as electronic compasses for heading (yaw) estimation, while barometric pressure sensors (barometers) measure air pressure to infer altitude. These sensors provide absolute orientation and elevation references that complement relative measurements from accelerometers and gyroscopes, thereby improving long-term accuracy.

Magnetometers and barometers are especially important in environments where GPS or external references are unavailable. Table 6.2 summarizes the roles of these environmental sensors in dead reckoning. Both sensors face unique challenges: magnetometers are prone to interference and require careful calibration, and barometers are affected by weather-induced pressure changes. In this section, we focus on the mathematical models and theoretical aspects of using magnetometers for heading estimation and barometers for altitude sensing in dead reckoning.

6.4.1 Magnetometers for Heading Estimation

The Earth's magnetic field provides a global reference for orientation. At any given location, the Earth's field can be modeled as a vector with known

Table 6.2
Key Environmental Sensors in Dead Reckoning and Their Roles

Sensor	Measured Quantity	Role in Dead Reckoning
Magnetometer	Earth's magnetic field (3-axis)	Heading (yaw) estimation
Barometer	Atmospheric pressure	Altitude estimation

inclination and declination angles. In a local North-East-Down (NED) reference frame, the magnetic field vector can be represented as $\mathbf{B}_M = [B_N, B_E, B_D]^T$, where B_N and B_E are the horizontal components (pointing toward magnetic North and East, respectively) and B_D is the vertical (downward) component. A three-axis magnetometer measures the Earth's field in the sensor's body frame. The ideal magnetometer measurement model can be written as:

$$\tilde{\mathbf{m}}^B = \mathbf{C}_M^B \mathbf{B}_M + \mathbf{b}_{\text{bias}} + \epsilon \tag{6.51}$$

where $\tilde{\mathbf{m}}^B$ is the magnetometer measurement vector in the body frame, $\mathbf{C}_M^B$ is the rotation matrix from the navigation frame to the body frame (dependent on the sensor's roll, pitch, yaw), $\mathbf{b}_{\text{bias}}$ represents biases (hard-iron offsets), and $\boldsymbol{\epsilon}$ represents measurement noise (and small distortions). Equation (6.51) indicates that, ignoring noise and bias, the measured field is just the Earth's field rotated into the body frame.

To estimate heading from $\tilde{\mathbf{m}}^B$, the influence of roll and pitch must be removed so that only the horizontal component of $\mathbf{B}_M$ is considered. If the roll ϕ and pitch θ of the sensor (with respect to the horizontal plane) are known (e.g., from an accelerometer), the magnetometer reading can be tilt-compensated. Let $[\tilde{m}_x^B, \tilde{m}_y^B, \tilde{m}_z^B]^T$ denote the magnetometer readings in the body frame axes. The horizontal components of the Earth's field in the body frame, (M_x, M_y), can be obtained by rotating the magnetometer measurement back to a level orientation:

$$\begin{aligned} M_x &= \tilde{m}_x^B \cos(\theta) + \tilde{m}_z^B \sin(\theta) \\ M_y &= \tilde{m}_y^B \cos(\phi) + \tilde{m}_x^B \sin(\theta)\sin(\phi) - \tilde{m}_z^B \cos(\theta)\sin(\phi) \end{aligned} \tag{6.52}$$

The heading relative to magnetic North is then given by:

$$\psi = \text{atan2}\left(M_y, M_x\right) \tag{6.53}$$

which yields the magnetometer-based heading. In practice, a known local magnetic declination can be added to ψ to convert a magnetic heading to a true heading (relative to geographic North).

Heading estimation via magnetometers faces several challenges. First, magnetometer readings require calibration to correct for sensor biases and scale factors. Hard-iron effects (constant additive biases caused by permanent magnets or direct current (DC) in the vicinity) and soft-iron effects (scale distortions caused by nearby ferrous materials) can skew $\tilde{\mathbf{m}}^B$. Calibration is typically performed by rotating the sensor in a clean magnetic environment and fitting the measurements to a sphere or ellipsoid to solve for biases and misalignments [35]. Second, external magnetic disturbances can cause temporary heading errors. Nearby ferromagnetic objects or electromagnetic fields can distort the local field, leading to heading errors if interpreted as changes in $\mathbf{B}_M$. Theoretically, one can improve accuracy by modeling and filtering out such disturbances. For example, algorithms can estimate time-varying bias terms $\mathbf{b}_{\text{bias}}$ as part of the state in a filter or reject outlier measurements when sudden magnetic distortions are detected. Another approach is to use multiple magnetometers or reference sensors to detect and cancel interference, although this adds complexity.

Various theoretical methods have been proposed to enhance magnetometer heading accuracy. One approach is to integrate magnetometer data with gyroscope data in a Kalman filter or complementary filter so that gyro-derived heading (which is smooth but drifts) is continuously corrected by magnetometer measurements (which are absolute but noisy) [36, 37]. Another approach is to apply optimal estimation or smoothing techniques that account for sensor noise characteristics. In summary, magnetometers provide a crucial absolute heading reference for dead reckoning, but careful calibration and filtering are required to mitigate errors from sensor imperfections and environmental disturbances.

6.4.2 Barometer for Altitude Sensing and Integration

Barometric pressure sensors measure ambient air pressure, which decreases with increasing altitude. This physical principle allows barometers to function as altimeters. The relationship between pressure and altitude can be derived from the barometric formula under standard atmospheric conditions. Assuming a standard tropospheric lapse rate (temperature decreases linearly with altitude), the pressure altitude h as a function of the current atmospheric pressure P is given by:

$$h = \frac{T_0}{L} \times \left(1 - \left(\frac{P}{P_0} \right)^{\frac{R \times L}{g \times M}} \right) \tag{6.54}$$

where T_0 is the temperature at sea level (standard atmosphere: 288.15K), and L is the temperature lapse rate (standard: 0.0065 K/m). P_0 represents the pressure at standard ground level (101,325 Pa).

Depending on the desired altitude reference, the atmospheric pressure P used in the above equation can be defined in two ways:

- Local pressure is used if the height above the ground level is desired.
- Mean sea level (MSL) pressure is used if geodetic height above sea level is desired.

R is the gas constant (8.314 J/(mol·K)), g is the gravitational acceleration, and M is the molar mass of air (0.029 kg/mol). Figure 6.13 illustrates the barometer solution, which generally applies when using MSL pressure with QNH correction (QNH refers to adjusting the local pressure to the MSL pressure and using it to set the altimeter so that it displays the altitude relative to sea level (i.e., geodetic height)).

Barometric altimeters are often integrated into dead reckoning systems to provide the vertical position component. A barometer on its own can measure changes in altitude with good relative accuracy (often resolving altitude differences on the order of meters). However, for absolute altitude, the sensor must be calibrated to a reference pressure. In a dead reckoning context, barometric altitude can be combined with other sensor data (such as vertical acceleration from an IMU) to improve overall altitude estimation [38]. For example, one can use the barometer to correct low-frequency drift in altitude that comes from double-integrating accelerometer data, while relying on accelerometers to capture rapid vertical motions.

Several challenges arise in using barometers for altitude sensing. One major issue is that atmospheric pressure is influenced not only by altitude but also by weather systems (e.g., high-pressure and low-pressure regions). This means that, over time, P_0 can effectively change, causing the inferred altitude to drift even if the actual altitude is constant. Another challenge is sensor noise

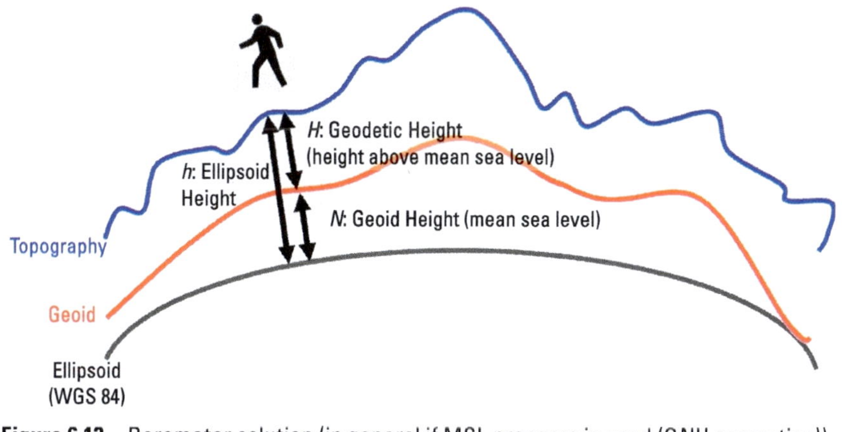

Figure 6.13 Barometer solution (in general if MSL pressure is used (QNH correction)). Google Earth used the Geoid.

and resolution; barometric readings can fluctuate due to turbulence or sensor quantization, introducing noise in altitude estimates. Additionally, in indoor environments or sealed areas, pressure readings might be affected by heating, ventilation, and air conditioning (HVAC) systems or other factors unrelated to true altitude changes.

Theoretical solutions have been developed to address these challenges. To handle slow pressure drift due to weather, one method is to periodically recalibrate P_0 using known altitude references when available. In a filter framework, one can also treat the reference pressure or bias as part of the state to be estimated, thus correcting for it in real time. To reduce noise, optimal filtering (e.g., a Kalman filter) can be applied to smooth the altitude estimates, using a process model for altitude. For instance, a simple process model might assume that altitude changes only when there is a corresponding acceleration, so that sudden pressure changes without corroborating acceleration can be discounted as noise. Another theoretical approach is to fuse the barometric data with other environmental data such as temperature and humidity to adjust the model in (6.54) (since T_0 and L can vary with weather), thereby improving the accuracy of the altitude calculation. In summary, barometers are effective for sensing altitude in dead reckoning, but they must be carefully integrated and calibrated to mitigate errors from atmospheric variations and sensor noise.

6.4.3 Sensor Fusion Strategies for Environmental Sensors

A key theoretical aspect of using environmental sensors in dead reckoning is how to fuse them with other sensors (such as IMU sensors) to obtain a more accurate and robust estimate of the navigation state. Sensor fusion algorithms combine multiple sensor inputs by weighting them according to their uncertainty and dynamics. Common frameworks include the Kalman filter (and its variants such as the EKF) and particle filters. These frameworks differ in complexity and optimality, as summarized in Table 6.3.

Table 6.3
Comparison of Sensor Fusion Strategies

Method	**Advantages**	**Limitations**
EKF	Optimal (minimum variance) estimate if models are accurate	Requires linearization and precise noise models
Particle filter	Handles nonlinear/non-Gaussian cases	High computational cost, complexity in implementation

In a Kalman filter framework, one defines a state vector (e.g., including position, velocity, orientation, sensor biases) and uses a dynamic model to predict state evolution and measurement models for sensor readings. For example, a simplified state might include heading ψ and altitude h as components. The process (state transition) and measurement equations can be expressed as:

$$\begin{aligned} \mathbf{x}_{k+1} &= f\left(\mathbf{x}_k, \mathbf{u}_k\right) + \mathbf{w}_k \\ \mathbf{z}_k &= h\left(\mathbf{x}_k\right) + \mathbf{v}_k \end{aligned} \tag{6.55}$$

where $\mathbf{x}_k$ is the state at time t_k, $\mathbf{u}_k$ represents control inputs or inertial measurements (e.g., gyro turn rate or acceleration), $\mathbf{z}_k$ is the measurement vector (e.g., magnetometer heading and barometric altitude readings), and $\mathbf{w}_k$, $\mathbf{v}_k$ are process and measurement noise, respectively. The functions $f(\ )$ and $h(\ )$ represent the system's dynamic model and measurement model. For instance, $f(\ )$ might propagate the heading by adding the integrated gyro turn rate to ψ, and $h(\ )$ might extract the expected magnetometer reading. The Kalman filter (or EKF for nonlinear $f(\ )$, $h(\ )$) uses these equations to predict the state and then update it when new measurements arrive, weighting the uncertainty of the prediction versus the observation.

The theoretical advantage of combining magnetometer and barometer data with inertial sensors is error mitigation. Gyroscope integration alone would let heading drift unbounded, but a magnetometer measurement provides an absolute reference to clamp that drift. Likewise, integrating accelerometer data for vertical position is prone to drift due to biases, but barometric altitude can serve as a reference to correct that drift. By weighting these appropriately (for instance, a Kalman filter automatically assigns higher weight to the barometer reading when accelerometer-derived altitude is uncertain, and vice versa), the fused estimate can be much more accurate than what each sensor would achieve on its own. Sensor fusion also enables estimation of sensor biases (e.g., magnetometer bias or accelerometer bias) as additional state variables, thus improving long-term accuracy by correcting those biases online.

6.5 Error Modeling and Calibrations

As described previously, each sensor type exhibits unique error characteristics that must be modeled and calibrated to achieve accurate navigation solutions. In this section, we discuss the error sources and calibration techniques for four key sensors: the IMU, wheel odometry sensor, magnetometer, and barometric altimeter.

6.5.1 Error Sources

In this section, we review the error sources associated with various sensors. The main error sources are summarized in Table 6.4.

6.5.1.1 IMUs

IMUs (accelerometers and gyroscopes) have errors such as bias, noise, scale factor, and misalignment. A constant gyroscope bias causes orientation drift over time, leading to position errors that grow cubically if uncorrected. Similarly, accelerometer biases result in errors in velocity and position estimates. Additionally, inertial sensor outputs contain noise, causing drift when raw angular velocity and acceleration data are integrated to obtain angles and velocities. For gyroscopes and accelerometers, this drift is characterized as angle random walk (ARW) and velocity random walk (VRW), respectively. Stochastic noise in IMU readings is commonly analyzed using the Allan variance method [39], which helps to identify specific noise types, including random walk and bias instability. Here we provide an example demonstrating how Allan variance can be applied to estimate these parameters. We utilize gyroscope data and compute the Allan deviation, as shown in Figures 6.14 and 6.15. The vertical

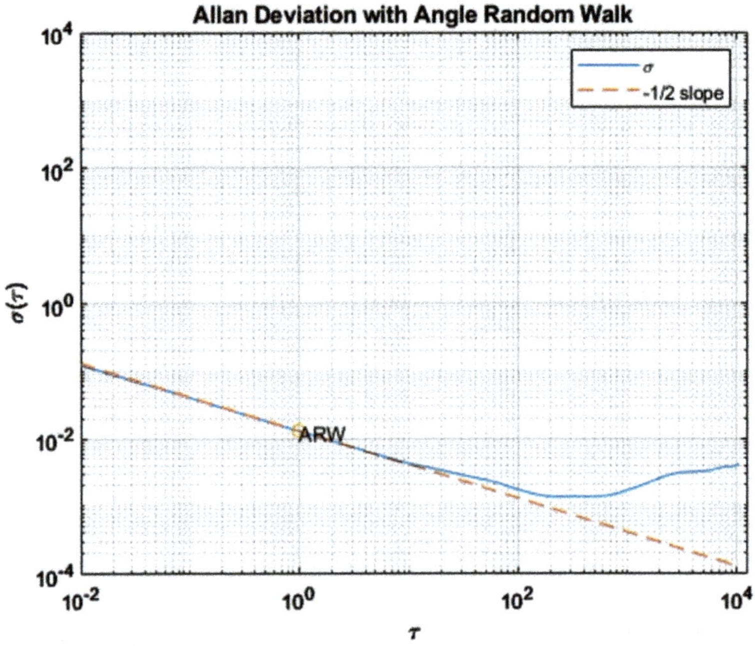

Figure 6.14 Allan deviation analysis with ARW for gyroscope data.

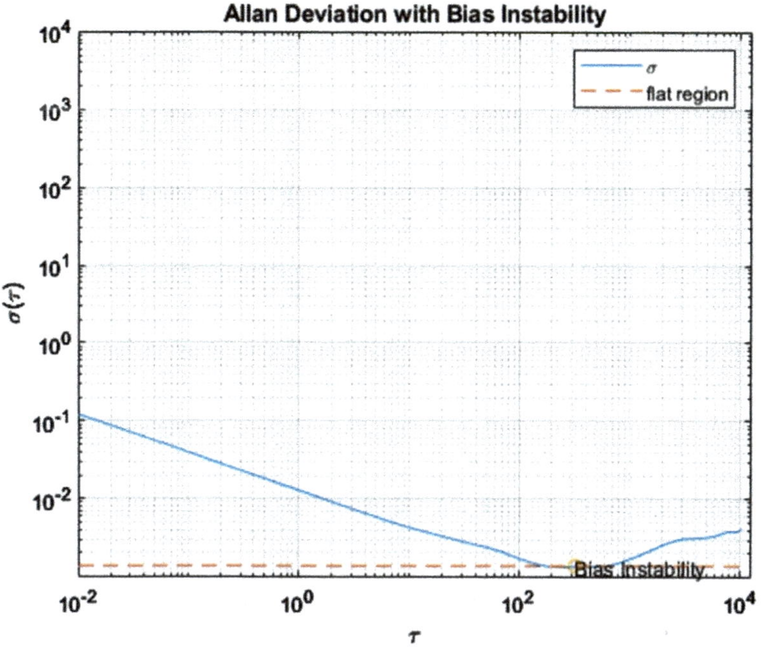

Figure 6.15 Allan deviation analysis with bias instability for gyroscope data.

axis represents the Allan deviation, $\sigma(\tau)$, while the horizontal axis denotes the averaging time. The ARW can be expressed as follows:

$$\sigma(\tau) = \frac{\text{ARW}}{\sqrt{\tau}} \tag{6.56}$$

Taking the logarithm of the equation, we obtain:

$$\log(\sigma(\tau)) = \log(\text{ARW}) - \frac{1}{2}\log(\tau) \tag{6.57}$$

This means that, by performing a log-linear fit on the Allan deviation, where the slope is approximately $-1/2$, the exponential of the fitted intercept gives the ARW value. Once the ARW is determined, the standard deviation of the white noise can be calculated as follows:

$$\sigma_{\omega} = \text{ARW} \times \sqrt{\Delta t} \tag{6.58}$$

where Δt represents the sampling interval of the gyroscope data.

For bias instability, locate the lowest point (flat region) of the Allan deviation curve and use it as the estimated value of bias instability. The code for computing the Allan variance is available in the Sensor Models section of the MATLAB Navigation Toolbox.

In addition, scale factor errors (gain errors) and axis misalignment lead to each axis measuring a slightly incorrect or coupled value of the true motion. Thereby, IMU error models often take the form:

$$\begin{aligned}\tilde{\mathbf{a}} &= \left(\mathbf{I} + \mathbf{S}_a + \Delta_a\right)\mathbf{a}_{\text{true}} + \mathbf{b}_a + \mathbf{n}_a \\ \tilde{\boldsymbol{\omega}} &= \left(\mathbf{I} + \mathbf{S}_\omega + \Delta_\omega\right)\boldsymbol{\omega}_{\text{true}} + \mathbf{b}_\omega + \mathbf{n}_\omega\end{aligned} \tag{6.59}$$

For an accelerometer measuring specific force **a** and a gyroscope measuring rotation rate $\boldsymbol{\omega}$, errors include scale factor errors **S**, misalignments Δ, biases **b**, and noises **n**. Environmental factors, such as temperature changes, can cause these biases to drift over time.

6.5.1.2 Wheel Odometry

Wheel odometry errors are typically classified as either systematic errors (such as scale factors or misalignments) or nonsystematic errors (random or terrain-induced). A common systematic error arises from the misestimation of wheel circumference or wheel base, resulting in consistent scale-factor errors in distance and turning calculations, which lead to cumulative position drift. Nonsystematic errors include wheel slip (e.g., wheels spinning without corresponding forward motion on low-friction surfaces) and uneven contact with the ground, both of which cause unpredictable jumps in odometry. Additionally, wheel encoder measurements introduce quantization noise due to finite encoder resolution, although this is typically minor compared to other error sources. In a simplified odometry error model, deterministic errors can be effectively captured using parameters representing left/right wheel scale factors and a fixed angular misalignment, whereas random slip events are modeled as process noise within a navigation filter.

6.5.1.3 Magnetometer

Magnetometers have bias, scale factor, misalignment and environment-induced errors. Noise in magnetometers is typically low (sensor resolution on the order of microteslas), but environmental interference inside buildings can be severe. Nearby ferrous metals or electrical currents produce local magnetic fields effectively introducing a time-varying bias. These environment-induced errors can lead to large heading inaccuracies if not accounted for.

6.5.1.4 Barometer

Barometric altimeters measure air pressure to infer altitude. A pressure sensor's primary error is a bias in the pressure reading, which translates to an altitude offset. Noise in pressure readings causes short-term altitude jitter (usually a few centimeters after filtering). A scale (sensitivity) error is usually small due to factory calibration. However, barometer readings are strongly affected by environmental changes: weather-induced pressure variation or indoor HVAC systems can cause the measured pressure to drift independent of actual altitude changes (e.g., a mere 0.1% change in pressure can induce several meters of apparent altitude change). We can model the measured pressure as $\tilde{h} = h_{\text{true}} + b_h + n_h$, where b_h may slowly vary with the environment. Without compensation, such drift b_h would appear as a false change in altitude.

6.5.2 Error Calibrations

To mitigate these errors, calibrations are performed both in controlled settings and during operation. Broadly, we have: (1) offline calibration in laboratories or at initialization to identify biases, scale factors, and misalignments; (2) online mitigation using algorithms (filtering or real-time estimation) to compensate for errors during use; and (3) hardware approaches that reduce errors via physical design or redundancy. Table 6.5 summarizes examples of each.

6.5.2.1 IMU Calibration

For IMUs, offline calibration involves measuring sensor output under known conditions, for example, static orientations (to solve for accelerometer biases) and constant rotation on a precision turntable (to determine gyro scale factor error) [47]. Temperature chamber (Figure 6.16) tests can map bias variation versus temperature, and Allan variance analysis of static data yields noise parameters. The identified calibration parameters are then applied to correct raw measurements. Online methods refine this calibration: for instance, foot-mounted IMUs use zero-velocity updates (when a foot is on the ground, true velocity is zero) to reset accumulated error and estimate bias in real time [19]. Additionally, an EKF in an INS can include bias states that are continuously estimated by integrating other sensors or known position updates. As a hardware approach, using multiple IMUs and averaging their outputs can reduce random noise (since uncorrelated noise tends to cancel out) [48]. Higher-grade sensors or mechanically isolating the IMU from vibrations are other hardware strategies to improve performance.

Table 6.4
Error Sources in Different Sensors

Error Source	IMU	Wheel Odometry	Magnetometer	Barometer
Bias (offset) [2, 40, 41]	Yes: Small constant biases in accelerometers and gyroscopes (e.g., nonzero outputs at rest) cause errors that accumulate over time. In IPIN, bias drift significantly impacts accuracy, requiring regular calibration or zero-velocity updates to reduce drift.	No (negligible): An encoder measures incremental rotation; it generally produces no output when the wheel is still, so there is not an additive bias term.	Yes: It typically exhibits constant bias due to hard-iron effects (permanent magnetization of the device or nearby ferrous materials). This appears as an additive offset in the magnetic field reading that persists until calibrated out. Indoor devices often require magnetometer calibration to remove hard-iron bias. If uncorrected, the bias skews heading estimates.	Yes: It often exhibits baseline offsets varying between devices and over time, causing altitude biases of several meters. Barometers are usually calibrated to known reference pressures or altitudes. Periodic recalibration is essential for accurate elevation estimates indoors.
Noise (random error) [2, 40, 42, 43]	Yes: Their readings contain significant random noise (thermal and electrical). In low-cost MEMS ones, this noise causes orientation and velocity errors to grow stochastically (often modeled as angular random walk and velocity random walk).	Yes: It has random errors due to encoder resolution limits and wheel slip variations. Encoder quantization (discrete wheel tick counts) introduces small rounding errors in distance, and uneven floor surfaces or slight slippage cause nonrepeatable deviations.	Yes: The measurements include sensor noise and short-term field fluctuations. The MEMS ones exhibit inherent noise in the measured field (e.g., on the order of tens of microteslas resolution), and indoor magnetic readings can fluctuate due to transient electromagnetic	Yes: Their readings are prone to small random fluctuations. Sensor thermal noise and air turbulence cause the pressure to vary by a few tenths of a hectopascal even in a steady environment, translating to noise of decimeters in altitude.

Table 6.4
(Cont.)

Error Source	IMU	Wheel Odometry	Magnetometer	Barometer
	In IPIN, this high-frequency sensor noise is a key contributor to inertial drift. Advanced filtering (e.g., Kalman filters) is used to partially mitigate IMU noise.	These noise-like errors are zero-mean over time but cause the robot's computed position to wander as they accumulate. Probabilistic models often treat such odometry uncertainty as random noise in localization algorithms.	disturbances (appliances, electrical currents). This noise is usually modeled as zero-mean Gaussian noise in the magnetometer error model. Averaging or lowpass filtering is often applied to reduce random magnetic noise.	IPIN systems often apply averaging or Kalman filtering to smooth out barometric noise while tracking floor levels.
Scale factor error [40, 44, 45]	Yes: They have scale factor errors, meaning the sensor's output gain deviates from the ideal. Without proper factory calibration, each axis can have a sensitivity error (ppm or % of full-scale). In indoor use, these errors lead to distortion in the computed trajectory unless the IMU is calibrated. Modern IMUs often undergo multiaxis calibration to correct scale factors before use.	Yes: A wheel sensor's distance per tick is based on an assumed wheel circumference and wheel base. Any miscalibration here yields a scale factor error. For example, if the actual wheel diameter is larger than assumed, every rotation covers more distance than the encoder calculation, introducing a proportional error. Such scale errors are systematic and common if the odometry is not calibrated for the specific robot and floor conditions.	Yes: Their scale factor errors occur when the sensor axes have unequal gains or nonlinearity, often due to manufacturing tolerances (soft-iron effects can also appear as scaling on axes). This causes the true geomagnetic field, which should plot as a sphere when the device is rotated, to appear as an ellipsoid. Indoor magnetometers frequently exhibit this error, requiring calibration by fitting an ellipsoid to measured data and scaling axes accordingly.	No (minimal): The sensors are factory-calibrated and generally have negligible scale factor errors. Height-pressure relationships follow the barometric formula, with deviations mainly due to environmental assumptions (e.g., temperature, humidity), not sensor gain. As long as outputs are in proper units, errors are dominated by bias and corrected through offset calibration, not scaling.

Table 6.4
(Cont.)

Error Source	IMU	Wheel Odometry	Magnetometer	Barometer
Misalignment (nonorthogonality) [40, 45, 46]	Yes: Axis misalignment is common in MEMS IMUs, where sensor axes may not be perfectly orthogonal or aligned with the device frame. This causes cross-axis coupling and inaccurate readings. Misalignment must be corrected using a calibration matrix. High-end IMUs are precalibrated, but low-cost units typically require alignment during initialization.	Yes: In differential drive odometry, misalignment refers to physical alignment errors like wheels not perfectly parallel or an unknown effective wheel base length. A slight skew in wheel mounting can cause one wheel to travel a different path, leading to a gradual heading bias (systematic turning error). Even a tiny angular misalignment between the wheel encoder axis and the robot frame can accumulate into significant position error over distance. This is one of the two dominant systematic odometry errors that are routinely calibrated out.	Yes: The axis misalignment arises from assembly tolerances and misalignment with the device's heading frame, leading to heading errors. Calibration typically estimates a misalignment matrix along with scale factors to align the sensor to the navigation frame. Indoors, this is handled during commissioning or in real-time via sensor fusion.	No: It measures ambient air pressure, a scalar quantity, so orientation or alignment has no effect on its reading. Unlike directional sensors, there are no axes to misalign in a pressure sensor; it outputs the same pressure regardless of how the device is rotated. Indoor barometric readings depend on location (and airflow conditions) but not on the sensor orientation, so misalignment is not applicable for barometers.

Table 6.4
(Cont.)

Error Source	IMU	Wheel Odometry	Magnetometer	Barometer
Specific environment-induced error [2, 40–42]	Yes (temperature and vibration): Indoor environments indirectly affect IMU accuracy. MEMS IMU biases are temperature-sensitive. Vibrations from machinery or walking impacts can also introduce errors. While IMUs are not affected by electromagnetic or pressure factors, the lack of external references indoors allows bias and noise to accumulate.	Yes: Indoor surfaces impact wheel sensor accuracy. Slippery floors cause wheel slip—rotation without actual movement—leading to odometry errors. Soft carpets can alter effective wheel radius, and floor bumps or cables disrupt normal motion. These nonsystematic errors accumulate unpredictably. To mitigate them, robots often use slip detection, recalibration, or fuse data from IMUs or vision.	Yes: Indoor environments often distort magnetometer readings due to steel structures, wiring, and appliances. Elevators, power lines, or reinforced walls can shift the magnetic heading by tens of degrees. Magnetic anomalies vary across rooms, making headings unreliable without frequent calibration. Many systems either avoid magnetometers indoors or fuse them with gyros to filter out bad data.	Yes: Indoor barometer readings can be affected by HVAC systems, ventilation, and door or elevator movements, causing pressure shifts unrelated to altitude. Weather changes also cause slow drift. As a result, barometric height estimates indoors can be unreliable without compensation. Systems often rely on relative changes, floor-based references, or filtering to separate true elevation from environmental effects.

Table 6.5
Examples of Calibration and Compensation Methods for Each Sensor

Sensor	Offline Calibration	Online Calibration	Hardware Improvements
IMU	Multiposition static tests; turntable rotation and Allan variance	Filter bias estimation; ZUPT	Redundant IMUs (IMU array); thermal isolation
Wheel odometry	Drive known patterns (e.g., squares)	Adaptive slip correction	High-resolution encoders; extra wheel sensor
Magnetometer	Rotate in all axes (fit ellipsoid)	Auto bias update; outlier rejection	Isolate from motors; dual units
Barometer	Set reference offset	Periodic reference update; filter	Multiple sensors (averaging)

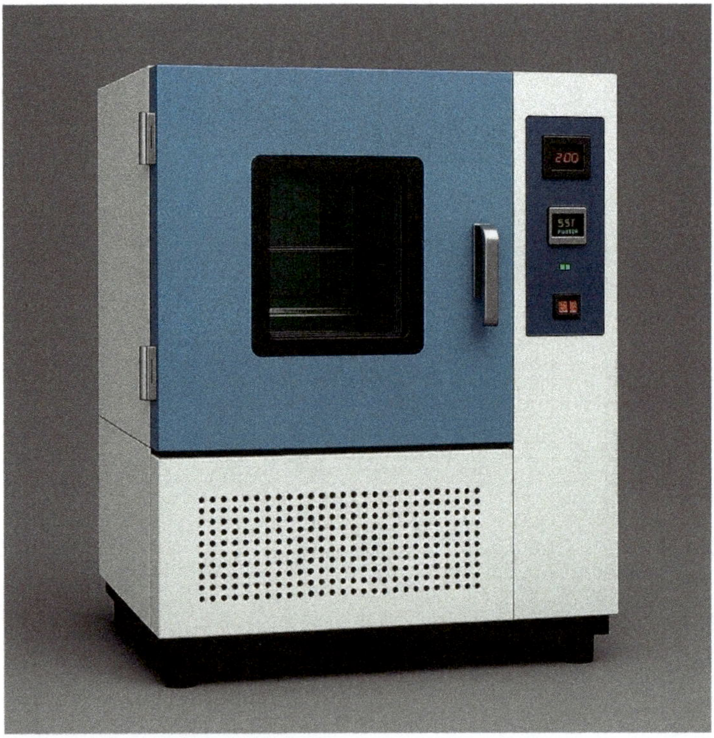

Figure 6.16 The temperature chamber.

6.5.2.2 Wheel Odometry Calibration

Wheel odometry offline calibration is often done by driving the robot in known patterns and adjusting parameters. A classic example is the UMBmark test (driving in a square and returning to the start) to quantify and correct systematic odometry errors such as unequal wheel diameters [49]. By analyzing the final pose error, one can estimate wheel radius and wheel base corrections. Online odometry calibration can be achieved by sensor fusion: for instance, if the robot has an IMU or vision system, it can detect when odometry diverges from other sensors and adjust wheel slip or scale factors on-the-fly [50]. Some robots use adaptive algorithms to reduce reliance on wheel data when slip is detected (e.g., if wheel encoders report motion but an IMU indicates the robot has stopped, the system recognizes slippage). Hardware improvements include using high-resolution encoders to reduce the quantization error and add auxiliary sensing. For example, an extra free-rolling wheel with its own encoder can provide ground-truth displacement to detect slipping of the drive wheels. Ensuring good wheel traction (tire material, weight distribution) is also important to minimize slip-induced errors.

6.5.2.3 Magnetometer Calibration

Magnetometer offline calibration is performed by collecting data in many orientations and fitting an ellipsoid (in high-end setups, a three-axis Helmholtz coil can apply known fields for calibration) to find the hard-iron bias and soft-iron scaling. This yields a bias vector and calibration matrix so that the output can be corrected to true field values. Many devices ask users to perform a “figure-8” motion that effectively gathers such calibration data. Online, systems can monitor the magnetic field magnitude and, if a deviation from the expected Earth field is detected over time, slowly adjust the bias estimate. Also, if a sudden magnetic disturbance is sensed (e.g., the magnetometer reading changes rapidly while the IMU indicates no rotation), the system can temporarily ignore or down-weight magnetometer data. In terms of hardware, the magnetometer should be placed away from current-carrying wires or ferrous components in the device. Some designs use dual magnetometers at different locations; if one is disturbed, the other might provide a cleaner signal. Using magnetically “quiet” materials and adding shielding can further reduce biases.

6.5.2.4 Barometer Calibration

Barometer calibration is straightforward. Offline, one can set an initial offset so that the barometer reads the correct altitude at a known reference (for example, zero at a known ground level or a specific floor altitude). High-end systems might calibrate the pressure sensor at multiple known pressures and temperatures to refine its accuracy curve. Online, since pressure can drift with weather and indoor conditions, the system may regularly recalibrate using known altitude references. For instance, if an indoor map provides floor heights, the barometric altitude reading can be reset when the user is detected to reach a new floor (e.g., upon taking an elevator). Filters such as Kalman filters are often used to blend barometric altitude with other vertical measurements (accelerometer integration or other altimetric sensors) to correct drift. As a hardware solution, some devices incorporate multiple barometric sensors and average their readings to reduce noise and detect anomalies (if one sensor diverges, it can be flagged). Ensuring the sensor is properly vented to ambient air is also important for accuracy. Overall, thorough calibration of these sensors, combined with robust error modeling, significantly improves the accuracy and robustness of indoor navigation systems. In practice, these methods are often used in combination (e.g., applying a factory calibration and then running an online bias estimator) to ensure long-term stability.

6.6 Drift Mitigation and Corrections

This section introduces theoretical approaches to mitigate drift, built on three pillars: (1) robust initialization, including initial alignment and proper covariance setup for the Kalman filter; (2) exploitation of motion constraints, such as zero-velocity updates (ZUPT), zero-angular-rate updates (ZARU), and nonholonomic constraints (NHC); and (3) external aiding via sensor fusion with additional sensors. We provide a detailed overview of the algorithms and the underlying theory for each method, focusing on how they constrain error growth and enhance indoor positioning performance.

6.6.1 Initialization

Proper initialization of an INS is critical to prevent large initial errors from seeding future drift. Initialization involves two main components: (1) initial alignment, which determines the orientation of the INS with respect to a reference frame, and (2) initialization of the state error covariance matrix **P** (e.g., for an EKF), which reflects confidence in the initial state estimates.

6.6.1.1 Initial Alignment

Initial alignment is the process of finding the orientation (attitude) of the sensor platform relative to the navigation frame before navigation begins. For land-based or indoor INS, this often means determining heading (yaw) with respect to north and leveling the pitch and roll with respect to gravity [11]. A common method for static initial alignment is to use accelerometer readings to define the gravity vector (thus setting the roll and pitch angles by assuming the device is stationary and gravity points downwards), and to use either a magnetometer or known reference direction to establish yaw. In the absence of a magnetometer, if the device remains static for a period, one can also utilize the gyroscope to detect Earth's rotation (for example, in high-precision systems, the gyroscope can sense the Earth's rotation rate projected in the sensor axes to aid in finding North). Once coarse alignment is done (perhaps yielding attitude within a few degrees), a fine alignment can be performed by an attitude Kalman filter that further refines the orientation using the INS equations themselves, treating the initial position as known and velocity as zero. The fine alignment stage often runs the INS in a feedback loop for a short duration while stationary, estimating residual orientation errors and gyroscope biases.

Mathematically, if we denote the true orientation (as a rotation matrix or quaternion) as $\mathbf{C}_B^M$ (body to nav frame rotation) and our initial guess as $\mathbf{C}_{B,0}^M$, the alignment error can be represented by a small rotation $\tilde{\boldsymbol{\theta}}$ such that $\mathbf{C}_B^M \approx \exp([\tilde{\boldsymbol{\theta}}]_\times)\mathbf{C}_{B,0}^M$, where $[\ ]_\times$ is the skew-symmetric matrix of a vector. Fine alignment algorithms attempt to drive $\tilde{\boldsymbol{\theta}}$ to zero by observing inertial sensor outputs under the assumption of known true motion (typically that the system is at rest, so true specific force is just $-\mathbf{g}$ in the vertical and true angular rate is Earth's rotation in nav-frame coordinates). Any discrepancy between measured acceleration/rotation and the expected values given the current state estimate is attributed to misalignment or sensor bias, and a Kalman filter can estimate those errors.

6.6.1.2 Initial Covariance Setting

In an EKF-based INS, we maintain a state error covariance matrix $\mathbf{P}$ that represents the uncertainty in the navigation states (position, velocity, attitude) and any included sensor error states (e.g., gyroscope bias, accelerometer bias). The initialization of $\mathbf{P}$ is crucial for filter performance; if the values are too small (overconfident), the filter will underestimate initial uncertainty and might not properly correct initial errors, while if too large, the filter may be overly sensitive and respond erratically to measurements.

Typically, one sets the initial variances for position and velocity states based on knowledge of the starting condition. For example, if the starting position is known precisely (e.g., at the origin of a local coordinate system), position variance can be set very low (almost zero). Velocity can be assumed zero initially for a stationary start, but one might still assign a small variance to account for any slight movement or sensor noise. Attitude uncertainty depends on the alignment process; after coarse leveling and heading initialization, one might assume a few degrees of uncertainty in yaw and smaller in pitch/roll. Gyroscope and accelerometer biases often have very large initial uncertainty if uncalibrated (since they could be anywhere within the sensor's specified bias error bounds).

During the initial alignment procedure, the filter may also be employed to reduce these uncertainties. For example, as the fine alignment runs while the device is static, the filter will update and typically drive down the attitude and bias uncertainty rapidly as it converges. After successful initialization, the INS is ready to navigate with a consistent reference frame and a well-characterized error covariance, which is the foundation for applying further drift mitigation techniques.

6.6.2 Motion Constraints for Drift Mitigation

Table 6.6 summarizes these motion constraints. Each provides a form of "ground truth" for certain state components (zero velocity or zero turn-rate) under known conditions, thereby stopping the unchecked growth of those errors. By integrating such updates into the navigation filter, the dead reckoning can operate for longer periods without external position fixes while keeping drift to manageable levels. Next we will briefly introduce the motion constraints.

6.6.2.1 Zero Velocity Update

As introduced in the previous section, zero velocity update (ZUPT) [18] is a technique primarily used in pedestrian (foot-mounted) INS or other

Table 6.6
Summary of Motion Constraints for Drift Mitigation

Constraint	Quantity Forced to Zero	Applicable Condition	Primary Effect
ZUPT	Linear velocity	Come to a complete stop or let your foot touch the ground	Corrects velocity and position drift
ZARU	Angular turn-rate	No rotation (often at stop)	Calibrates gyro bias, corrects orientation
NHC	Lateral and vertical velocity	Normal wheeled motion	Prevents sideways/vertical drift, improves heading

applications where the sensor periodically comes to rest (i.e., has zero velocity) for short durations. The basic idea is to detect moments when the IMU is stationary relative to the navigation frame and, at those moments, inject a pseudo-measurement indicating that the velocity is zero. This approach can significantly correct accumulated velocity errors and, indirectly, improve position accuracy.

For example, in a foot-mounted INS (often used for indoor pedestrian tracking), each time that the foot is on the ground flat during the stance phase of a step, its velocity in the global frame is effectively zero for a brief period. At the detected zero-velocity instants, a measurement update is performed in the EKF. Essentially, the filter is told that the true velocity is zero with some measurement noise $\mathbf{r}_k$. The Kalman gain will then adjust the state estimates: velocity errors are set to near zero, and accordingly, position error (which is correlated with velocity error in the state covariance) is also pulled back. Additionally, the filter can estimate and correct accelerometer biases that would otherwise cause a nonzero velocity drift if no external frame measurement is given.

ZUPT has been shown to effectively bound the positional drift in foot-mounted INS to a few percent of distance traveled, even without external sensors. However, it relies on the presence of frequent zero-velocity periods. In cases of wheeled robots or other platforms that may not come to a full stop regularly, ZUPT is less directly applicable (although one can sometimes use ZUPTs at deliberate stops or if the platform is known to stop occasionally).

6.6.2.2 Zero Angular Rate Update

The zero angular rate update (ZARU) is conceptually similar to ZUPT but applies to the rotational motion: it assumes that, at certain times, the device has (approximately) zero angular velocity. This is often the case during the stance phase of a foot in pedestrian dead reckoning, when the foot is flat on the ground and not twisting significantly. If a zero-angular-rate detector identifies that the rotational motion is below a threshold (i.e., any measured gyro output is predominantly bias, with true rotation negligibly small), one can perform a pseudomeasurement that the angular rate is zero [51].

In an EKF context, a ZARU can be implemented by observing the gyroscope measurements:

$$\mathbf{z}_{\mathrm{zaru}} = 0,\ h(\mathbf{x}) = \boldsymbol{\omega}_b \tag{6.60}$$

for each axis of the gyroscope (or all three simultaneously as a vector measurement), where $\boldsymbol{\omega}_b$ is the angular rate of the body relative to the inertial frame (as measured by the gyro, in body axes). By updating the filter with the information that true rotation is zero (within some noise tolerance), the filter primarily corrects the gyroscope bias states. Essentially, any observed gyro output is interpreted as bias. Over time, ZARU can significantly reduce gyro bias drift, which, in turn, stabilizes the heading (yaw) estimate that would otherwise drift. This is especially useful as yaw in an INS is otherwise unobservable without external references (no component of gravity provides yaw information, unlike roll and pitch).

ZARU is complementary to ZUPT in foot-mounted systems [51]. However, its effectiveness depends on the sensor and motion characteristics. If the threshold for "zero rotation" is too high, we might treat actual motion as bias, causing errors; if too low, we rarely apply ZARU. In normal walking, feet often have slight rotations even during stance (people can swivel their feet), limiting opportunities for true zero-rate updates. Therefore, some implementations combine ZARU with heuristic heading drift reduction methods that use partial observations of heading change when foot motion is low, rather than a strict zero update [52]. In summary, ZARU is a useful constraint to curtail gyro bias accumulation and heading drift in systems where the device experiences intervals of almost no rotation.

6.6.2.3 Nonholonomic Constraints

Nonholonomic constraints (NHC) refer to constraints on motion that are nonintegrable as position constraints but restrict the velocities of a system. In the context of land vehicle navigation (and sometimes applied to pedestrian or other systems in a heuristic way), NHC usually implies that certain velocity components are zero in the vehicle's body frame. For example, a typical wheeled vehicle cannot slip sideways or jump, so its velocity in the lateral (sideways) and vertical directions relative to the vehicle frame should be zero (assuming no wheel slip and a level ground). This provides two constraints:

$$\tilde{\mathbf{v}}_k^S = \left[0,\ \tilde{v}_k^S, 0\right]^{\mathrm{T}} \tag{6.61}$$

Here, $\tilde{\mathbf{v}}_k^S$ represents the measurement in the speed coordinate system, assuming the x, y, and z axes point to the right, forward, and upward, respectively. This equation can be used as a pseudomeasurement in the EKF, similar to ZUPT, but specifically influencing certain components of the velocity.

By applying NHC updates, the filter can constrain lateral and vertical velocity errors to zero whenever the vehicle model is assumed valid. The primary effect is to prevent spurious growth of velocity in those unpowered directions and to correct small velocity errors before they lead to large position errors.

It also aids in estimating accelerometer biases: for instance, a slight bias that would cause a gradual buildup of a nonzero lateral velocity will be corrected when the NHC measurement forces the lateral velocity estimate back to zero.

The NHC are very powerful for vehicle INS during external information is lost (indoor or in tunnels), as they essentially keep the vehicle on track by not allowing the inertial solution to wander sideways. However, one must be cautious: NHC assumes no sideslip or vertical movement. In reality, vehicles can have minor wheel slip or suspension movements. If those are significant, treating them as zero can introduce errors. Still, with appropriate measurement noise tuning, the EKF can accommodate slight violations of the constraint without instability.

It is worth noting that similar constraints can be applied in other contexts: for example, a person walking might have a constraint that average velocity in the vertical direction is zero over a gait cycle (they come back to ground level), or robotic systems with specific kinematic constraints can incorporate them into the filter. The general idea is to incorporate any known nonholonomic behavior of the system as additional information to arrest drift.

6.6.3 External Aiding by Sensor Fusion

While motion constraints significantly mitigate drift, most navigation systems also incorporate external aiding sensors to provide absolute references. Sensor fusion combines the relative motion sensing of an IMU with other sensors that either do not drift or drift much less, ensuring bounded-error performance.

Ultimately, the most robust approach to drift mitigation is a fully integrated multisensor system. This can include an IMU, wheel odometer, magnetometer, barometer, RF receiver, and even cameras or lidar. Each sensor has different error characteristics and failure modes. The filter combines all available measurements to estimate the navigation state. For example, a Wi-Fi or BLE position will reset any residual drift in all INS degrees of freedom. Visual or lidar odometry can provide relative motion that bounds drift in between absolute positions. The INS in turn bridges gaps when those external signals are unavailable, providing continuous estimates. In such fusion, the INS essentially acts as the core integrator of motion, and all other sensors either constrain its errors or correct its output.

External aiding sensors significantly enhance the long-term accuracy of navigation systems by preventing unbounded error growth. Designing the filter involves accounting for the noise and bias characteristics of each sensor (process and measurement noise tuning) and possibly time synchronization and alignment between sensors. The considerations of sensor fusion are detailed in Chapter 8.

6.7 Conclusions

Positioning based on proprioceptive and environmental sensors remains a critical area of research with substantial potential for improvement in accuracy and reliability. Future research directions are oriented around advancements in both hardware and software:

- *Advancements in sensor technology:* Efforts continue towards developing next-generation inertial sensors with significantly enhanced performance characteristics. Quantum-based inertial sensors, such as cold-atom interferometry gyroscopes and accelerometers, promise to deliver drastically reduced drift and higher precision, potentially achieving strategic-grade navigation performance. Although currently limited by size, complexity, and cost, ongoing advancements are expected to gradually make these technologies more accessible for practical applications.
- *Machine learning and data-driven approaches:* Recent developments in machine learning, particularly deep learning techniques, have shown promise in leveraging patterns within inertial data. Methods involving recurrent neural networks (RNNs) or transformer-based architectures can implicitly model complex human or vehicular motion dynamics, reducing positional drift significantly. These approaches, particularly suitable for pedestrian tracking and smartphone-based inertial odometry, still face challenges in terms of generalization and interpretability. Nevertheless, hybrid models integrating traditional Kalman filtering with learning-based anomaly detection and corrections represent an active and rapidly expanding research frontier.

Overall, combining cutting-edge sensor technology with sophisticated data-driven approaches represents the most promising pathway forward in proprioceptive and environmental sensor-based positioning systems.

References

[1] Tao, Z., et al., "Mapping and Localization Using GPS, Lane Markings and Proprioceptive Sensors," *2013 IEEE/RSJ International Conference on Intelligent Robots and Systems*, 2013.

[2] Woodman, O. J. *An Introduction to Inertial Navigation*, No. UCAM-CL-TR-696, University of Cambridge, Computer Laboratory, 2007.

[3] El-Sheimy, N., and A. Youssef, "Inertial Sensors Technologies for Navigation Applications: State of the Art and Future Trends," *Satellite Navigation*, Vol. 1.1, 2020, pp. 1–21.

[4] El-Sheimy, N., and Y. Li, "Indoor Navigation: State of the Art and Future Trends," *Satellite Navigation*, Vol. 2.1, 2021, pp. 1–23.

[5] Roetenberg, D., "Inertial and Magnetic Sensing of Human Motion," These de doctorat, University of Twente, 2006.

[6] Titterton, D., and J. L. Weston, "Strapdown Inertial Navigation Technology," *IET*, Vol. 17, 2004.

[7] Brandt, A., and J. F. Gardner, "Constrained Navigation Algorithms for Strapdown Inertial Navigation Systems with Reduced Set of Sensors," *Proceedings of the 1998 American Control Conference. ACC (IEEE Cat. No. 98CH36207)*, Vol. 3, 1998.

[8] Friedland, B., "Analysis Strapdown Navigation Using Quaternions," *IEEE Transactions on Aerospace and Electronic Systems*, Vol. 5, 1978, pp. 764–768.

[9] Teslić, L., I. Škrjanc, and G. Klančar, "EKF-Based Localization of a Wheeled Mobile Robot on Structured Environments," *Journal of Intelligent & Robotic Systems*, Vol. 62, 2011, pp. 187–203.

[10] Cho, B. -S., et al., "A Dead Reckoning Localization System for Mobile Robots Using Inertial Sensors and Wheel Revolution Encoding," *Journal of Mechanical Science and Technology*, Vol. 25, 2011, pp. 2907–2917.

[11] Groves, P. D., "Principles of GNSS, Inertial, and Multisensor Integrated Navigation Systems [Book Review]," *IEEE Aerospace and Electronic Systems Magazine*, Vol. 30, No. 2, 2015, pp. 26–27.

[12] Wu, Y., C. Goodall, and N. El-Sheimy, "Self-Calibration for IMU/Odometer Land Navigation: Simulation and Test Results," *Proceedings of the 2010 International Technical Meeting of The Institute of Navigation*, 2010.

[13] Mikov, A., et al., "Sensor Fusion for Land Vehicle Localization Using Inertial MEMS and Odometry," *2019 IEEE International Symposium on Inertial Sensors and Systems (INERTIAL)*, 2019.

[14] Niu, X., S. Nassar, and N. El-Sheimy, "An Accurate Land-Vehicle MEMS IMU/GPS Navigation System Using 3D Auxiliary Velocity Updates," *Navigation*, Vol. 54, No. 3, 2007, pp. 177–188.

[15] Li, Y., et al., "Observability Analysis of Non-Holonomic Constraints for Land-Vehicle Navigation Systems," *Proceedings of the 25th International Technical Meeting of the Satellite Division of The Institute of Navigation (ION GNSS 2012)*, 2012.

[16] Wang, Y., A. Chernyshoff, and A. M. Shkel, "Error Analysis of ZUPT-Aided Pedestrian Inertial Navigation," *2018 IEEE International Conference on Indoor Positioning and Indoor Navigation (IPIN)*, 2018.

[17] Xiaofang, L., et al., "Applications of Zero-Velocity Detector and Kalman Filter in Zero Velocity Update for Inertial Navigation System," *Proceedings of 2014 IEEE Chinese Guidance, Navigation and Control Conference*, 2014.

[18] Wang, Y., and A. M. Shkel, "A Review on ZUPT-Aided Pedestrian Inertial Navigation," *2020 27th IEEE Saint Petersburg International Conference on Integrated Navigation Systems (ICINS)*, 2020.

[19] Skog, I., et al., "Zero-Velocity Detection—An Algorithm Evaluation," *IEEE Transactions on Biomedical Engineering*, Vol. 57, No. 11, 2010, pp. 2657–2666.

[20] Levi, R. W., and T. Judd, "Dead Reckoning Navigational System Using Accelerometer to Measure Foot Impacts," U.S. Patent No. 5,583,776, December 10, 1996.

[21] Yan, D., C. Shi, and T. Li, "An Improved PDR System with Accurate Heading and Step Length Estimation Using Handheld Smartphone," *The Journal of Navigation*, Vol. 75, No. 1, 2022, pp. 141–159.

[22] Chen, R., L. Pei, and Y. Chen, "A Smart Phone Based PDR Solution for Indoor Navigation," *Proceedings of the 24th International Technical Meeting of the Satellite Division of the Institute of Navigation (ION GNSS 2011)*, 2011.

[23] Díez, L. E., et al., "Step Length Estimation Methods Based on Inertial Sensors: A Review," *IEEE Sensors Journal*, Vol. 18, No. 17, 2018, pp. 6908–6926.

[24] Weinberg, H., "Using the ADXL202 in Pedometer and Personal Navigation Applications," *Analog Devices AN-602 Application Note*, Vol. 2, No. 2, 2002, pp. 1–6.

[25] Kim, J. W., et al., "A Step, Stride and Heading Determination for the Pedestrian Navigation System," *Journal of Global Positioning Systems*, Vol. 3, No. 1-2, 2004, pp. 273–279.

[26] Wang, Q., et al., "Personalized Stride-Length Estimation Based on Active Online Learning," *IEEE Internet of Things Journal*, Vol. 7, No. 6, 2020, pp. 4885–4897.

[27] Vandermeeren, S., and H. Steendam, "Deep-Learning-Based Step Detection and Step Length Estimation with a Handheld IMU," *IEEE Sensors Journal*, Vol. 22, No. 24, 2022, pp. 24205–24221.

[28] Ping, Z., et al., "Pedestrian Stride-Length Estimation Based on Bidirectional LSTM Network," *2020 Chinese Automation Congress (CAC)*, 2020.

[29] Cho, D. -K., et al., "Autogait: A Mobile Platform That Accurately Estimates the Distance Walked," *2010 IEEE International Conference on Pervasive Computing and Communications (PerCom)*, 2010.

[30] Wisth, D., M. Camurri, and M. Fallon, "Robust Legged Robot State Estimation Using Factor Graph Optimization," *IEEE Robotics and Automation Letters*, Vol. 4, No. 4, 2019, pp. 4507–4514.

[31] Rubio, F., F. Valero, and C. Llopis-Albert, "A Review of Mobile Robots: Concepts, Methods, Theoretical Framework, and Applications," *International Journal of Advanced Robotic Systems*, Vol. 16, No. 2, 2019, 1729881419839596.

[32] Wang, Q., et al., "Recent Advances in Pedestrian Inertial Navigation Based on Smartphone: A Review," *IEEE Sensors Journal*, 2022.

[33] Herath, S., H. Yan, and Y. Furukawa, "Ronin: Robust Neural Inertial Navigation in the Wild: Benchmark, Evaluations, & New Methods," *2020 IEEE International Conference on Robotics and Automation (ICRA)*, 2020, pp. 3146–3152.

[34] Bai, S., W. Wen, and C. Shi, "3DIO: Low-Drift 3D Deep-Inertial Odometry for Indoor Localization Using an IMU," *IEEE Internet of Things Journal*, 2024.

[35] Foster, C. C., and G. H. Elkaim, "Extension of a Two-Step Calibration Methodology to Include Nonorthogonal Sensor Axes," *IEEE Transactions on Aerospace and Electronic Systems*, Vol. 44, No. 3, 2008, pp. 1070–1078.

[36] Suh, Y. S., "Orientation Estimation Using a Quaternion-Based Indirect Kalman Filter with Adaptive Estimation of External Acceleration," *IEEE Transactions on Instrumentation and Measurement*, Vol. 59, No. 12, 2010, pp. 3296–3305.

[37] Madgwick, S. O. H., A. J. L. Harrison, and R. Vaidyanathan, "Estimation of IMU and MARG Orientation Using a Gradient Descent Algorithm," *2011 IEEE International Conference on Rehabilitation Robotics*, 2011, pp. 1–7.

[38] Sabatini, A. M., and V. Genovese, "A Sensor Fusion Method for Tracking Vertical Velocity and Height Based on Inertial and Barometric Altimeter Measurements," *Sensors*, Vol. 14, No. 8, 2014, pp. 13324–13347.

[39] El-Sheimy, N., H. Hou, and X. Niu, "Analysis and Modeling of Inertial Sensors Using Allan Variance," *IEEE Transactions on Instrumentation and Measurement*, Vol. 57, No. 1, 2007, pp. 140–149.

[40] Ouyang, G., and K. Abed-Meraim, "Analysis of Magnetic Field Measurements for Indoor Positioning," *Sensors*, Vol. 22, No. 11, 2022, p. 4014.

[41] Fetzer, T., et al., "Using Barometer for Floor Assignation Within Statistical Indoor Localization," *Sensors*, Vol. 23, No. 1, 2022, p. 80.

[42] Chong, K. S., and L. Kleeman, "Accurate Odometry and Error Modelling for a Mobile Robot," *Proceedings of IEEE International Conference on Robotics and Automation*, Vol. 4, 1997, pp. 2783–2788.

[43] Li, B., B. Harvey, and T. Gallagher, "Using Barometers to Determine the Height for Indoor Positioning," *International IEEE Conference on Indoor Positioning and Indoor Navigation*, 2013, pp. 1–7.

[44] Pan, J., C. Zhang, and Q. Cai, "An Accurate Calibration Method for Accelerometer Nonlinear Scale Factor on a Low-Cost Three-Axis Turntable," *Measurement Science and Technology*, Vol. 25, No. 2, 2014, 025102.

[45] Bai, S., et al., "Improved Preintegration Method for GNSS/IMU/In-Vehicle Sensors Navigation Using Graph Optimization," *IEEE Transactions on Vehicular Technology*, Vol. 70, No. 11, 2021, pp. 11446–11457.

[46] Bai, S., et al., "A System-Level Self-Calibration Method for Installation Errors in a Dual-Axis Rotational Inertial Navigation System," *Sensors*, Vol. 19, No. 18, 2019, p. 4005.

[47] Syed, Z. F., et al., "A New Multi-Position Calibration Method for MEMS Inertial Navigation Systems," *Measurement Science and Technology*, Vol. 18, No. 7, 2007, p. 1897.

[48] Wang, L., et al., "Improving the Navigation Performance of the MEMS IMU Array by Precise Calibration," *IEEE Sensors Journal*, Vol. 21, No. 22, 2021, pp. 26050–26058.

[49] Borenstein, J., and L. Feng, "UMBmark: A Method for Measuring, Comparing, and Correcting Dead-Reckoning Errors in Mobile Robots, " SPIE, Vol. 2591, December 1995, pp. 113–124.

[50] Martinelli, A., N. Tomatis, and R. Siegwart, "Simultaneous Localization and Odometry Self Calibration for Mobile Robot," *Autonomous Robots*, Vol. 22, 2007, pp. 75–85.

[51] Rajagopal, S., "Personal Dead Reckoning System with Shoe Mounted Inertial Sensors," Master's Degree Project, KTH Royal Institue of Technology, Stockholm 2008.

[52] Zampella, F., et al., "Unscented Kalman Filter and Magnetic Angular Rate Update (MARU) for an Improved Pedestrian Dead-Reckoning," *Proceedings of the 2012 IEEE/ION Position, Location and Navigation Symposium*, 2012, pp. 129–139.

7

Indoor Simultaneous Localization and Mapping

7.1 Introduction to Simultaneous Localization and Mapping

A map is a key element for indoor navigation. According to the *Cambridge Dictionary*, a map is "a drawing that gives you a particular type of information about a particular area." In other words, the composition of features associated with coordinate positions can be regarded as a map. Currently, most indoor mapping methods use simultaneous localization and mapping (SLAM) to build the first indoor map. This chapter will first introduce mapping methods using lidar,, which can provide decimeter-to-centimeter accuracy in depth measurements to generate accurate 3-D representations of indoor areas. However, the cost of lidar could hinder its applications in indoor navigation. In general, a camera is a low-cost option visually similar to the perception of our eyes. The second half of this chapter introduces visual SLAM for both monocular and depth cameras. The integration of lidar and camera is briefly discussed as a professional indoor mapping method that provides realistic 3-D indoor maps for augmented reality applications. Finally, the roles of IMU in SLAM are discussed to provide readers with the core benefits of this integration. A flowchart of a general SLAM, which includes odometry and mapping, is shown in Figure 7.1.

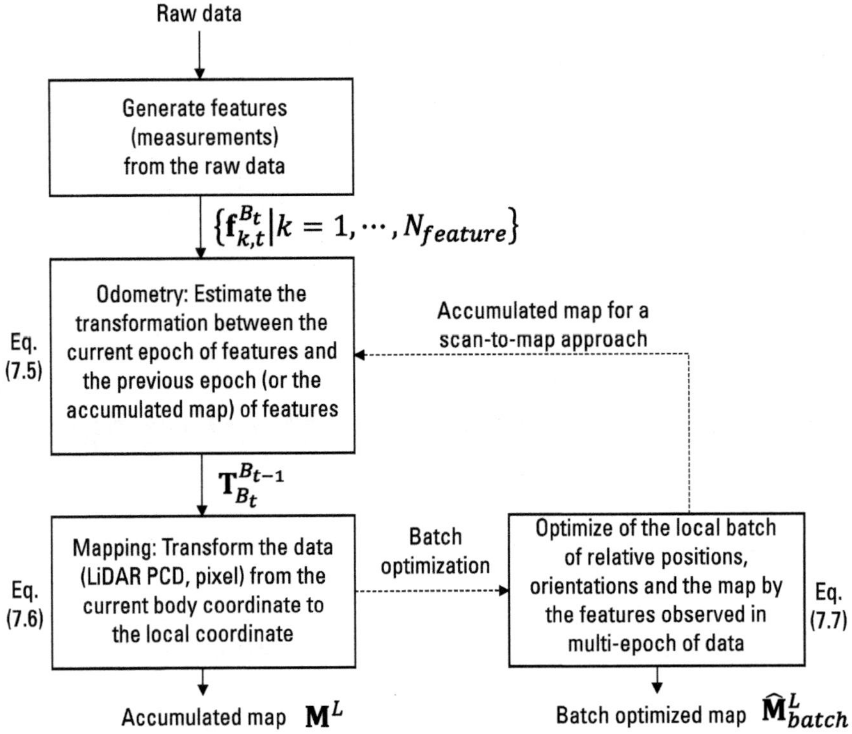

Figure 7.1 The generalized flowchart of SLAM.

7.1.1 The Historical Development of SLAM

SLAM is a computational problem that involves constructing or updating a map of an unknown environment while simultaneously keeping track of an agent's location within it. SLAM is fundamental to autonomous systems navigating in previously unexplored environments.

The SLAM problem was first conceptualized in the 1980s, with significant early contributions coming from researchers such as Hugh Durrant-Whyte and John J. Leonard, who built on the work of Smith and Cheeseman [1]. The initial theoretical foundations were established using EKF, marking the beginning of what we now call the classical era of SLAM research.

The development of SLAM can be broadly categorized into several eras:

1. *Classical era (1986–2004):* This period focused on establishing the theoretical foundations of SLAM. The research was dominated by filter-based approaches, particularly EKF-SLAM [1] and particle

filters. A significant breakthrough came with FastSLAM [2], which used Rao-Blackwellized particle filters to efficiently solve SLAM by factoring the joint posterior into a product of a robot path posterior and landmark posteriors conditioned on the path.

2. *Algorithmic-analysis era (2004–2015):* During this period, researchers focused on understanding computational complexity, consistency, and convergence properties. Graph-based optimization methods emerged, and researchers began exploring how to handle large-scale environments and long-term operation. Key developments included Square Root SAM [3] and GraphSLAM [4].
3. *Robust-perception era (2015–present):* The current era is characterized by focusing on real-world robustness and adaptability. Modern SLAM systems are designed to work in dynamic, changing environments, with research emphasizing semantic understanding, object-level mapping, and deep learning integration. Key developments in this era include Oriented FAST and Rotated BRIEF (ORB)-SLAM [5], Large-Scale Direct Monocular (LSD)-SLAM [6], and lidar-based methods such as lidar Odometry and Mapping (LOAM) [7] and lidar Inertial Odometry-Smoothing and Mapping (LIO-SAM) [8].

Figure 7.2 provides a visual timeline of the evolution of SLAM technologies from 1990 to 2020. This figure illustrates the progression through three distinct eras, each represented by a different colored line. The classical era begins with EKF-SLAM in 1990 and extends to FastSLAM in 2002. The

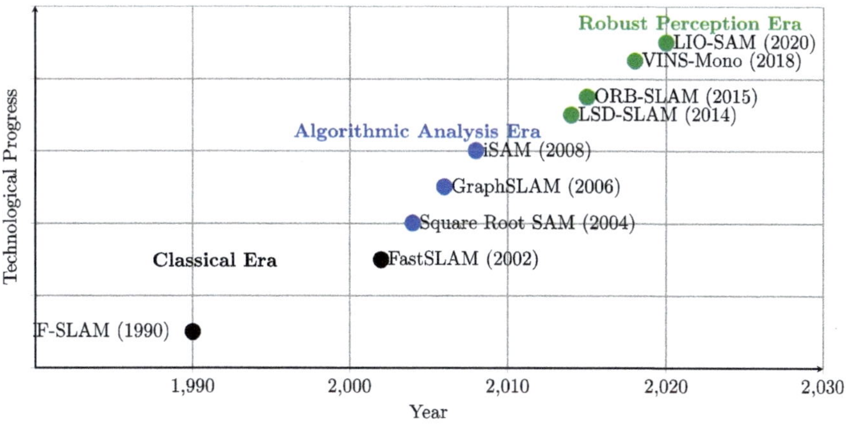

Figure 7.2 Evolution of SLAM technologies over time, showing key algorithms and their approximate relative impact.

algorithmic-analysis era spans from Square Root SAM in 2004 to various developments including GraphSLAM and iSAM. The most recent robust perception era includes modern approaches like LSD-SLAM, ORB-SLAM, VINS-Mono, and LIO-SAM. The vertical axis represents technological progress, showing how each era has built upon the foundations laid by previous research while introducing new paradigms and capabilities.

7.1.2 SLAM Frameworks and Evolution

The evolution of SLAM frameworks closely follows the hardware and algorithmic advancements in robotics and computer vision. Here, we provide an overview of major SLAM frameworks that have shaped the field.

- *GraphSLAM (2006):* This approach formulated SLAM as a graph optimization problem. Poses and landmarks form the nodes of a graph, while measurements create edges (constraints) between nodes. The SLAM problem is then solved by finding the configuration of nodes that best satisfies all constraints [4]. GraphSLAM represented a significant paradigm shift from filter-based methods by recasting the entire SLAM problem into a sparse graph structure. By exploiting this sparsity, GraphSLAM achieves better computational efficiency for large-scale environments compared to earlier approaches. The method constructs information matrices where nonzero elements correspond to directly related poses and landmarks, allowing for efficient optimization using sparse linear algebra techniques. This graph-based representation also provides a natural framework for incorporating loop closures and handling uncertainty in a principled manner. GraphSLAM's formulation has become foundational for many modern SLAM systems, as it scales well to large environments and complex scenarios by focusing computational resources on the most informative constraints.
- *Parallel tracking and mapping (PTAM) (2007):* This introduced the idea of splitting tracking and mapping into separate threads, allowing real-time operation even with more computationally intensive mapping processes. PTAM was primarily designed for augmented reality applications but influenced many subsequent visual SLAM systems [9]. The key innovation of PTAM was recognizing that tracking camera motion could operate at a frame rate with minimal computational resources, while the more complex task of map building and refinement could run asynchronously at lower frequencies. This

parallel architecture enabled the use of computationally expensive bundle adjustment techniques in real-time applications for the first time. PTAM also introduced several practical innovations including a keyframe-based mapping approach, a map initialization procedure using a stereo initialization from a five-point algorithm, and an efficient relocalization system based on a modified version of randomized trees. By demonstrating that accurate camera tracking could be achieved in real time on consumer hardware, PTAM sparked a revolution in visual SLAM research and applications, particularly in the AR/VR domain where low latency is critical.

- *LSD-SLAM (2014):* Unlike feature-based methods, LSD-SLAM directly works with image intensities, creating semidense depth maps. It efficiently handles scale drift and loop closures, making it suitable for large-scale environments [6]. As a direct method, LSD-SLAM represented a significant departure from the feature-based paradigm by operating directly on image intensities rather than extracted features. This approach allows it to exploit more of the available image information, particularly in texture-poor regions where feature detection often fails. LSD-SLAM builds semidense depth maps by estimating depths only at pixels with sufficient image gradient, striking a balance between computational efficiency and information utilization. One of its major contributions is a novel scale-aware image alignment algorithm that addresses the inherent scale ambiguity in monocular vision. LSD-SLAM also incorporates a pose graph optimization framework for global consistency and loop closure, allowing it to create large-scale maps with reduced drift. By demonstrating that direct methods could operate efficiently at scale, LSD-SLAM inspired subsequent research in direct and semidirect SLAM approaches that seek to maximize the use of available visual information.
- *ORB-SLAM (2015):* This feature-based system uses ORB features for all SLAM tasks: tracking, mapping, relocalization, and loop closing. Its robust design has made it one of the most popular visual SLAM systems, with versions for monocular, stereo, and RGB-D cameras [5]. ORB-SLAM represents the culmination of years of research in feature-based SLAM, integrating best practices and novel solutions into a comprehensive framework. The system employs a carefully designed three-thread architecture separating tracking, local mapping, and loop closing tasks. Its use of ORB features provides a good balance of computational efficiency and descriptive power, enabling real-time

performance on modest hardware while maintaining robust feature matching across different viewpoints and lighting conditions. ORB-SLAM implements an efficient keyframe-based bundle adjustment approach for local map optimization, combined with a pose graph optimization for global consistency. The system also incorporates an innovative map initialization method that automatically detects either planar or general scenes, adapting its initialization strategy accordingly. These capabilities, combined with careful engineering for robustness, have made ORB-SLAM a standard baseline in the field and the foundation for numerous subsequent systems.

- *LOAM (2014):* This algorithm optimizes for low-drift and real-time performance without using GPS or IMU data. It extracts plane and edge features from point clouds and uses them for fast scan-to-scan odometry estimation and scan-to-map matching [7]. LOAM introduced a novel approach to lidar SLAM by decoupling the odometry and mapping processes into two parallel algorithms running at different frequencies. The high-frequency odometry algorithm extracts edge and planar features from the point cloud and matches them to the previous scan to estimate motion, while the lower-frequency mapping algorithm registers these features to a global map for refinement and loop closure. This dual-rate approach enables LOAM to achieve both real-time performance and high accuracy. LOAM's feature extraction methodology is particularly innovative, using the distinctiveness of local surface geometry to classify points as edge or planar features, which reduces the computational burden of point cloud registration while maintaining sufficient constraints for accurate pose estimation. The system also incorporates distortion correction to account for sensor motion during scanning, which is crucial for accurate mapping during continuous movement. LOAM's exceptional performance in the KITTI odometry benchmark demonstrated that lidar-only SLAM could achieve accuracy comparable to or exceeding systems using additional sensors such as IMU or GPS, establishing it as a foundational approach for subsequent lidar SLAM research.
- *VINS-Mono (2018):* This is a robust monocular visual-inertial state estimator. It tightly couples visual and inertial measurements to handle challenging scenarios like dynamic environments and fast motion [10]. VINS-Mono represents a significant advancement in visual-inertial fusion, addressing the limitations of vision-only systems while maintaining computational efficiency. The system employs a sliding

window-based optimization framework that tightly integrates visual features and pre-integrated IMU measurements to estimate the full state including position, velocity, orientation, and sensor biases. VINS-Mono incorporates a robust initialization procedure that can reliably estimate initial states without requiring specific motions or known patterns. One of its key innovations is the incorporation of a 4-DOF pose graph for loop closure that maintains the gravity direction constraint from IMU, enabling more accurate global consistency. The system also features online spatial and temporal calibration capabilities, automatically determining the transformation between camera and IMU as well as time offsets between sensor measurements. VINS-Mono's robust outlier rejection strategies and failure detection mechanisms make it particularly suitable for challenging real-world deployments where lighting conditions vary and dynamic objects are common. Its modular design and open-source implementation have made it a popular foundation for autonomous drone navigation, AR applications, and research in visual-inertial fusion.

- *LIO-SAM (2020):* This builds on LOAM by tightly coupling lidar and IMU data using factor graph optimization. It incorporates IMU pre-integration and efficient loop closure to achieve high accuracy with real-time performance [8]. LIO-SAM addresses several key limitations of previous lidar SLAM systems through its innovative factor graph formulation and sensor fusion approach. Unlike loosely coupled methods that use IMU only for initial state estimation, LIO-SAM fully integrates IMU constraints throughout the optimization process, enabling robust performance during rapid motion and featureless environments. The system employs a factor graph architecture with four types of factors: IMU pre-integration factors, odometry factors, GPS factors (when available), and loop closure factors. A key innovation is its dual-model design, maintaining two separate factor graphs: one for global mapping that preserves all constraints, and another for real-time odometry that is periodically reset to ensure computational efficiency. LIO-SAM also introduces scan context-based loop closure detection that leverages the 360° field of view of mechanical lidars for robust place recognition. The system's ability to incorporate GNSS factors when available allows for georeferenced mapping, while still maintaining high accuracy in GNSS-denied environments. By achieving both high accuracy and real-time performance on modest hardware, LIO-SAM has enabled new applications in autonomous navigation, particularly for robots operating in complex 3-D environments where robust localization is essential.

The progression from early filter-based approaches to modern optimization-based methods has been driven by several factors:

- *Computational resources:* Modern computers allow for more complex algorithms and real-time processing of dense sensor data. The exponential growth in computing power, particularly in embedded systems and mobile processors, has enabled the deployment of sophisticated SLAM algorithms. The development of specialized hardware accelerators for visual processing and machine learning tasks has further expanded the capabilities of real-time SLAM systems.
- *Sensor technology:* The development of affordable, high-quality sensors such as RGB-D cameras and 3-D lidars has enabled more accurate environment perception. The dramatic reduction in cost and size of 3-D sensing technologies has been crucial for widespread SLAM adoption. Solid-state lidars have made high-precision depth sensing more accessible, while the mass production of RGB-D cameras initially driven by the gaming industry has provided affordable depth sensing for indoor applications. Time-of-flight cameras have improved depth accuracy in close-range applications, and event-based cameras with microsecond temporal resolution have opened new possibilities for high-speed motion tracking. Additionally, improvements in MEMS-based inertial sensors have made high-quality IMUs available at consumer price points, facilitating visual-inertial fusion approaches that enhance robustness and accuracy.
- *Algorithmic innovations:* New optimization techniques, feature extraction methods, and loop closure strategies have improved SLAM robustness and accuracy. Mathematical breakthroughs in sparse optimization, manifold representation of rotations, and factor graph formulations have enabled more efficient and accurate state estimation. Learned feature descriptors leveraging deep neural networks have significantly improved feature-matching robustness across varying conditions. Semantic SLAM approaches that incorporate object recognition and scene understanding are bridging the gap between geometric mapping and higher-level environmental representation. Novel outlier rejection techniques and robust estimation methods have enhanced performance in dynamic environments with moving

objects. The incorporation of probabilistic approaches for handling uncertainty continues to improve reliability in challenging conditions, while recent advances in differentiable programming enable end-to-end optimization of entire SLAM pipelines.

- *Multisensor fusion:* Modern SLAM systems often integrate data from multiple sensors (e.g., camera, lidar, IMU) to overcome the limitations of individual sensors. Complementary sensor characteristics can be leveraged to enhance robustness; cameras provide rich texture and color information, lidars offer precise depth measurements regardless of lighting, and IMUs provide high-frequency motion estimates resistant to visual degradation. Sophisticated fusion frameworks using factor graphs, Kalman filter variants, or deep learning approaches can optimally combine these heterogeneous data sources. Tight coupling of sensors at the measurement level, rather than simply fusing outputs, has proven particularly effective for handling challenging scenarios such as rapid motion or feature-poor environments. Recent research has explored adaptive fusion strategies that dynamically adjust sensor weighting based on environmental conditions and sensor health, further improving reliability across diverse operating conditions. The integration of additional sensors such as wheel encoders, barometers, magnetometers, and GNSS receivers provides even more constraints for robust state estimation in specific application domains.

7.1.3 Comparison of Popular SLAM Implementations

Different SLAM implementations have their strengths and weaknesses, making them suitable for different applications. Table 7.1 provides a comparison of some popular SLAM frameworks based on various aspects.

Table 7.2 compares the pipeline components of lidar-based and vision-based SLAM systems.

Currently, the evolution of SLAM continues with active research in several directions:

1. *Semantic SLAM:* Incorporating object-level and semantic understanding into SLAM systems [11].
2. *Deep learning integration:* Using neural networks for feature extraction, depth estimation, and loop closure detection [12].
3. *Dynamic environment handling:* Developing methods to handle moving objects and changing environments [13].

Table 7.1
Basic Comparison of Popular SLAM Implementations

Aspect	ORB-SLAM	LSD-SLAM	LOAM	VINS-Mono	LIO-SAM
Sensor type	Camera (Mono/ Stereo/ RGB-D)	Monocular camera	Lidar	Monocular camera + IMU	Lidar + IMU
Feature type	Sparse features (ORB)	Semidense depth maps	Edge and planar features	Visual features (optical flow)	Edge and planar features
Map representation	Sparse point cloud	Semidense depth maps	Point cloud	Sparse point cloud	Point cloud
Localization accuracy	High (with loop closure)	Medium	High	Very high	Very high
Computational cost	Medium	Medium-high	Medium-high	High	Medium-high
Loop closure	Yes	Yes	No (original version)	Yes	Yes
Scale recovery	No (monocular)	No (monocular)	Yes	Yes (using IMU)	Yes
Real-time performance	Yes	Yes (with GPU)	Yes	Yes	Yes
Environment types	Well-textured	Well-textured	Structured environments	Various	Various
Major advantages	Accurate, robust loop closure	Works in large environments	Good for unmanned vehicles	Robust to dynamic objects	High accuracy, leverages IMU
Major limitations	Fails in textureless areas	Requires GPU for real-time	Motion distortion issues	Initialization challenges	Sensor synchronization requirements
Reference	[5]	[6]	[7]	[10]	[8]

4. *Resource-constrained SLAM:* Creating efficient algorithms for mobile and embedded devices [14].
5. *Event-based SLAM:* Leveraging neuromorphic cameras (event cameras) for high-speed, high-dynamic range visual SLAM [15].

Table 7.2
Comparison of lidar-Based and Vision-Based SLAM Pipelines

Pipeline Component	Lidar SLAM	Visual SLAM
Data preprocessing	Point cloud downsampling Motion distortion correction Ground/outlier removal	Image rectification Lens distortion correction Exposure adjustment
Feature extraction	Edge features Planar features Corner points Normal distributions (NDT)	Corner features (FAST, Harris) Blob features (SIFT, SURF, ORB) Direct methods (photometric error)
Data association	Nearest neighbor matching ICP matching Feature-to-feature matching	Feature descriptor matching Optical flow tracking Epipolar constraints
Motion estimation	Point-to-point registration Point-to-plane registration NDT registration	Essential/fundamental matrix Perspective-n-Point (PnP) Homography (for planar scenes)
Back-end optimization	Factor graph optimization Pose graph optimization Bundle adjustment (with IMU)	Bundle adjustment Pose graph optimization Key frame-based optimization
Loop closure	ICP-based scan matching Place recognition using point cloud Scan context matching	Bag of Words (BoW) DBoW2/DBoW3 Visual place recognition
Map representation	Point cloud map Surfel map Occupancy grid/voxel grid	Sparse point cloud Semidense depth maps Dense volumetric (TSDF)
Multisensor fusion	Lidar-IMU tight coupling Lidar-visual integration GNSS integration	Visual-inertial fusion Multicamera systems Visual-GNSS integration

These advancements are pushing SLAM systems toward greater autonomy, robustness, and applicability across diverse scenarios, from autonomous driving to AR/VR applications.

7.2 General Mathematical Model of SLAM

First, the features (which are the measurements used for the coming estimation, denoted by **f**) are extracted from the raw data. If this is the first epoch, the device's (i.e., camera and lidar) body local coordinate is used as the local map coordinate for the upcoming estimation, denoted by B in the superscript. Note that the features extracted at the first epoch are used as the initial map. The movement of an agent can be estimated by the transformation between features from adjacent epochs, incorporating a rotation **C** and a translation $\Delta\mathbf{x}$ in 3-D space (from the subscript to the superscript coordinate systems). The feature transformation can be represented by the coordinate transformation, using

$$\mathbf{f}_{k,t}^{B_{t-1}} = \mathbf{C}_{B_t}^{B_{t-1}}\mathbf{f}_{k,t}^{B_t} + \Delta\mathbf{x}_{B_t}^{B_{t-1}} \tag{7.1}$$

where the superscripts B_t and B_{t-1} denote corresponding to the local body coordinate system at the epochs t and $t-1$, respectively. For multiple epochs, the transformation is usually described by matrix multiplication in a compact manner, as follows:

$$\begin{bmatrix} \mathbf{f}_{k,t}^{B_{t-1}} \\ 1 \end{bmatrix} = \begin{bmatrix} \mathbf{C}_{B_t}^{B_{t-1}} & \Delta\mathbf{x}_{B_t}^{B_{t-1}} \\ 0 & 1 \end{bmatrix} \begin{bmatrix} \mathbf{f}_{k,t}^{B_t} \\ 1 \end{bmatrix} \tag{7.2}$$

For convenience, the transformation is simply represented by the following multiplication, where the feature vector will be automatically extended with a row of 1 at the bottom during the multiplication.

$$\mathbf{f}_{k,t}^{B_{t-1}} = \mathbf{T}_{B_t}^{B_{t-1}}\mathbf{f}_{k,t}^{B_t} \tag{7.3}$$

$$\mathbf{T}_{B_t}^{B_{t-1}} = \begin{bmatrix} \mathbf{C}_{B_t}^{B_{t-1}} & \Delta\mathbf{x}_{B_t}^{B_{t-1}} \\ 0 & 1 \end{bmatrix} \tag{7.4}$$

where **T** is the transformation matrix between the previous and current epoch of features is estimated, also known as a pose in robotics field. It is then used to transform a kth feature from the current epoch of the device's body coordinate, $\mathbf{f}_{k,t}^{B_t}$, back to the kth feature of the previous epoch of device's body coordinate, $\mathbf{f}_{k,t}^{B_{t-1}}$. Thus, the odometry can be obtained by optimizing $\mathbf{T}_{B_t}^{B_{t-1}}$ to align the features from the previous epoch with those from the current one after the transformation, as below.

$$\hat{\mathbf{T}}_{B_t}^{B_{t-1}} = \arg\min_{\mathbf{T}_{B_t}^{B_{t-1}}} \sum_{k=1}^{N_{\text{feature}}} \left\| \mathbf{f}_{k,t-1}^{B_{t-1}} - \mathbf{T}_{B_t}^{B_{t-1}} \mathbf{f}_{k,t}^{B_t} \right\|^2 \tag{7.5}$$

For the mapping stage, the features at different epochs are all transformed to the local map coordinate to form the map $\mathbf{M}^L$ as below.

$$\mathbf{M}^L = \left\{ \mathbf{M}_t^L | \forall t \right\} \tag{7.6}$$

with $\mathbf{M}_t^L = \{\mathbf{f}_{k,t}^L | k = 1, \cdots, N_{\text{feature}}\}$ and $\mathbf{f}_{k,t}^L = \mathbf{T}_{B_0}^L \left(\prod_1^t \mathbf{T}_{B_t}^{B_{t-1}}\right) \mathbf{f}_{k,t}^{B_t}$.

Then both odometry and mapping are completed, meaning the SLAM is achieved. SLAM will make use of the fact that all the features are mutually observed in multi-epoch of the data. It means that SLAM can further apply a batch optimization considering the covisibility in a window containing several scans. The estimated transformation matrix and the map earlier are used as initial guesses for this batch optimization. The transformations in a window are batch optimized as follows.

$$\begin{aligned} \hat{\mathbf{T}}_{\text{window}} &= \arg\min_{\mathbf{T}_{\text{window}}} \sum_{t=i}^{i+w} \sum_{k=1}^{N_{\text{feature}}} \left\| \mathbf{M}_{\text{window}}^L - \mathbf{T}_{B_0}^L \left(\prod_1^t \mathbf{T}_{B_t}^{B_{t-1}} \right) \mathbf{f}_{k,t}^{B_t} \right\|^2 \\ \mathbf{T}_{\text{window}} &= \left\{ \mathbf{T}_{B_t}^{B_{t-1}} | \forall t \in [i, i+w] \right\} \end{aligned} \tag{7.7}$$

where i and $i + w$ denote the index of the first scan and the last scan in the window, respectively. $\mathbf{M}_{\text{window}}^L$ denotes the window map defined as $\{\mathbf{M}_t^L | \forall t \in [i, i + w]\}$. It is worth mentioning that for the constraint construction of the certain pose $\mathbf{T}_{B_t}^{B_{t-1}}$, the map points contributed by the corresponding scan $\mathbf{M}_t^L$ should be excluded from $\mathbf{M}_{\text{window}}^L$ during the match.

Finally, the batch optimized map can be obtained as $\hat{\mathbf{M}}_{\text{batch}}^L = \{\hat{\mathbf{M}}_t^L | \forall t\}$. $\hat{\mathbf{M}}_t^L$ is registered by the batch optimized transformation from (7.7). It is different from (7.6) where the map is accumulated by the pose from odometry (7.5).

The maps built by these methods are mostly in a local coordinate. The connection between indoor maps to outdoor maps requires the coordination transformation from the local map coordinate to the world geodetic coordinate. If the transformation between the world coordinate and the local map coordinate is available (which is usually given by a GNSS receiver), the SLAM map can be transformed to the world coordinate to provide absolute map information.

$$\mathbf{M}^W = \mathbf{T}_L^W \mathbf{M}^L \tag{7.8}$$

Table 7.3 provides an overview of the relationship between measurements, model, and the state to be estimated. Table 7.3 focuses on lidar and visual odometry, which are the primary methods discussed in this chapter. The formulas listed in Table 7.3 will be detailed in the following sections.

7.3 Lidar SLAM

A lidar sensor transmits pulsed light wave from a channel and receives a reflected wave that hits an object. There are two types of popular lidar: mechanical lidar, and solid-state lidar. In general, the former one has a sparser point cloud data (PCD) but wider coverage of field of view, while the latter one has a dense PCD but narrower coverage of field of view. The distance between the lidar and object can therefore be determined by a two-way TOA ranging method as shown in the left of Figure 7.3. If there are multiple channels, it is important to distinguish the reflection is coming from a specific channel, resulting in a sequence of binary codes (pulses) that has a characteristic of a low cross-correlation is usually used. Since it requires a few pulses to for the correlation, the lidar ranger usually has a requirement on the minimum distance between the sensor and an object. For a mechanical 3-D lidar, a servomotor will rotate all the channels in 360° to receive the reflections from its surrounding environment in 360°, forming a point cloud as shown in the right of Figure 7.3. The point cloud is to describe as the distances between all the objects (which cause the reflections) and the lidar sensor and it is defined as follows.

$$\mathbf{P}_t = \begin{bmatrix} \mathbf{p}_1 & \cdots & \mathbf{p}_i \end{bmatrix} = \begin{bmatrix} x_1 & \cdots & x_i \\ y_1 & \cdots & y_i \\ z_1 & \cdots & z_i \end{bmatrix}, i \in \text{the received reflections} \quad (7.9)$$

Table 7.3

Comparison of lidar and Visual Odometry in Terms of Measurement, Model, and State

Methods	Measurement	Model	State
Lidar	$\mathbf{p}_{i,t}^{B_t}$	$\mathbf{p}_{j,t-1}^{B_{t-1}} = \mathbf{C}_{B_t}^{B_{t-1}}\mathbf{p}_{i,t}^{B_t} + \Delta\mathbf{x}_{B_t}^{B_{t-1}}$	$\mathbf{T}_{B_t}^{B_{t-1}} = \begin{bmatrix} \mathbf{C}_{B_t}^{B_{t-1}} & \Delta\mathbf{x}_{B_t}^{B_{t-1}} \\ 0 & 1 \end{bmatrix}$
Monocular camera	$\mathbf{p}_{k,t+\Delta t}^{I}$ $\mathbf{p}_{k,t}^{I}$	$\left(\mathbf{K}^{-1}\mathbf{p}_{k,t+\Delta t}^{I}\right)^T \mathbf{t}^\wedge \mathbf{R}\mathbf{K}^{-1}\mathbf{p}_{k,t}^{I} = 0$	$\mathbf{T} = \begin{bmatrix} \mathbf{R} & \mathbf{t} \\ 0 & 1 \end{bmatrix}$ and depth of features

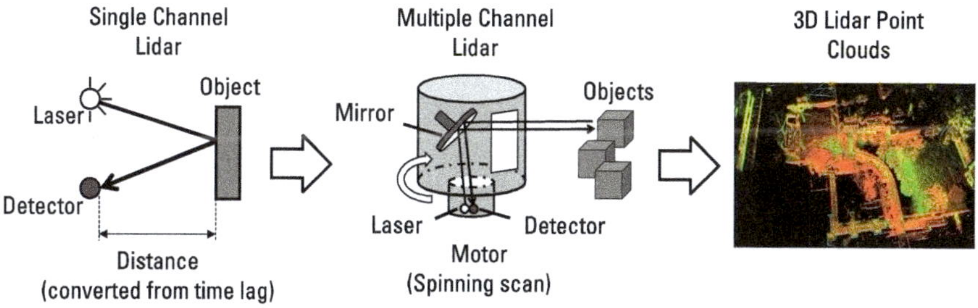

Figure 7.3 Mechanical lidar sensing principle.

where the subscript i denotes the index of the point scanned in epoch t and (x_i, y_i, z_i) denotes the location of a point in the local body coordinate of the lidar at epoch t. The point cloud, $\mathbf{P}_t$, is therefore the measurement of lidar. The model is the surrounding environment, as shown in Figure 7.4. In the following section, a conventional point-based and two state-of-the-art feature-based lidar models are introduced. The challenges of lidar SLAM are also discussed [16].

7.3.1 Point Cloud-Based Lidar SLAM: ICP

The most well-known conventional method is iterative closest point (ICP). It minimizes the point-to-point distance between two scans of lidar PCD. The previous scan can be regarded as a model of surrounding environment and the current scan is regarded as the measurement. ICP can estimate the transformation matrix between the two scans $\mathbf{T}_{B_t}^{B_{t-1}}$ when the minimum distance is achieved by iterations. It is described as follows.

$$\hat{\mathbf{T}}_{B_t}^{B_{t-1}} = \arg\min_{\mathbf{T}_{B_t}^{B_{t-1}}} \sum_{i=1}^{N_{\text{model}}} \sum_{j=1}^{N_{\text{measurement}}} b_{i,j} \left\| \mathbf{p}_{i,t-1}^{B_{t-1}} - \mathbf{T}_{B_t}^{B_{t-1}} \mathbf{p}_{j,t}^{B_t} \right\|^2 \tag{7.10}$$

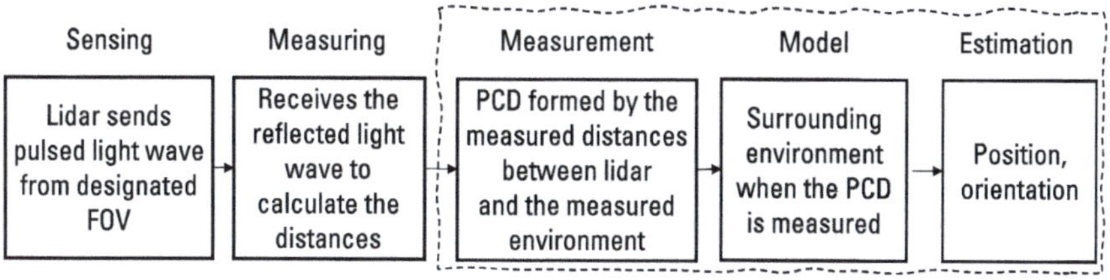

Figure 7.4 The measurements, models, and estimation of using lidar point cloud data for odometry.

where N_{model} and $N_{\text{measurement}}$ denote the number of points in the previous and current scans, respectively. $\mathbf{p}_{i,t-1}^{B_{t-1}}$ and $\mathbf{p}_{j,t}^{B_t}$ are the points in the previous and current scans, respectively. $b_{i,j}$ is a binary selector of the paired points, meaning that, if $\mathbf{p}_{i,t-1}^{B_{t-1}}$ and $\mathbf{p}_{j,t}^{B_t}$ is a pair, then $b_{i,j} = 1$ and otherwise $b_{i,j} = 0$ To determine whether the two points in the two scans is a pair, generally speaking, a threshold, $d_{\text{threshold}}$, of the distance of the two points is used to make the judgment.

$$b_{i,j} = \begin{cases} 0, \text{ if } \left\| \mathbf{p}_{i,t-1}^{B_{t-1}} - \left(\mathbf{T}_{B_t}^{B_{t-1}}\right)_0 \mathbf{p}_{j,t}^{B_t} \right\|_2 > d_{\text{threshold}} \\ 1, \text{ otherwise} \end{cases} \tag{7.11}$$

where the subscript 0 for $\left(\mathbf{T}_{B_t}^{B_{t-1}}\right)$ denotes initial guesses. The value of the threshold is based on applications, for example, 1m is used for the application of lidar in an indoor parking lot. For a high (depending on the velocity of the lidar) scan rate of lidar, the null initial guess is fine. One way to obtain a better initial guess is to apply a constant translational and rotational velocities assumption to predict the transformation matrix of the next epoch of time to solve (7.10), a nonlinear optimizer solver as mentioned Section 3.4.1. To reduce the computational load, the rotation matrix can be solved first by singular value decomposition (SVD), and then the translation matrix can be solved by substituting $\hat{\mathbf{C}}_{B_t}^{B_{t-1}}$ into $\Delta\hat{\mathbf{x}}_{B_t}^{B_{t-1}} = \overline{\mathbf{p}}_{t-1}^{B_{t-1}} - \left(\hat{\mathbf{C}}_{B_t}^{B_{t-1}}\overline{\mathbf{p}}_t^{B_t}\right)$ [17]. The $\overline{\mathbf{p}}_{t-1}^{B_{t-1}}$ and $\overline{\mathbf{p}}_t^{B_t}$ denotes the geometric center of the points in the previous scan and the current scan, respectively. As a scan of lidar PCD can easily contain more than 10,000 points using a typical mechanical 3-D lidar, the computational load of the conventional ICP is high, meaning it could not be used in the applications that have limited computing resources. A straightforward idea to maintain the SLAM performance but to reduce the computational load is to maintain the observability of the measurement model but to reduce the number of points used. A voxel grid filter can be used to sampling the dense raw points [18]. Another solution is to exploit the representative features from the dense raw points where only partial of the distinct features are used for the point registration.

Finally, either a direct assembly of the PCD, for example, (7.6), or a batch estimation considering all the PCD, for example, (7.7), can be used to obtain the accumulated PCD map $\mathbf{M} = \{\mathbf{p}_1^L \cdots \mathbf{p}_{N\text{point}}^L\}$ where N_{point} denotes the total number of the registered points.

To maintain the observability of the measurement model of the lidar SLAM, an idea is to extract "features" from the PCD. Two popular methods are: (1) to represent the raw points inside a voxel by a normal distribution, and (2) to extract 2-D planes and 1-D edges from the PCD.

7.3.2 Feature-Based Lidar SLAM: NDT

The normal distribution transform (NDT) method represents the points inside a voxel as a 3-D normal distribution as shown in Figure 7.5 [19]. In other words, it describes the probability of finding a point in a certain position by a set of normal distributions [20].

For kth voxel, the average $\overline{\mathbf{p}}_{k,t-1}^{B_{t-1}}$ and covariance matrix $\boldsymbol{\Sigma}_{k,t-1}^{B_{t-1}}$ of the points inside the voxel are calculated:

$$\overline{\mathbf{p}}_{k,t-1}^{B_{t-1}} = \frac{1}{n_k}\sum_{i=1}^{n_k} \mathbf{p}_{i,t-1}^{B_{t-1}} \tag{7.12}$$

$$\boldsymbol{\Sigma}_{k,t-1}^{B_{t-1}} = \frac{1}{n_k - 1}\sum_{i=1}^{n_k}\left(\mathbf{p}_{i,t-1}^{B_{t-1}} - \overline{\mathbf{p}}_{k,t-1}^{B_{t-1}}\right)\left(\mathbf{p}_{i,t-1}^{B_{t-1}} - \overline{\mathbf{p}}_{k,t-1}^{B_{t-1}}\right)^T \tag{7.13}$$

where n_k denotes the total number of points inside the kth voxel. To apply these normal distribution features in SLAM, the transformation matrix $\mathbf{T}_{B_t}^{B_{t-1}}$ is associated with the NDT measurement model. Comparing to ICP, the NDT method models the surrounding environment by voxels that represented by normal distributions. If the points in the current scan can achieve a maximum joint probability with a given transformation matrix $\hat{\mathbf{T}}_{B_t}^{B_{t-1}}$, then the given $\hat{\mathbf{T}}_{B_t}^{B_{t-1}}$ is solution with the maximum likelihood. The likelihood (which can be regarded as a score) of voxel k referring to $\mathbf{T}_{B_t}^{B_{t-1}}$ can be described as:

$$\text{likelihood}_k\left(\mathbf{T}_{B_t}^{B_{t-1}}\right) = \prod_{j=1}^{N_{\text{measurement}}} \exp\left(-\frac{\left\|\mathbf{T}_{B_t}^{B_{t-1}}\mathbf{p}_{j,t}^{B_t} - \overline{\mathbf{p}}_{k,t-1}^{B_{t-1}}\right\|_{\boldsymbol{\Sigma}_{k,t-1}^{B_{t-1}}}^2}{2}\right) \tag{7.14}$$

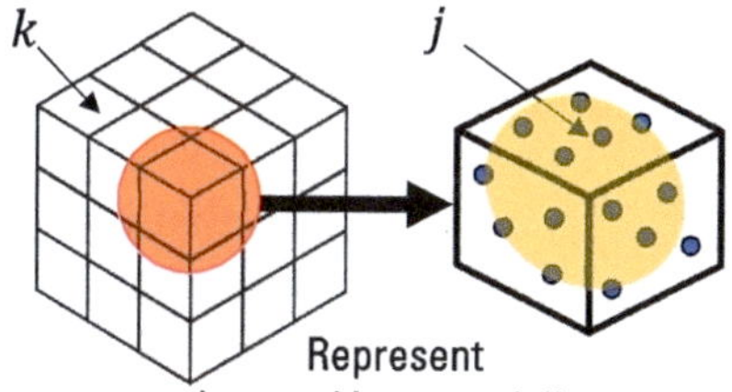

Figure 7.5 Illustration of representing points as normal distributions.

where $\|\mathbf{r}\|_{\Sigma}^{2}$ denotes $\mathbf{r}^{T}\Sigma^{-1}\mathbf{r}$. Considering all the voxel, the joint likelihood becomes:

$$\text{likelihood}\left(\mathbf{T}_{B_t}^{B_{t-1}}\right)=\prod_{k=1}^{N_{\text{voxel}}}\prod_{j=1}^{N_{\text{measurement}}}\exp\left(-\frac{\left\|\mathbf{T}_{B_t}^{B_{t-1}}\mathbf{p}_{j,t}^{B_t}-\overline{\mathbf{p}}_{k,t-1}^{B_{t-1}}\right\|_{\Sigma_{k,t-1}^{B_{t-1}}}^{2}}{2}\right) \tag{7.15}$$

Since the likelihood is Gaussian distributed, the MLE of (7.15) can be estimated by a nonlinear LS estimation.

$$\hat{\mathbf{T}}_{B_t}^{B_{t-1}}=\underset{\mathbf{T}_{B_t}^{B_{t-1}}}{\arg\min}\frac{1}{2}\sum_{k=1}^{N_{\text{voxel}}}\sum_{j=1}^{N_{\text{measurement}}}\left\|\mathbf{T}_{B_t}^{B_{t-1}}\mathbf{p}_{j,t}^{B_t}-\overline{\mathbf{p}}_{k,t-1}^{B_{t-1}}\right\|_{\Sigma_{k,t-1}^{B_{t-1}}}^{2} \tag{7.16}$$

The NDT method divides the surrounding environment into fix-sized cells. This may lead to discontinuities in the model representation since some voxel may contain very little points. The insufficient constraints could be problematic in calculating the Jacobian matrix of the NDT measurement model [21]. A method that combines the ideas of ICP and NDT is proposed and called generalized ICP [22]. It first associates a group of points between the previous and current scans and then models the group of points by normal distributions and finally does the distribution-to-distribution matching.

7.3.3 Feature-Based Lidar SLAM: LOAM (Plane and Edge)

A state-of-the-art lidar method is LOAM [7]. Its aims are: (1) to maintain a low-drift of the mapped PCD map, and (2) to reduce the computational load of SLAM. In the previously mentioned ICP and NDT, the two SLAMs are matching the previous and current scans (i.e., a scan-to-scan matching). These scan-to-scan matchings only consider the points in the vicinity of the two scans, meaning that the points (features) in a long distance are not really used in the calculation of the transformation; see (7.10) and (7.16). This results in drifting to the conventional ICP and NDT maps. LOAM proposed a scan-to-map approach (refer to the dashed line in the upper right of Figure 7.1). The map means a locally registered map during the SLAM process. This local map is the aggregated PCD of all previous scans so that it also considers the points (features) in a long distance. This scan-to-map approach can effectively reduce the drift. However, this approach drastically increases the computational load; hence, the LOAM has to reduce the computational load at the same time. To achieve this aim, LOAM extracts representative edge and planar features from PCD to reduce the redundancy of the dense point clouds while maintaining the model observability. The extraction accords to the curvature of a few segmented points in a current scan $\mathbf{P}_t$. A point is selected as an edge point if its curvature value is larger than a predetermined threshold, or classified as a planar point by a smaller curvature, as Figure 7.6 shows.

To further reduce the computation load of the SLAM, the lower update rate of the scan-to-map approach is desired. This reduction of the update rate results in another challenge; a more stringent requirement on the quality of the initial guess comparing it to a scan-to-scan approach. In LOAM, a scan-to-scan odometry is used to provide that initial guess for the later scan-to-map matching. Thus, it is a two-step approach. Equation (7.17) describes the rough transformation estimated by the odometry.

$$\hat{\mathbf{T}}_{\text{odom}} = \arg\min_{\mathbf{T}_{\text{odom}}} \frac{1}{2}\left\{ \sum_{k=1}^{N^{\text{edge}}} \left\| \omega_{(k)} f\left(\mathbf{p}_{(k,t)}^{\text{edge}}, \mathbf{T}_{\text{odom}}\right)\right\|^2 + \sum_{k=1}^{N^{\text{plane}}} \left\| \omega_{(k)} f\left(\mathbf{p}_{(k,t)}^{\text{plane}}, \mathbf{T}_{\text{odom}}\right)\right\|^2 \right\} \tag{7.17}$$

Figure 7.6 Illustration of extracted edge (black points) and planar points (gray boxes) for scan-to-map concerning a frame of lidar point clouds (gray points).

where $\omega_{(k)}$ denotes the bisquare weight, setting as $1 - 1.8 \times f(\mathbf{p}_{(k,t)}, \mathbf{T}_{\text{odom}})$ [7]. $f\left(\mathbf{p}_{(k,t)}^{\text{edge}}, \mathbf{T}_{\text{odom}}\right)$ denotes a point-to-line distance, which is calculated as follows:

$$f\left(\mathbf{p}_{(k,t)}^{\text{edge}}, \mathbf{T}_{\text{odom}}\right) = \frac{\left\| \left(\mathbf{T}_{\text{odom}}\mathbf{p}_{(k,t)}^{\text{edge}} - \mathbf{p}_{(j,t-1)}^{\text{edge}}\right) \times \left(\mathbf{T}_{\text{odom}}\mathbf{p}_{(k,t)}^{\text{edge}} - \mathbf{p}_{(l,t-1)}^{\text{edge}}\right) \right\|_2}{\left\| \mathbf{p}_{(j,t-1)}^{\text{edge}} - \mathbf{p}_{(l,t-1)}^{edge} \right\|_2} \tag{7.18}$$

where $\mathbf{p}_{(k,t)}^{\text{edge}}$ denotes the kth edge feature points on the current scan. $\mathbf{p}_{(j,t-1)}^{\text{edge}}$ and $\mathbf{p}_{(l,t-1)}^{\text{edge}}$ denote two closest edge features of $\mathbf{p}_{(k,t)}^{\text{edge}}$ in the previous scan that are assumed to be on the corresponding line segment of $\mathbf{p}_{(k,t)}^{\text{edge}}$. As shown in Figure 7.7, $\mathbf{p}_{(j,t-1)}^{\text{edge}}$ and $\mathbf{p}_{(l,t-1)}^{\text{edge}}$ are selected from adjacent rings so that vertical lines orthogonal to the laser beam are exploited. Such lines are free from motion distortion compared with that parallel to the laser beam.

In (7.19), $f\left(\mathbf{p}_{(k,t)}^{\text{plane}}, \mathbf{T}_{\text{odom}}\right)$ is formulated as the point-to-plane distance for planar feature points as follows:

$$f\left(\mathbf{p}_{(k,t)}^{\text{plane}}, \mathbf{T}_{\text{odom}}\right) = \frac{\left\| \begin{array}{c} \left(\mathbf{T}_{\text{odom}}\mathbf{p}_{(k,t)}^{\text{plane}} - \mathbf{p}_{(j,t-1)}^{\text{plane}}\right) \\ \left(\left(\mathbf{p}_{(j,t-1)}^{\text{plane}} - \mathbf{p}_{(l,t-1)}^{\text{plane}}\right) \times \left(\mathbf{p}_{(j,t-1)}^{\text{plane}} - \mathbf{p}_{(m,t-1)}^{\text{plane}}\right)\right) \end{array} \right\|_2}{\left\| \left(\mathbf{p}_{(j,t-1)}^{\text{plane}} - \mathbf{p}_{(l,t-1)}^{\text{plane}}\right) \times \left(\mathbf{p}_{(j,t-1)}^{\text{plane}} - \mathbf{p}_{(m,t-1)}^{\text{plane}}\right) \right\|_2} \tag{7.19}$$

where $\mathbf{p}_{(k,t)}^{\text{plane}}$ denotes the kth planar feature points on the current scan. $\mathbf{p}_{(j,t-1)}^{\text{plane}}$, $\mathbf{p}_{(l,t-1)}^{\text{plane}}$, and $\mathbf{p}_{(m,t-1)}^{\text{plane}}$ denotes the three closest planar feature points of $\mathbf{p}_{(k,t)}^{\text{plane}}$ in the previous scan which are assumed be on the corresponding planar patch of $\mathbf{p}_{(k,t)}^{\text{plane}}$. As shown in Figure 7.7, $\mathbf{p}_{(j,t-1)}^{\text{plane}}$ and $\mathbf{p}_{(l,t-1)}^{\text{plane}}$ are selected from the

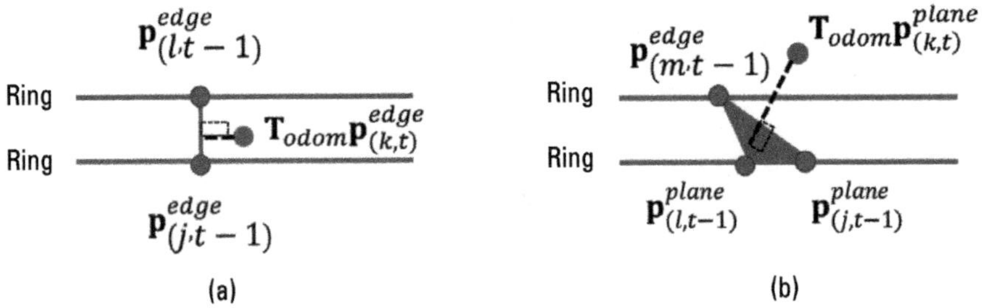

Figure 7.7 Illustration of point-to-line distance (a) and point-to-plane (b) distance for scan-to-scan odometry.

same ring while $\mathbf{p}_{(m,t-1)}^{\text{plane}}$ are selected from an adjacent ring. Consequently, the three points are not colinear to form a planar patch.

Then the rough transformation $\mathbf{T}_{\text{odometry}}$ is used to estimate a refined transformation by the scan-to-map approach. The objective function is formulated as (7.17). However, the point-to-line (plane) distance is that between $\mathbf{p}_{(k,t)}^{\text{edge}}\left(\mathbf{p}_{(k,t)}^{\text{plane}}\right)$ and the line (plane) fitted by its closest points selected from the map rather the previous scan.

Finally, the refined transformation will be used in the batch optimization as mentioned in (7.7).

7.3.4 Challenges of Lidar SLAM

For IPIN applications, adverse weather conditions are not a problem for lidar SLAM. The two major challenges are motion distortion and the presence of dynamic objects.

7.3.4.1 Motion Distortion

Ideally speaking, lidar SLAM assumes that all the points are scanned at the same location of the ego-vehicle. In reality, the movement of the ego-vehicle while the lidar is scanning will cause a motion distortion to the scanned lidar PCD. This phenomenon is illustrated in Figures 7.8 and 7.9. As can be seen when lidar stays static as the gray dot in the left part of Figure 7.8, the scanning pattern remains as a circle. While when lidar moves from the left dot to the right dot as in the right part of Figure 7.8, the scanning pattern is distorted. This is a factor resulting in the drift of accumulated lidar PCD map. This distortion can be compensated for if the rotation rate and acceleration of the ego-vehicle are roughly known. A popular method is to integrate with IMU to provide such measurements to compensate the distortion [23]. The poses of the lidar during the scan sweep are predicted. As shown in the following equation, each point is transformed to the starting moment of the scan so that the motion distortion is compensated for:

$$\tilde{\mathbf{p}}_i = \mathbf{T}_i \mathbf{p}_i \tag{7.20}$$

where $\mathbf{p}_i$ denotes the ith point produced by the lidar during a scan sweep at the lidar body frame. $\mathbf{T}_i$ is the lidar pose currently related to scan starting time. $\tilde{\mathbf{p}}_i$ denotes the point with distortion compensation. $\mathbf{T}_i$ is predicted as:

$$\mathbf{T}_i = \begin{bmatrix} \mathbf{R}_i & \mathbf{t}_i \\ 0 & 1 \end{bmatrix} = \begin{bmatrix} \exp\left(\boldsymbol{\omega} \cdot \Delta t_i\right)^{\wedge} & \mathbf{v} \cdot \Delta t_i \\ 0 & 1 \end{bmatrix} \tag{7.21}$$

where Δt_i denotes the time increment when $\mathbf{p}_i$ is achieved relative to the scan starting point. $\boldsymbol{\omega}$ and $\mathbf{v}$ are the average rotation rate and average velocity provided by IMU during Δt_i, respectively. $(\ )^\wedge$ represents the skew symmetry matrix of a vector which transforms the rotation vector to the rotation matrix with further exponential transformation.

7.3.4.2 Dynamic Objects

A major assumption made in lidar SLAM is that the surrounding environment is static, meaning that there are only still objects. In other words, lidar SLAM assumes that the change of the distances between the ego-vehicle to the surrounding objects is due to the relative transformation of ego-vehicle the between two epochs. As Figure 7.10 shows, the position change of the dynamic object is not due to the position change of the ego-vehicle, resulting in the estimated relative transformation of ego-vehicle not being accurate. This is particularly a problem when lidar SLAM is used in urban areas where the number of dynamic objects is excessive [24].

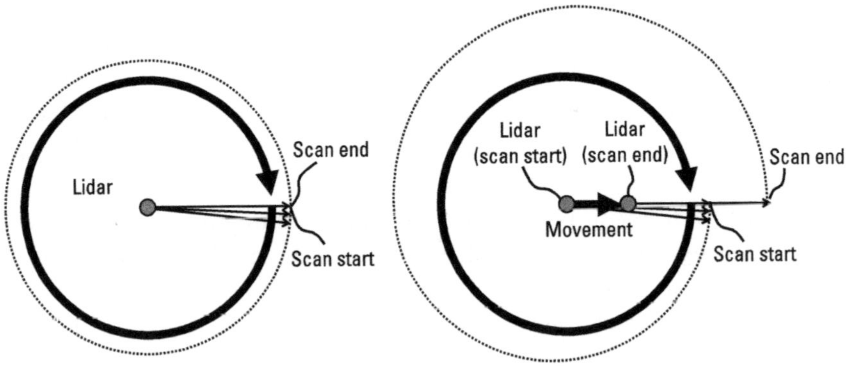

Figure 7.8 Illustration of motion distortion due to lidar movement.

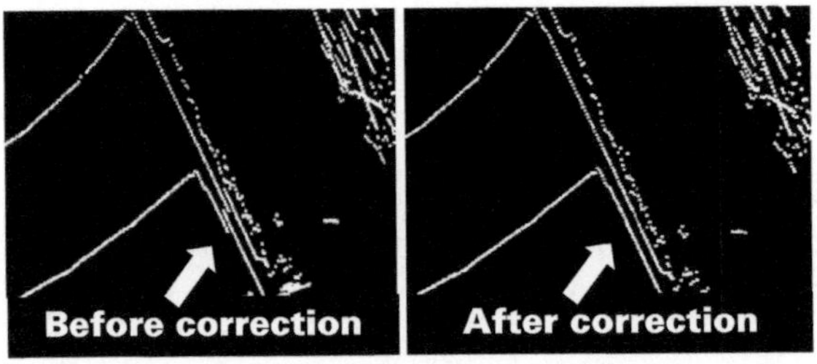

(a) (b)

Figure 7.9 Motion distortion before (a) and after (b) applying correction.

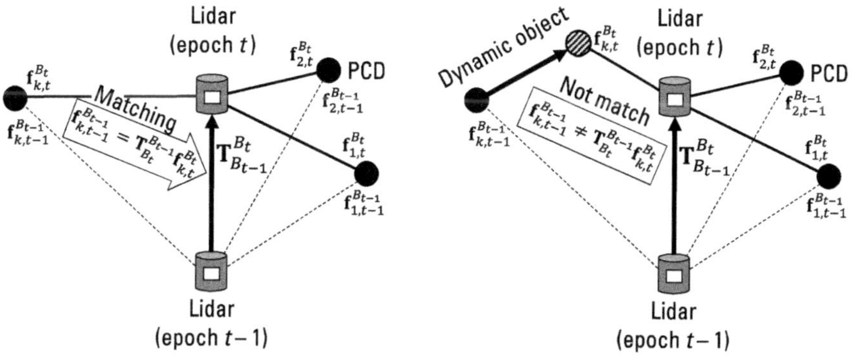

Figure 7.10 Lidar odometry without (a) and with (b) the impact of a dynamic object.

There are several approaches to mitigate the effect. One of the intuitive approaches is to detect the dynamic object using advanced object recognition approach [25] to exclude the point cloud that came from the dynamic object before its use in lidar SLAM. Another approach is to apply a robust estimator (such as the ones introduced in Section 3.1.1) to mitigate the impact of these outlier measurements [26].

7.3.5 Closed-Loop Constraints

A loop closure should be detected once the sensor carrier arrives at the same place it has been. A Euclidean distance-based loop closure detection approach is proposed [8]. For the latest estimated pose $\mathbf{T}_j$, the closest prior pose $\mathbf{T}_i$ is searched in the surrounding Euclidean space with a certain size. $\mathbf{T}_i$ is selected as the loop pair for $\mathbf{T}_j$ if the Euclidean distance between them is smaller than a predetermined threshold. The transformation $\Delta\mathbf{T}'_{ij}$ between the loop pair is achieved by matching the i scan with the sub-map composed of scans around the j one. The closed-loop constraint is formulated as the residual between the observed pose increment $\Delta\mathbf{T}'_{ij}$ and the practical one derived from the state as follows:

$$\mathbf{f}\left(\mathbf{T}_i,\mathbf{T}_j,\Delta\mathbf{T}'_{ij}\right) = \ln\left(\Delta\mathbf{T}'^{-1}_{ij}\mathbf{T}_i^{-1}\mathbf{T}_j\right)^{\vee} \tag{7.22}$$

For implementation feasibility (e.g., derivative calculation of the residual relative to the state), the residual is transformed from the matrix format to the

Lie algebra format by $\ln(\ )^{\vee}$. $\ln(\ )$ transforms the transformation matrix to the skew-symmetric matrix constructed from its corresponding Lie algebra. $(\ )^{\vee}$ recover the Lie algebra from its skew symmetric matrix.

7.4 Visual SLAM

There are two popular types of cameras that are used for visual SLAM for IPIN applications, monocular and RGB-D cameras, which will be introduced next. Stereo camera-based SLAM is not introduced in the book for the following reasons. The stereo camera-based approach uses the baseline between the left and right cameras to calculate the disparity to reconstruct the depth of the images. However, the baseline is usually very short (e.g., the stereo camera on smartphones), resulting in the reconstructed depth being inaccurate for objects at long distances. The poor depth measurements limit its performance in odometry and SLAM. As a result, a monocular camera-based SLAM is usually used even when a stereo camera is available. In other words, only the left or right camera is used. As a result, the monocular camera-based SLAM is still the most popular approach.

7.4.1 Monocular Camera

Compared with lidar, a monocular camera cannot directly measure the distance (also known as "depth" in image processing) between the sensed objects and the camera. To use a monocular camera in a visual odometry, it has to use a sequence of 2-D images (t and $t + 1$) to estimate the transformation matrix of the camera pose between two epochs of time. In other words, the change of a same feature in two images can be used to infer the change of the camera pose as shown in Figure 7.11. The overall flow of a visual SLAM can be simplified as in Figure 7.12, for which each of the functions will be introduced next.

7.4.1.1 Feature Detection and Tracking

As the visual odometry makes use of the change of a same feature in two images to infer the change of the camera pose, it is utterly important to detect distinctive features from an image. Obviously, corner points are much more distinctive than lines since it can be easily identified in a 2-D image by our eyes (see Figure 7.13).

The corner point can be detected by a widely applied method, the Shi-Tomasi method [27]. Its idea is to check whether there are two lines in a searched area of the image. First, the image is transformed from red-green-blue

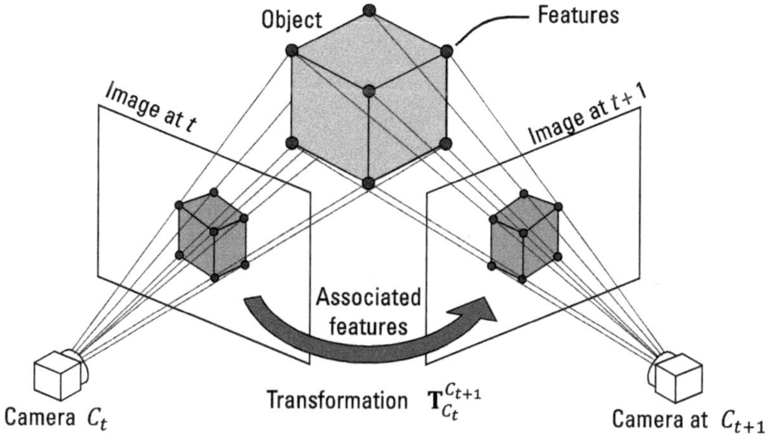

Figure 7.11 Illustration of the idea of estimating a transformation matrix using the detected and tracking features on a sequence of images.

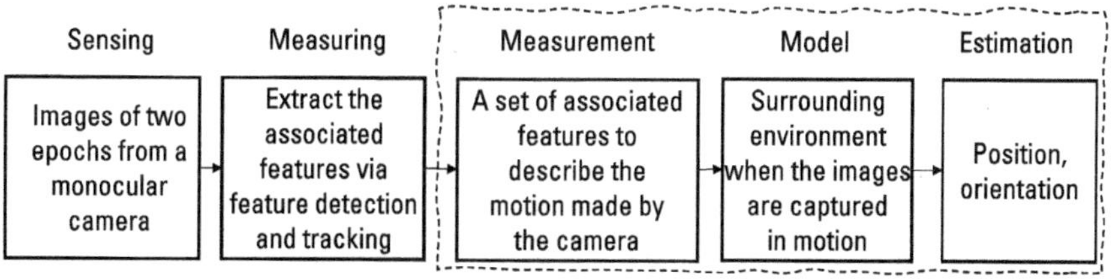

Figure 7.12 The measurements, models and estimation of using an image for a monocular camera based odometry.

(RGB) to a grayscale image for making only one value in each pixel. Then a 2-D window with the size of w_u and w_v in u- and v-axes is used as a region of interest (ROI) to find corner points, as shown in Figure 7.14.

The goal here is to find the corner point, meaning the intersection of two lines. Thus, the representation of the lines using the changes of grayscale

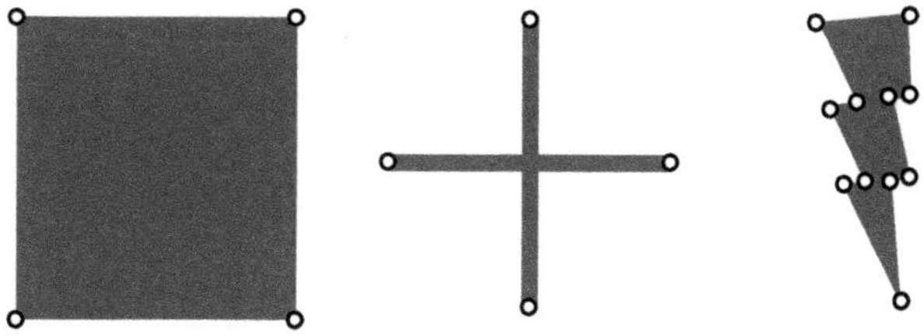

Figure 7.13 Illustration of corner points are distinctive in a 2-D plane.

Figure 7.14 (a) Illustration of a 2-D window used as a ROI to the searched corner points, and (b) associated eigenvalues when there is either a line or a corner in the window.

values insides the window is important. The grayscale change $E_p(u_p, v_p)$ within the ROI of the searched pixel can be calculated as:

$$E_p\left(u_p, v_p\right) = \sum_{w_u}\sum_{w_v} W\left(w_u, w_v\right)\left[I\left(u_p + w_u, v_p + w_v\right) - I\left(u_p, v_p\right)\right]^2 \quad (7.23)$$

where I denotes a grayscale image. u_p and v_p denote the location of the pixel index p used to search the corner points, which is the center of the searching window. $I(u_p, v_p)$ denotes a specific grayscale value. $w(w_u, w_v)$ is the weighting of the corresponding pixel within this window. The calculation of (7.23) can be regarded as a weighted convolution process between the $I(u_p, v_p)$ and the surrounding grayscale value in this window. In general, the window will be set as a quadratic Gaussian distribution with the window center as the origin.

By linearizing the change of grayscale values in the window using a first-order Taylor series expansion, the following equation can be derived:

$$I\left(u_p + w_u, v_p + w_v\right) \approx I\left(u_p, v_p\right) + I_{w_u} w_u + I_{w_v} w_v \quad (7.24)$$

where $I_{w_u} = \partial I(u_p, v_p)/\partial u_p$ and $I_{w_v} = \partial I(u_p, v_p)/\partial v_p$. I_{w_u} and I_{w_v} are the gradient of the gray value changes according to pixel location directions in u and v, respectively (i.e., the partial differential of I). By substituting (7.24) into (7.23), the grayscale change within the window can be modeled as:

$$E_p\left(u_p, v_p\right) \approx \begin{bmatrix} u_p & v_p \end{bmatrix} \mathbf{G} \begin{bmatrix} u_p \\ v_p \end{bmatrix} \quad (7.25)$$

$$\mathbf{G} = \sum_{w_u}\sum_{w_v} W(w_u, w_v)\begin{bmatrix} I_{w_u}^2 & I_{w_u} I_{w_v} \\ I_{w_v} I_{w_u} & I_{w_v}^2 \end{bmatrix} = \mathbf{U}^{-1}\begin{bmatrix} \lambda_{p,1} & 0 \\ 0 & \lambda_{p,2} \end{bmatrix}\mathbf{U} \quad (7.26)$$

If there is a corner point in the window, both the eigenvalues in the derived 2-D model should be larger than a certain value as shown in the right of Figure 7.14. Note that an eigenvector of a model is a nonzero vector that changes at most when the model is applied to by a state value. Eigenvalues are then used to indicate how much the vectors are scaled. In this example, the vectors can be understood as the major lines in the window. Thus, an SVD is applied in the model to obtain the eigenvalues, $\lambda_{p,1}$ and $\lambda_{p,2}$. If both the eigenvalues are larger than a predefined threshold, $\lambda_{\min}$, then it is concluded that there is a corner point in this window. All the pixels p in the image are scanned to find all the detected corner points $\mathbf{f}$. For the example in Figure 7.14 with a uniform weighting W within the window, the detection of features (i.e., corner points) is achieved by

$$\sum_{w_u}\sum_{w_v}\begin{bmatrix} I_{w_u}^2 & I_{w_u} I_{w_v} \\ I_{w_v} I_{w_u} & I_{w_v}^2 \end{bmatrix} = \mathbf{U}^{-1}\begin{bmatrix} \lambda_{p,1} & 0 \\ 0 & \lambda_{p,2} \end{bmatrix}\mathbf{U} \quad (7.27)$$

$$\mathbf{f} = \left\{ p \mid \lambda_{p,1} > \lambda_{\min}, \& \lambda_{p,2} > \lambda_{\min} \right\} \quad (7.28)$$

As mentioned in Figure 7.11, the key of a visual odometry to use a 2-D correspondence of images to infer the change of the 3-D pose (3-D position and 3-D orientation) of a camera. A popular and computational efficient method to obtain the 2-D correspondence of the detected features is Optical Flow with the Lucas-Kanade method [28]. Its idea is as follows and is shown in Figure 7.15.

For each kth detected feature in the image of the current epoch, located at $[u_k \quad v_k]^{\mathrm{T}} \in \mathbf{f}$, a window in the image of the current epoch is used to check

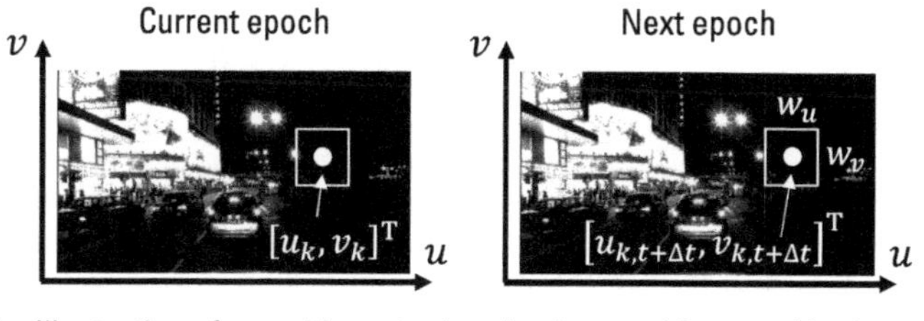

Figure 7.15 Illustration of searching whether the detected feature k in the current epoch exists in the image of the next epoch.

whether the feature still exists in the image of the next epoch. Note that the window here is used to track feature movements, which is different from the ROI window in (7.23) for detecting features. If the grayscale value changes dramatically over a threshold $\varepsilon_{\text{grayscale}}$, it is assumed that the detected feature no longer exists in the current image. It should be noted if the brightness of the image change may cause the loss of detected features, which is one of the reasons that monocular camera-based visual odometry is less robust.

$$\varepsilon_k = \sum_{w_u}\sum_{w_v}\left(I_t\left(u_k, v_k\right) - I_{t+\Delta t}\left(u_k + w_u, v_k + w_v\right)\right)^2 \tag{7.29}$$

$$\mathbf{f}_{\text{track}} = \left\{\begin{bmatrix} u_k & v_k \end{bmatrix}^{\mathrm{T}} \middle| \varepsilon_k < \varepsilon_{\text{grayscale}}\right\} \tag{7.30}$$

The next step is to derive the pixel change Δu_k and Δv_k of a tracked feature k in $\mathbf{f}_{\text{track}}$. Note that the pixel change is the 2-D correspondence that is used as a measurement mentioned in Figure 7.12. If the constant brightness assumption can be made, then, for each k, the grayscale value of the tracked features in the images of the previous and current epochs are remained the same.

$$I\left(u_k, v_k, t\right) = I\left(u_k + \Delta u_k, v_k + \Delta v_k, t + \Delta t\right) \tag{7.31}$$

By applying the first order of Taylor series expansion, the following equation can be derived. Here, the change of grayscale value is assumed to be linear.

$$I\left(u_k + \Delta u_k, v_k + \Delta v_k, t + \Delta t\right) \approx I\left(u_k, v_k, t\right) + \frac{\partial I}{\partial u_k}\Delta u_k + \frac{\partial I}{\partial v_k}\Delta v_k + \frac{\partial I}{\partial t}\Delta t \tag{7.32}$$

If another assumption that the detected feature did not move in the real world is made (static feature), where the movement of Δu_k and Δv_k are due to camera movement, the following equations can be obtained.

$$\frac{\partial I}{\partial u_k}\Delta u_k + \frac{\partial I}{\partial v_k}\Delta v_k + \frac{\partial I}{\partial t}\Delta t = 0 \tag{7.33}$$

$$\frac{\partial I}{\partial u_k}\frac{\Delta u_k}{\Delta t} + \frac{\partial I}{\partial v_k}\frac{\Delta v_k}{\Delta t} = \frac{\partial I}{\partial t} \tag{7.34}$$

$$\begin{bmatrix} I_{u_k} & I_{v_k} \end{bmatrix}\begin{bmatrix} \dot{u}_k \\ \dot{v}_k \end{bmatrix} = \begin{bmatrix} -I_t \end{bmatrix} \tag{7.35}$$

where $\dot{u}_k$ and $\dot{v}_k$ are the velocity of the tracked feature at the u and v directions, respectively.

Finally, another assumption is made that all the pixels in a searched window in the current epoch of the image are spatially consistent. For example, if a static object is captured inside the window, then this assumption is valid. This assumption means that all the pixels in the search window based on the kth tracked feature are used to calculate the velocity of the pixel of the kth feature tracked. The benefit is to provide more measurements to estimate the velocity of a tracked feature compared with just using a tracked pixel to a tracked pixel of the two epochs. In other words, the estimation of the pixel change of a tracked feature k becomes an overdetermined system. The matrix form below can be obtained as shown in the equation:

$$\begin{bmatrix} I_{u,1} & I_{v,1} \\ I_{u,2} & I_{v,2} \\ \vdots & \vdots \\ I_{u,(w_u \times w_v)} & I_{v,(w_u \times w_v)} \end{bmatrix} \begin{bmatrix} \dot{u}_k \\ \dot{v}_k \end{bmatrix} = - \begin{bmatrix} I_{t,1} \\ I_{t,2} \\ \vdots \\ I_{t,(w_u \times w_v)} \end{bmatrix}$$

$$\begin{gathered} \mathbf{H}_k \mathbf{x}_k = -\mathbf{x}_k \\ \hat{\mathbf{x}}_k = \left(\mathbf{H}_k^{\mathrm{T}} \mathbf{H}_k \right)^{-1} \mathbf{H}_k^{\mathrm{T}} \left(-\mathbf{z}_k \right) \\ k \in \mathbf{f}_{\text{track}}, \text{ feature tracked} \end{gathered} \tag{7.36}$$

An example of detected and tracked visual features by the methods mentioned above is shown in Figure 7.16. The white lines are the measured 2-D correspondences for the estimation of the camera pose. It should be noted that the accuracy and robustness of feature detections and tracking would greatly determine the performance of visual odometry and SLAM. ORB is a popular and a patent-free method to combine the features as a descriptor to improve the robustness and accuracy of the estimation of the 2-D correspondences [29]. A comparison of the different feature-tracking methods can be found [30].

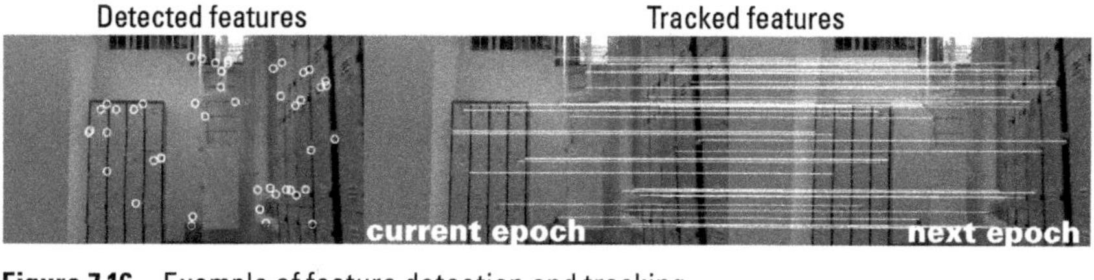

Figure 7.16 Example of feature detection and tracking.

Deep learning networks have been used to detect and track the features [31], which may relax the three assumptions: (1) the constant brightness between two epochs, (2) the track features remaining static in two epochs, and (3) a spatial consistency of the window searched, made in the Lucas-Kanade method.

7.4.1.2 Camera Model

There are several examples of a camera model:

- *Pinhole model for generating an image:* To estimate the pose of the camera, three coordinate systems are involved: the pixel plane (2-D), the imaging plane (2-D), and the camera-body coordinate (3-D). Figure 7.17 illustrates a pinhole imaging model. If there is an object located at $\mathbf{p}^C = [x^C \quad y^C \quad z^C]^T$ in the camera-body coordinate, by this model, its position in imaging plane becomes $[x^I \quad y^I]^T$. According to the triangular similarity, they have a relationship,

$$\frac{z^C}{f} = \frac{x^C}{x^I} = \frac{y^C}{y^I} \tag{7.37}$$

where f denotes the focal length of the camera. As previously mentioned, an image is usually described by pixels, meaning a pixel plane. The position of a pixel $\mathbf{p}^{\text{pixel}} = [u \quad v]^T$ can be obtained by applying scaling (θ_x and θ_y) and translation (c_x and c_y) to the object in the image plane, which is obtained by another scaling with the focal length from the camera-body coordinate, as below.

$$u = \theta_x x^I + c_x = \frac{\theta_x f x^C}{z^C} + c_x = \frac{f_x x^C}{z^C} + c_x \tag{7.38}$$

$$v = \theta_y y^I + c_y = \frac{\theta_y f x^C}{z^C} + c_x = \frac{f_y y^C}{z^C} + c_y \tag{7.39}$$

By reorganizing (7.38) and (7.39) in a matrix form related to the camera-body coordinate,

$$z^C \begin{bmatrix} u \\ v \\ 1 \end{bmatrix} = \begin{bmatrix} f_x & 0 & c_x \\ 0 & f_y & c_y \\ 0 & 0 & 1 \end{bmatrix} \begin{bmatrix} x^C \\ y^C \\ z^C \end{bmatrix} \tag{7.40}$$

$$s\mathbf{p}^{\text{pixel}} = \mathbf{K}\mathbf{p}^{\text{C}}$$

where **K** denotes an intrinsic parameter matrix. f_x and f_y are the focal length in x- and y-axes of the imaging plane incorporating the pixel-image resolution scaling, respectively. $[c_x, c_y]^T$ is the principal point, and $s = z^C$ denotes the depth of the object. It is clear that the object in the physical world became mathematically related to the pixel in an image. Generally speaking, the intrinsic parameter matrix only needs to be calibrated once before the camera is used to apply visual odometry.

- *Distortion:* In visual SLAM, the camera distortion is not negligible. As shown in Figure 7.18, it is caused by both the lens distortion and the assembly angle between the camera sensor and vertical plane. It can divide into radial and tangential distortions (Figure 7.18). These distortions can be corrected by the following model [32],

$$\begin{cases} \hat{x}^I = x^I\left(1 + k_1 r^2 + k_2 r^4 + k_3 r^6 + 2p_1 x^I y^I + p_2\left(r^2 + 2{x^I}^2\right)\right) \\ \hat{y}^I = y^I\left(1 + k_1 r^2 + k_2 r^4 + k_3 r^6 + 2p_2 x^I y^I + p_1\left(r^2 + 2{y^I}^2\right)\right) \end{cases} \tag{7.41}$$

where, (k_1, k_2, k_3) and (d_1, d_2) are the coefficients of the radial and tangential distortions, respectively. $(x^I \quad y^I)$ is the ideal position on the imaging plane (without distortion), and $\left(\hat{x}^I, \hat{y}^I\right)$ is the real distorted position. With the intrinsic parameters, the distortion parameters can be obtained by camera calibration. Based on this undistorted model, the true projection position $\mathbf{p}^{\text{pixel}} = [u \quad v]$ of a point on the pixel plane is

$$u = f_{x^I}\hat{x}^I + c_{x^I} \tag{7.42}$$

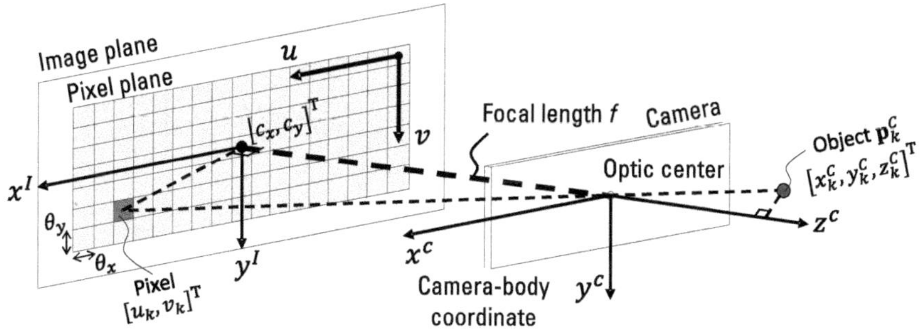

Figure 7.17 The planes and coordinate related to a pinhole imaging model.

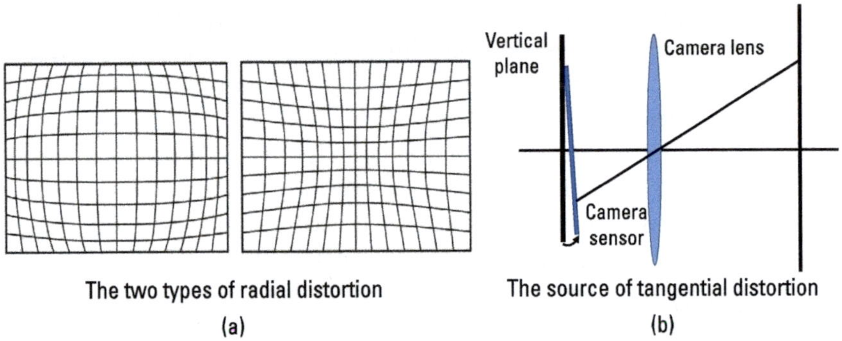

Figure 7.18 Illustration of the camera radial and tangential distortions: (a) the two types of radial distortion and (b) the source of tangential distortion.

$$y = f_{y^I}\hat{y}^I + c_{y^I} \tag{7.43}$$

7.4.1.3 Visual Odometry

The relationship between the two consecutive camera poses (two epochs) and detected features in the images can be described using Figure 7.19. This is known as an epipolar geometry [33].

The feature points $\mathbf{p}_{k,t}^{\text{pixel}_t}$ and $\mathbf{p}_{k,t+\Delta t}^{\text{pixel}_{t+\Delta t}}$ are a matching pair in the two images, and they correspond to the same 3-D point, $\mathbf{p}_{k,t}^{C_t}$. The camera centers at two adjacent epochs and the 3-D point $\mathbf{p}_{k,t}^{C_t}$ can determine a plane, called the epipolar plane. The line between these two camera centers is the baseline. The intersections of the baseline and the image planes at two adjacent epochs are epipoles, $e_{k,t}$ and $e_{k,t+\Delta t}$. The lines $l_{k,t}$ and $l_{k,t+\Delta t}$ between a pixel and the epipole are called the epipolar lines. The pose of the kth detected feature in

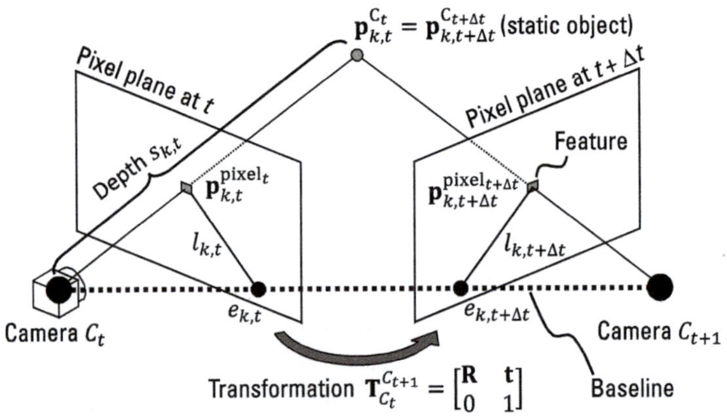

Figure 7.19 Illustration of epipolar geometry with features between adjacent epochs.

the camera-body frame at time $t + \Delta t$ can be obtained from the pose at t by applying a rotation matrix $\mathbf{R}$ and a translation vector $\mathbf{t}$ from the camera-body coordinate C_t to $C_{t+\Delta t}$, as follows:

$$\mathbf{p}_{k,t+\Delta t}^{C_{t+\Delta t}} = \mathbf{R}\mathbf{p}_{k,t}^{C_t} + \mathbf{t} \tag{7.44}$$

where the rotation and translation can be described by a transformation matrix as (7.4). In the pixel frame, the detected feature k at time t and $t + \Delta t$ can be expressed as below.

$$\mathbf{p}_{k,t+\Delta t}^{\text{pixel}} = \mathbf{p}_{k,t}^{\text{pixel}} + \begin{bmatrix} \dot{u}_k \\ \dot{v}_k \end{bmatrix} \Delta t,\ k \in \mathbf{f}_{\text{track}},\ \text{feature tracked} \tag{7.45}$$

where $\begin{bmatrix} \dot{u}_k \\ \dot{v}_k \end{bmatrix}$ is the measurement that is obtained from features detection and tracking by (7.36). Then (7.40) is substituted with (7.44).

$$s_{k,t+\Delta t}\mathbf{p}_{k,t+\Delta t}^{\text{pixel}} = \mathbf{K}\left(\mathbf{R}\mathbf{p}_{k,t}^{C_t} + \mathbf{t}\right) \tag{7.46}$$

Note that the depth of the kth feature is changing with time. By using an epipolar constraint, $\mathbf{p}_{k,t}^{C_t} \cdot \left(\mathbf{t} \times \mathbf{p}_{k,t}^{C_t}\right) = 0$ which means that the inner product of the vector, $\mathbf{p}_{k,t}^{C_t}$, and the normal vector of an epipolar plane for the detected feature k, $\left(\mathbf{t} \times \mathbf{p}_{k,t}^{C_t}\right)$, are zero, the following equation can be obtained [34]. The derivation can be found in Appendix 7A.

$$\left(\mathbf{K}^{-1}\mathbf{p}_{k,t+\Delta t}^{\text{pixel}}\right)^T \mathbf{t}^{\wedge}\mathbf{R}\mathbf{K}^{-1}\mathbf{p}_{k,t}^{\text{pixel}} = 0 \tag{7.47}$$

where the operator ^ denotes the skew-symmetric matrix of a vector, as follows:

$$\mathbf{t} = \begin{bmatrix} t_x \\ t_y \\ t_z \end{bmatrix},\ \mathbf{t}^{\wedge} = \begin{bmatrix} 0 & -t_z & t_y \\ t_z & 0 & -t_x \\ -t_y & t_x & 0 \end{bmatrix} \tag{7.48}$$

Then an essential matrix $\mathbf{E}$ is defined as following:

$$\mathbf{E} = \mathbf{t}^{\wedge}\mathbf{R} = \begin{bmatrix} e_1 & e_2 & e_3 \\ e_4 & e_5 & e_6 \\ e_7 & e_8 & e_9 \end{bmatrix} \tag{7.49}$$

$$\left[\mathbf{K}^{-1}\mathbf{p}_{k,t+\Delta t}^{\text{pixel}}\right]^{\mathrm{T}}\begin{bmatrix} e_1 & e_2 & e_3 \\ e_4 & e_5 & e_6 \\ e_7 & e_8 & e_9 \end{bmatrix}\left[\mathbf{K}^{-1}\mathbf{p}_{k,t}^{\text{pixel}}\right]=0 \tag{7.50}$$

As can be seen here, by finding eight pairs of matching feature points between two frames [35], the exact solution of the fundamental matrix can be solved. Since the tracked feature might contain random noise, it is common to use more than eight pairs of features to find the elements in the fundamental matrix, meaning that an LS estimation can be used. Since the translation matrix is $\mathbf{t}^{\wedge}$ is a skew-symmetric matrix and $\mathbf{R}$ is an orthogonal matrix, the SVD of the essential matrix $\mathbf{E}$ can decouple $\mathbf{t}^{\wedge}$ and $\mathbf{R}$ as shown in the equations below [36]:

$$\mathbf{E}=\mathbf{U}\boldsymbol{\Sigma}\mathbf{V}^{T} \tag{7.51}$$

$$\mathbf{t}^{\wedge}=\left[\mathbf{t}_1^{\wedge}\ \ \mathbf{t}_2^{\wedge}\right],\ \mathbf{R}=\left[\mathbf{R}_1\ \ \mathbf{R}_2\right] \tag{7.52}$$

$$\mathbf{t}_1^{\wedge}=\mathbf{U}\mathbf{R}_z(+90°)\boldsymbol{\Sigma}\mathbf{U}^{\mathrm{T}} \tag{7.53}$$

$$\mathbf{t}_2^{\wedge}=\mathbf{U}\mathbf{R}_z(-90°)\boldsymbol{\Sigma}\mathbf{U}^{\mathrm{T}} \tag{7.54}$$

$$\mathbf{R}_1=\mathbf{U}\mathbf{R}_z^{\mathrm{T}}(+90°)\mathbf{V}^{\mathrm{T}} \tag{7.55}$$

$$\mathbf{R}_2=\mathbf{U}\mathbf{R}_z^{\mathrm{T}}(-90°)\mathbf{V}^{\mathrm{T}} \tag{7.56}$$

$$\mathbf{R}_z(\theta)=\begin{bmatrix} \cos\theta & -\sin\theta & 0 \\ \sin\theta & \cos\theta & 0 \\ 0 & 0 & 1 \end{bmatrix} \tag{7.57}$$

Finally, the transformation can be obtained, meaning that the visual odometry is achieved. After solving $\mathbf{R}_1$, $\mathbf{R}_2$, $\mathbf{t}_1^{\wedge}$, and $\mathbf{t}_2^{\wedge}$, four combinations of transformation solution $\mathbf{R}\mathbf{p}_{k,t}^{C_t}+\mathbf{t}$ can be obtained. Note that the solution is valid only when the associated depth of $\mathbf{p}^{C}$ is positive in the camera-body frame at time t and $t+\Delta t$. Thus, the unique solution of the transformation matrix can be selected from the above four solutions.

7.4.2 Monocular SLAM

In monocular SLAM, an inherent challenge called scale uncertainty is always present. This problem refers to the real size of the object in the world frame being unknown. Briefly, if we scale down or up the motion of camera as well as the scene size by a certain multiple, we will capture the same images. Thus, the trajectory estimated by the monocular SLAM will differ from the real trajectory by a scale.

According to the visual odometry, the transformation **T** between the two consecutive camera poses can be obtained by (7.51) to (7.57). Then the positions of the features in the 3-D space need to be calculated by the pose estimation results of camera. For the point with the 3-D position $\mathbf{p}_{k,t}^{C_t}$ in Figure 7.19, according to the camera projection model in (7.44), the two projected pixel positions $\mathbf{p}_{k,t}^{I}$, $\mathbf{p}_{k,t+\Delta t}^{I}$ in the two images can be obtained by feature detection, and we have

$$s_{k,t}\mathbf{p}_{k,t}^{I} = \mathbf{K}\mathbf{p}_{k,t}^{C_t} \tag{7.58}$$

$$s_{k,t+\Delta t}\mathbf{p}_{k,t+\Delta t}^{I} = \mathbf{K}\left(\mathbf{R}\mathbf{p}_{k,t}^{C_t} + \mathbf{t}\right) \tag{7.59}$$

By rearranging the equation:

$$\mathbf{K}^{-1}s_{k,t}\mathbf{p}_{k,t}^{I} = \mathbf{p}_{k,t}^{C_t} \tag{7.60}$$

$$\mathbf{K}^{-1}s_{k,t+\Delta t}\mathbf{p}_{k,t+\Delta t}^{I} = \mathbf{R}\mathbf{p}_{k,t}^{C_t} + \mathbf{t} \tag{7.61}$$

By substituting (7.60) into (7.61):

$$s_{k,t+\Delta t}\mathbf{K}^{-1}\mathbf{p}_{k,t+\Delta t}^{I} = s_{k,t}\mathbf{R}\mathbf{K}^{-1}\mathbf{p}_{k,t}^{I} + \mathbf{t} \tag{7.62}$$

$$s_{k,t+\Delta t}\mathbf{q}_{k,t+\Delta t}^{C_t} = s_{k,t}\mathbf{R}\mathbf{q}_{k,t}^{C_t} + \mathbf{t} \tag{7.63}$$

where $\mathbf{q}_{k,t+\Delta t}^{C_t} = \mathbf{K}^{-1}\mathbf{p}_{k,t+\Delta t}^{I}$ and $\mathbf{q}_{k,t}^{C_t} = \mathbf{K}^{-1}\mathbf{p}_{k,t}^{I}$ are the normalized coordinates of the 3-D point $\mathbf{p}_{k,t}^{C_t}$ with $Z = 1$ in the two camera frames.

In (7.63), there are two unknowns, $s_{t,t+\Delta t}$ and $s_{k,t}$. To solve this, we can use the cross-product with $\mathbf{q}_{k,t+\Delta t}^{C_t}$ on both sides of the equation:

$$\mathbf{q}_{k,t+\Delta t}^{C_t} \times \left(s_{k,t+\Delta t}\mathbf{q}_{k,t+\Delta t}^{C_t}\right) = \mathbf{q}_{k,t+\Delta t}^{C_t} \times \left(s_{k,t}\mathbf{R}\mathbf{q}_{k,t}^{C_t} + \mathbf{t}\right) \tag{7.64}$$

Since any vector crossed with itself equals zero, the left side becomes zero:

$$0 = \mathbf{q}_{k,t+\Delta t}^{C_t} \times \left(s_{k,t+\Delta t}\mathbf{q}_{k,t+\Delta t}^{C_t}\right) \tag{7.65}$$

$$0 = s_{k,t}\left(\mathbf{q}_{k,t+\Delta t}^{C_t} \times \mathbf{R}\mathbf{q}_{k,t}^{C_t}\right) + \mathbf{q}_{k,t+\Delta t}^{C_t} \times \mathbf{t} \tag{7.66}$$

Rearranging to isolate $s_{k,t}$:

$$s_{k,t}\left(\mathbf{q}_{k,t+\Delta t}^{C_t} \times \mathbf{R}\mathbf{q}_{k,t}^{C_t}\right) = -\left(\mathbf{q}_{k,t+\Delta t}^{C_t} \times \mathbf{t}\right) \tag{7.67}$$

In (7.67), **R**, **t** can be solved by (7.51) through (7.57). $\mathbf{q}_{k,t}^{C_t}$ and $\mathbf{q}_{k,t+\Delta t}^{C_t}$ are obtained by their definition. Thus, $s_{k,t}$ can be solved, and then substitute the solved $s_{k,t}$ into $s_{k,t}\mathbf{p}_{k,t}^{I} = \mathbf{K}\mathbf{p}_{k,t}^{C_t}$ to obtain $\mathbf{p}_{k,t}^{C_t}$, meaning the pose of the feature k in the camera-body frame at the time t. However, the solved $s_{k,t}$ is not the real depth of the point $\mathbf{p}_{k,t}^{C_t}$ in the world frame because of the scale uncertainty in the monocular vision. This is because multiplying a nonzero constant in (7.67) will not change the result of $s_{k,t}$.

A very naive SLAM can therefore be achieved using the equations below.

$$\mathbf{p}_k^L = \mathbf{T}_{C_0}^L \prod_t^0 \mathbf{T}_{C_t}^{C_{t-1}} \mathbf{p}_{k,t}^{C_t} \tag{7.68}$$

$$\mathbf{M}^L = \left\{\mathbf{p}_1^L \cdots \mathbf{p}_{N_{\text{feature}}}^L\right\} \tag{7.69}$$

However, its performance is not reliable and the results are prone to drift quickly. The fact that a feature can be detected in multiple epochs is not considered in this naive SLAM. Bundle adjustment is usually applied to tackle this quickly drifting problem.

7.4.2.1 Bundle Adjustment for Multi-Epoch SLAM

Finally, a batch optimization is applied to achieve a bundle adjustment of the map. Here, instead of using the time epoch to index the image, an image index i is used.

$$\left\{\mathbf{R}_i, \mathbf{t}_i, \mathbf{p}_k^{C_i}\right\} = \underset{\mathbf{R}_i, \mathbf{t}_i, \mathbf{p}_k^{C_i}}{\arg\min} \sum_{i \in \text{Img}_l} \sum_{k \in \text{Feature}_m} \left\| \mathbf{p}_{i,k}^{\text{pixel}_i} - \pi\left(\mathbf{R}_i \mathbf{p}_k^{C_i} + \mathbf{t}_i\right) \right\|^2 \tag{7.70}$$

where Img_l are the set of covisible images and Feature_m are all the feature points can be seen in these images I_{lmg}. $\pi(\ )$ is the projection function from the camera-body frame to the pixel frame, meaning (7.40), $\mathbf{p}^{C_i}$ is the 3-D point of the kth feature in the camera-body frame when the ith image is taken, $\mathbf{p}_k^{C_i}$ is the 3-D point of the kth feature in the camera-body frame when the ith image is taken, and $\mathbf{p}_{i,k}^{\text{pixel}_i}$ is the detected feature point in image I_i corresponding with the point $\mathbf{p}_k^{C_i}$. By reprojecting feature k to image i, considering the rotation matrix $\mathbf{R}_i$ and the translation vector $\mathbf{t}_i$, the predicted pixel position can be found. Then the transformation matrix that can minimize the Euclidean distance between the predicted and measured pixel position of the feature is the optimized relative positioning. The map with the consideration of bundle adjustment can be finally obtained using (7.68). Note that the $\mathbf{R}_i$, $\mathbf{t}_i$ estimated using the bundle adjustment are much more reliable than that calculated using (7.51) to (7.57). The rough estimation by (7.51) to (7.57) is usually only used as an initial guess for (7.70).

7.4.3 RGB-D Camera

A popular RGB-D camera (such as Microsoft Kinect) is to integrate an RGB camera with an infrared (IR) sensor. The IR sensor is a radiation-sensitive optoelectronic component, which can measure the distance of an object. It uses an IR LED to send an IR light and a photodiode to sense the reflected light. Similar to a lidar, the two-way TOF can therefore be used to measure the distance (or usually called "depth" in the image processing field) between the camera and an object. Due to the integration of an RGB camera and an IR sensor, each pixel also includes its depth information. The principle of integration is similar to Section 7.2.4. Since the depths are provided by an RGB-"D" camera, its visual odometry and SLAM are very similar to that of the lidar ones mentioned in the previous section. The RGB part is similar to that of monocular camera. The idea of ICP, NDT, and LOAM can be directly applied. However, it is usually more challenging compared to that using lidar. It is because:

1. Higher noise in the depth measurement is generated by the IR sensor compared with that of a lidar. One of the popular solutions to improve the performance of the SLAM using the RGB-D camera is to combine the pixel feature with the depth information in the SLAM. It is a PCD SLAM that integrates with a monocular camera-based SLAM. Effective integration depends on accurate knowledge of the information matrices of both systems. An adaptive approach is preferred, but it is also one of the challenges for this integration.
2. There is a narrower field of view in a camera compared to a mechanical lidar, similar to a solid-state lidar. The narrower field of view usually causes a decrease of the observability of the model, meaning a larger positioning error or a higher chance of obtaining a local minimum solution. A common approach of dealing with the lower observability

problem is to provide a better initial guess to the solution for the nonlinear optimization problem. However, this is a chicken-and-egg problem. For SLAM, an integration with IMU is an efficient approach to serve this purpose because IMU can measure the propagation of the agent's pose.

7.4.4 Challenges of a Visual SLAM

For monocular SLAM, the biggest bottleneck is the inability to determine the true scale of the scene, which makes the SLAM system relying only on a single camera severely limited in practical applications. Stereo SLAM can compute the depth information, but its accuracy is greatly dependent on the resolution of cameras and the baseline length between cameras, and the calculation of depth of each pixel is computationally time-consuming and unreliable. The integration of camera and IMU can effectively solve the problem of scale [10]. The RGB-D camera can avoid the large computation required for depth acquisition, but the range is usually small, and they are noisy and weak against interference. Therefore, it is commonly used indoors rather than in outdoor environments.

For the monocular camera-based SLAM, feature tracking plays an important role in determining the performance of data association for the motion estimation. The objective of feature tracking is to find the correct feature matching between consecutive series of images. The following are the challenges for rich and stable feature detection and checking:

- *Illumination change:* Both the optical flow-based feature-tracking method (Lucas-Kanade method) and the descriptor-based ORB descriptor rely on the image brightness constancy. Specifically, the optical flow-based feature-tracking method relies heavily on the assumption of brightness constancy to find that the feature corresponds between two consecutive images. However, the assumption of the image brightness constancy in practical scene may not always be satisfied due to lighting conditions and environment structures.
- *Imaging blur:* The image blur caused by a large motion or a sharp turn (large accelerations) will degrade the quality of features. In other words, the image blur leads to incorrect feature detection and tracking.
- *Dynamic objects:* The dynamic objects violate the assumption that the surrounding features are static for the SLAM. Specifically, dynamic objects violate the assumption of the optical-flow tracking algorithm in that the pixels around the key points have same motion.

- *Low-textured environment:* If the environment captured is a white wall without any corner points to be detected, then no features can be detected and tracked.
- *Environment with repetitive features:* If the multiple images taken at different locations are similar, then the optic flow will mistakenly give a near-zero velocity. For example, the corridor of hotel rooms is a typical scenario that visual SLAM will drift by very quickly.

7.4.5 Integration of a Camera with Lidar SLAM

If the expected lidar SLAM accuracy is higher than that of a visual SLAM, it is popular to map the pixels of an image into the accumulated lidar PCD map. This requires very precise calibrations on both the intrinsic parameters of the camera and the extrinsic parameters between the lidar and the camera body coordinate systems. The time synchronization between the camera and lidar is also required to be carefully handled. Assuming that the calibrations are done precisely, the pixel (which the RGB values) of an image can be transformed into a global coordinate system using the lidar SLAM. The idea is to give an RGB color to the PCD map generated by lidar SLAM as described in Figure 7.20.

Once the pixels are transformed into the lidar frame, the RGB value of the pixel closest to a PCD point is used to color that PCD point. Note that the field of view of the camera and lidar may differ. To ensure accurate integration, only PCD points and pixels within the overlapping field of view are used for data fusion.

7.5 Roles of IMU in Lidar SLAMs

Most of the SLAM methods are challenged by a practical aspect, the computational loads. The size of the model means the size of a matrix used in the linear algebra calculation. As the size of the matrix increases, the computational loads of calculating the inverse, the Jacobian matrix, will increase drastically

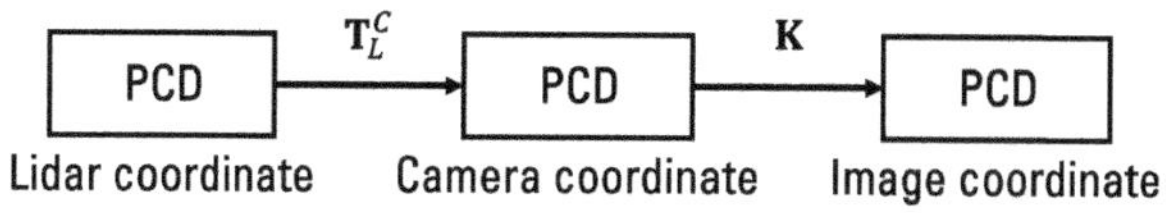

Figure 7.20 The flowchart of transforming pixels to a lidar frame.

[37]. Thus, a small-sized model that can maintain its observability is preferred. In a real system, a threshold on the numbers of the features used is usually predetermined to avoid the excessive computational load. The selection of the features used can be done using random sample consensus (RANSAC) [38]. However, there are cases in which the observability is compromised after the selection of the features. As mentioned in Section 7.2.2, IMU is a good candidate to provide a good initial guess to mitigate the challenge of the small model observability. IMUs measure the 3-D accelerations and the 3-D rotational rates, which can describe the propagation of the agent's pose in a short time. Here, how short it depends on the grade of the IMU used. For example, a MEMS-level IMU used in a smartphone can survive for a few seconds. In addition, if the SLAMs are calculated whenever the lidar PCD and images are received, the computational load could be too large for some IPIN applications. The rough output frequencies of lidar and the image could be 10 Hz and 15 Hz, respectively, meaning that SLAM calculation needs 10 or more times in a second, which may not be very meaningful if the movement of the agent is not fast. At this time, IMU can also be used to predict the motion of the agent, and the prediction can be used as an initial guess before the calculation of the odometry and SLAM. This can effectively reduce the output rate of the SLAM while maintaining accuracy. It is important to note that the update rate of the IMU is very high, which can reach 400 Hz. If the INS mechanization is used 400 Hz of the update rate, the computational load is high as well. Thus, a pre-integration method is popularly used to solve this problem.

For the lidar and INS integration, we recommend LIO-SAM [8], and its open-source code is available at https://github.com/TixiaoShan/LIO-SAM. For the visual odometry and INS integration, we recommend VINS [10], and its open-source code is available at https://github.com/HKUST- Aerial-Robotics/VINS-Mono.

7.6 Conclusions

This chapter has provided a comprehensive and systematic overview of indoor SLAM technologies, emphasizing their mathematical foundations, algorithmic advancements, and practical implementations. The chapter traces the historical development of SLAM, highlighting its evolution from classical filter-based approaches to modern optimization-based methods. Key milestones such as EKF-SLAM, FastSLAM, GraphSLAM, and recent robust perception frameworks have been discussed, illustrating how each era built upon the strengths of its predecessors while addressing their limitations.

The general mathematical model of SLAM has been detailed, outlining the core principles of feature extraction, transformation estimation, and batch optimization. This foundational framework has supported in-depth discussions on lidar SLAM, visual SLAM, and the integration of IMUs. Lidar SLAM has been explored through methods such as ICP, NDT, and LOAM, each offering distinct advantages in terms of accuracy, computational efficiency, and robustness. Visual SLAM has been examined in the context of monocular and RGB-D cameras, with a focus on feature detection, tracking, and the challenges posed by scale uncertainty and dynamic environments. The critical role of IMUs in enhancing SLAM performance has been emphasized, particularly in providing initial guesses for pose estimation and mitigating computational burdens.

The integration of lidar, cameras, and IMUs has been demonstrated as a powerful approach for achieving robust and accurate indoor mapping and localization. By leveraging the complementary strengths of these sensors, lidar's precise depth measurements, cameras' rich visual information, and IMUs' high-frequency motion estimates, modern SLAM systems can operate effectively in complex, dynamic environments. Comparative examples of lidar SLAM algorithms have further illustrated the practical implications of these technologies, demonstrating their potential for real-world applications.

Looking ahead, the next chapter will build upon the concepts introduced in this chapter by exploring advanced sensor fusion techniques. The integration of additional sensors and data sources to further enhance the reliability, accuracy, and adaptability of SLAM systems will be investigated. This will include discussions on adaptive fusion strategies and the handling of challenging scenarios such as dynamic objects and resource-constrained environments. The insights gained from this chapter help readers to grasp the fundamental concepts and methodologies underlying indoor SLAM technologies. These insights lay the groundwork for understanding the sophisticated sensor fusion frameworks that are currently being explored in the advancement of SLAM systems. By building on this foundation, readers will be better equipped to comprehend the complex sensor fusion techniques discussed in the next chapter, which aim to enhance the reliability, accuracy, and adaptability of SLAM systems for applications in indoor navigation and other relevant fields.

References

[1] Smith, R. C., and P. Cheeseman, "Estimating Uncertain Spatial Relationships In Robotics," *Autonomous Robot Vehicles*, 1990, pp. 167–193.

[2] Montemerlo, M., et al., "FastSLAM: A Factored Solution to the Simultaneous Localization and Mapping Problem," *Proceedings of the AAAI National Conference on Artificial Intelligence*, 2002, pp. 593–598.

[3] Dellaert, F., "Square Root SAM: Simultaneous Localization and Mapping Via Square Root Information Smoothing," *The International Journal of Robotics Research*, Vol. 25, 2006, pp. 1181–1203.

[4] Thrun, S., and M. Montemerlo, "The GraphSLAM Algorithm with Applications to Large-Scale Mapping of Urban Structures," *The International Journal of Robotics Research*, Vol. 25, 2006, pp. 403–429.

[5] Mur-Artal, R., J. M. M. Montiel, and J. D. Tardos, "ORB-SLAM: A Versatile and Accurate Monocular Slam System," *IEEE Transactions on Robotics*, Vol. 31, No. 5, 2015, pp. 1147–1163.

[6] Engel, J., T. Schöps, and D. Cremers, "LSD-SLAM: Large-Scale Direct Monocular SLAM," *European Conference on Computer Vision*, 2014, pp. 834–849.

[7] Zhang, J., and S. Singh, "LOAM: Lidar Odometry and Mapping in Real-Time," *Robotics: Science and Systems*, Vol. 2, 2014.

[8] Shan, T., et al., "LIO-SAM: Tightly-Coupled Lidar Inertial Odometry Via Smoothing and Mapping," *2020 IEEE/RSJ International Conference on Intelligent Robots and Systems (IROS)*, 2020, pp. 5135–5142.

[9] Klein, G., and D. Murray, "Parallel Tracking and Mapping for Small AR Workspaces," *2007 6th IEEE and ACM International Symposium on Mixed and Augmented Reality*, 2007, pp. 225–234.

[10] Qin, T., P. Li, and S. Shen, "VINS-Mono: A Robust and Versatile Monocular Visual-Inertial State Estimator," *IEEE Transactions on Robotics*, Vol. 34, No. 4, 2018, pp. 1004–1020.

[11] Salas-Moreno, R. F., et al., "SLAM++: Simultaneous Localisation and Mapping at the Level of Objects," *Proceedings of the IEEE Conference on Computer Vision and Pattern Recognition*, 2013, pp. 1352–1359.

[12] Tateno, K., et al., "CNN-SLAM: Real-Time Dense Monocular Slam with Learned Depth Prediction," *Proceedings of the IEEE Conference on Computer Vision and Pattern Recognition*, 2017, pp. 6243–6252.

[13] Bescos, B., et al., "DynaSLAM: Tracking, Mapping, and Inpainting in Dynamic Scenes," *IEEE Robotics and Automation Letters*, Vol. 3, 2018, pp. 4076–4083.

[14] Mur-Artal, R., and J. D. Tardós, "Orb-SLAM2: An Open-Source SLAM System for Monocular, Stereo, and RGB-D Cameras," *IEEE Transactions on Robotics*, Vol. 33, No. 5, 2017, pp. 1255–1262.

[15] Rebecq, H., T. Horstschaefer, and D. Scaramuzza, "Real-Time Visual-Inertial Odometry for Event Cameras Using Keyframe-Based Nonlinear Optimization," *British Machine Vision Conference (BMVC)*, 2017.

[16] Huang, Z., J. Rajasegaran, and A. Kanazawa, "Point-Slam: Dense Neural Point Cloud-Based Slam," *Proceedings Computer Vision–ECCV 2022: 17th European Conference*, Tel Aviv, Israel, October 23–27, 2022, pp. 597–614.

[17] Besl, P. J., and N. D. McKay, "Method for Registration of 3-D Shapes," *IEEE Transactions on Pattern Analysis & Machine Intelligence*, No. 2, 1992, pp. 239–256.

[18] Rusu, R. B., and S. Cousins, "3D Is Here: Point Cloud Library (PCL)," *2011 IEEE International Conference on Robotics and Automation*, 2011, pp. 1–4.

[19] Biber, P., and W. Straßer, "The Normal Distributions Transform: A New Approach to Laser Scan Matching," *Proceedings 2003 IEEE/RSJ International Conference on Intelligent Robots and Systems (IROS 2003)*, Vol. 3, 2003, pp. 2743–2748.

[20] Magnusson, M., A. Lilienthal, and T. Duckett, "Evaluation of 3D Registration Reliability and Speed—A Comparison of ICP and NDT," *2007 IEEE International Conference on Robotics and Automation*, 2009, pp. 3907–3912.

[21] Pang, J., et al., "3D NDT Transformation Estimation with Ground Segmentation Data," *IEEE Access*, Vol. 6, 2018, pp. 26838–26849.

[22] Segal, A., D. Haehnel, and S. Thrun, "Generalized-ICP," *Robotics: Science and Systems*, Vol. 2, 2009, p. 435.

[23] Ye, H., Y. Chen, and M. Liu, "Tightly Coupled 3D Lidar Inertial Odometry and Mapping," *2019 International Conference on Robotics and Automation (ICRA)*, 2019, pp. 3144–3150.

[24] Wen, P., et al., "Performance Evaluation of a Lidar Inertial Odometry and Mapping (LIOM) Method," *2018 IEEE International Geoscience and Remote Sensing Symposium (IGARSS 2018)*, 2018, pp. 7598–7601.

[25] Huang, P., et al., "Coarse-to-Fine Semantic Localization with HD Map for Autonomous Driving," *2021 IEEE International Conference on Robotics and Automation (ICRA)*, 2021, pp. 9664–9670.

[26] Wang, H., C. Wang, and L. Xie, "F-LOAM: Fast Lidar Odometry and Mapping," *2021 IEEE/RSJ International Conference on Intelligent Robots and Systems (IROS)*, 2021, pp. 4390–4396.

[27] Shi, J., and C. Tomasi, "Good Features to Track," *1994 Proceedings of IEEE Conference on Computer Vision and Pattern Recognition*, 1994, pp. 593–600.

[28] Lucas, B. D., et al., "An Iterative Image Registration Technique with an Application to Stereo Vision," *IJCAI*, Vol. 81, 1981, pp. 674–679.

[29] Rublee, E., et al., "Orb: An Efficient Alternative to Sift or Surf," *2011 International IEEE Conference on Computer Vision*, 2011, pp. 2564–2571.

[30] Bradski, G., and A. Kaehler, *Learning OpenCV: Computer Vision with the OpenCV Library*, Sebastapol, CA: O'Reilly Media, 2008.

[31] He, K., et al., "Deep Residual Learning for Image Recognition," *Proceedings of the IEEE Conference on Computer Vision and Pattern Recognition*, 2016, pp. 770–778.

[32] Zhang, Z., "A Flexible New Technique for Camera Calibration," *IEEE Transactions on Pattern Analysis and Machine Intelligence*, Vol. 22, 2000, pp. 1330–1334.

[33] Zhang, Z., "Determining Motion from 3D Line Segment Matches: A Comparative Study," *Image and Vision Computing*, Vol. 16, 1998, pp. 549–565.

[34] Gao, W., et al., *Lectures on Visual Perception*, New York: Springer, 2017.

[35] Hartley, R. I., "In Defense of the Eight-Point Algorithm," *IEEE Transactions on Pattern Analysis and Machine Intelligence*, Vol. 19, No. 6, 1997, pp. 580–593.

[36] Hartley, R., and A. Zisserman, *Multiple View Geometry in Computer Vision*, Cambridge, UK: Cambridge University Press, 2004.

Appendix 7A: Derivation of Epipolar Constraint

From (7.46), the transformation can be represented by

$$s_{k,t+\Delta t}\mathbf{p}_{k,t+\Delta t}^{\text{pixel}} = \mathbf{K}\left(\mathbf{R}\mathbf{p}_{k,t}^{\mathrm{C}_t} + \mathbf{t}\right) \tag{7A.1}$$

$$s_{k,t+\Delta t}\mathbf{p}_{k,t+\Delta t}^{\text{pixel}} = \mathbf{K}\mathbf{R}\mathbf{p}_{k,t}^{\mathrm{C}_t} + \mathbf{K}\mathbf{t} \tag{7A.2}$$

Then the inverse transformation is applied from the camera-body coordinate to the pixel plane:

$$s_{k,t+\Delta t}\mathbf{p}_{k,t+\Delta t}^{\text{pixel}} = \mathbf{K}\mathbf{R}\mathbf{K}^{-1}s_{k,t}\mathbf{p}_{k,t}^{\text{pixel}} + \mathbf{K}\mathbf{t} \tag{7A.3}$$

$$\mathbf{K}^{-1}s_{k,t+\Delta t}\mathbf{p}_{k,t+\Delta t}^{\text{pixel}} = \mathbf{R}\mathbf{K}^{-1}s_{k,t}\mathbf{p}_{k,t}^{\text{pixel}} + \mathbf{t} \tag{7A.4}$$

After that, we can use the property that for any vectors $\boldsymbol{a}$ and $\boldsymbol{b}$, $\boldsymbol{a} \times \boldsymbol{b} = \boldsymbol{a}^\wedge \boldsymbol{b}$, where $\boldsymbol{a}^\wedge$ is the skew-symmetric matrix of $\boldsymbol{a}$.

Multiplying both sides by $\boldsymbol{t}^\wedge$:

$$\mathbf{t}^\wedge\mathbf{K}^{-1}s_{k,t+\Delta t}\mathbf{p}_{k,t+\Delta t}^{\text{pixel}} = \mathbf{t}^\wedge\mathbf{R}\mathbf{K}^{-1}s_{k,t}\mathbf{p}_{k,t}^{\text{pixel}} + \mathbf{t}^\wedge\mathbf{t} \tag{7A.5}$$

Since $\boldsymbol{t}^\wedge\boldsymbol{t} = 0$ (the cross-product of a vector with itself is zero), we have:

$$\mathbf{t}^\wedge\mathbf{K}^{-1}s_{k,t+\Delta t}\mathbf{p}_{k,t+\Delta t}^{\text{pixel}} = \mathbf{t}^\wedge\mathbf{R}\mathbf{K}^{-1}s_{k,t}\mathbf{p}_{k,t}^{\text{pixel}} \tag{7A.6}$$

Taking the dot product with $\mathbf{K}^{-1}\mathbf{p}_{k,t}^{\text{pixel}}$ on both sides:

$$\left(\mathbf{K}^{-1}\mathbf{p}_{k,t}^{\text{pixel}}\right)^T \mathbf{t}^\wedge\mathbf{K}^{-1}s_{k,t+\Delta t}\mathbf{p}_{k,t+\Delta t}^{\text{pixel}} = \left(\mathbf{K}^{-1}\mathbf{p}_{k,t}^{\text{pixel}}\right)^T \mathbf{t}^\wedge\mathbf{R}\mathbf{K}^{-1}s_{k,t}\mathbf{p}_{k,t}^{\text{pixel}} \tag{7A.7}$$

The left side is zero because $\left(\mathbf{K}^{-1}\mathbf{p}_{k,t}^{\text{pixel}}\right)^T \mathbf{t}^\wedge \mathbf{K}^{-1}\mathbf{p}_{k,t+\Delta t}^{\text{pixel}}$ is the dot product of a vector with its own cross-product, which is always zero.

Therefore:

$$0 = \left(\mathbf{K}^{-1}\mathbf{p}_{k,t}^{\text{pixel}}\right)^T \mathbf{t}^\wedge \mathbf{R}\mathbf{K}^{-1} s_{k,t}\mathbf{p}_{k,t}^{\text{pixel}} \tag{7A.8}$$

Since $s_{k,t}$ is a scalar, we can rewrite:

$$0 = s_{k,t}\left(\mathbf{K}^{-1}\mathbf{p}_{k,t+\Delta t}^{\text{pixel}}\right)^T \mathbf{t}^\wedge \mathbf{R}\mathbf{K}^{-1}\mathbf{p}_{k,t}^{\text{pixel}} \tag{7A.9}$$

For this equation to be true for any nonzero $s_{k,t}$, we must have:

$$\left(\mathbf{K}^{-1}\mathbf{p}_{k,t+\Delta t}^{\text{pixel}}\right)^T \mathbf{t}^\wedge \mathbf{R}\mathbf{K}^{-1}\mathbf{p}_{k,t}^{\text{pixel}} = 0 \tag{7A.10}$$

This is the epipolar constraint equation.

8

Practical Aspects of Sensor Fusion

Sensor fusion is the process of combining data from multiple sensors to produce an estimate of the state of a system. By integrating diverse sensor measurements, a sensor fusion system [1] can improve positioning accuracy, increase robustness, and increase resilience against the effects of noise or uncertainty present in individual sensors. Indoor navigation systems heavily rely on sensor fusion. A pedestrian's smartphone can integrate inertial dead reckoning with signals from Wi-Fi access points or BLE beacons to overcome the lack of GPS indoors. The primary objective is to exploit complementary information. For example, one sensor may provide precise short-term motion cues while another gives absolute positioning, and their combination yields a more reliable and stable solution.

This chapter is organized as follows. Section 8.1 introduces the concepts of loosely coupled and tightly coupled sensor fusion frameworks, including their respective advantages and trade-offs, followed by a comparative summary. Section 8.2 examines the observability of sensor fusion systems, highlighting which states can and cannot be estimated reliably, along with techniques to address drift from unobservable states. Section 8.3 focuses on the tuning of sensor fusion systems, covering covariance estimation and the setting of thresholds. Section 8.4 examines calibration techniques, including intrinsic calibration for individual sensors such as cameras, lidars, and IMUs, as well as extrinsic calibration methods based on targets or motion. Section 8.5

addresses temporal calibration, discussing time synchronization, measurement timing, interpolation, and continuous-time methods. Section 8.6 discusses system efficiency, comparing real-time and offline processing strategies and considering memory usage. Finally, Section 8.7 summarizes key insights and practical takeaways.

8.1 Loosely Coupled and Tightly Coupled

Sensor fusion architectures are often categorized as loosely coupled (LC) or tightly coupled (TC) integration. These terms describe how deeply the sensor data are integrated with each other. In an LC approach, each sensor (or each subset of sensors) is first processed independently to produce its own state estimate, and then these estimates are combined at a higher level. In a TC approach, the raw measurements from all sensors feed into a single estimator that computes the fused state directly. The choice between LC or TC designs affects the system's accuracy, complexity, and robustness in different scenarios.

8.1.1 Loosely Coupled

LC fusion is characterized by a two-stage process. First, each sensor is solved on its own for the quantity of interest; then those results are fused. Each sensor can function (to some degree) on its own, and their outputs are combined at the navigation solution level. This makes design and implementation straightforward. One can swap out one sensor for a better one as long as it provides the same type (meaning the same format and unit) of output, and the fusion logic remains mostly unchanged. LC fusion is also computationally lighter in many cases because each sub-estimator deals with a smaller problem.

Figure 8.1 represents a typical IPIN example of LC fusion using Kalman filters based on IMU with auxiliary sensor updates. Imagine a smartphone tracking a user's steps (via IMU) to estimate relative movement, while a secondary sensor, such as Wi-Fi, occasionally provides absolute position fixes, for example, through fingerprinting. These intermittent updates are treated as observations in a high-level Kalman filter that corrects accumulated PDR drift. A practical example is a smartphone tracking a user's steps with its IMU while periodically receiving Wi-Fi-based location estimates. The PDR runs continuously, and each Wi-Fi fix anchors the trajectory. LC designs fuse only the processed outputs of each sensor, not raw data, enabling modularity and robustness. This architecture is widely adopted in real-world indoor navigation systems.

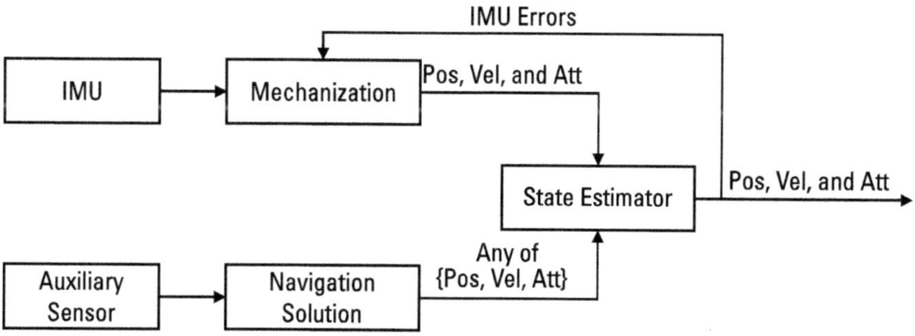

Figure 8.1 An example of a typical IMU/auxiliary sensor LC system.

However, LC systems have limitations. Because they rely on each sensor's independent solution, they may not fully utilize the information. There is information loss. For example, in an LC camera-IMU system, one might run visual SLAM to get a pose and integrate IMU for the pose, and then fuse poses. If the visual SLAM loses tracking (e.g., too few features), it provides no pose, and the fusion might fail to correct IMU drift. A tightly coupled (TC) visual-inertial odometry (VIO) would incorporate even a few tracked features with IMU data to still estimate pose. This underscores an LC drawback: it discards raw info that the individual subsystem could not solve alone. That said, LC systems often perform well in practice and are easier to validate. They remain very popular when the sensors individually can provide good solutions most of the time.

8.1.2 Tightly Coupled

TC fusion integrates sensor data at the measurement level, in a unified estimation process. Instead of first solving each sensor's outputs separately, a TC system feeds all raw measurements (or minimally processed measurements) into a single state estimator (such as a single EKF or a bundle adjustment based on FGO). This means that the estimator simultaneously considers, for example, IMU increments, visual features, and barometer readings all at once to estimate the state. TC fusion is generally more accurate and robust, especially in difficult conditions. Because it does not require each sensor to individually produce a full solution, it squeezes the maximum information out of every observation. For example, Figure 8.2 represents a TC visual-inertial (VI) system that will treat tracked feature pixel coordinates from the camera as measurements and the IMU's readings as helping to predict motion between image frames. The estimator (often a bundle adjustment based on FGO) will solve for the camera pose and scene feature positions along with IMU biases altogether. If at some frame only a few features are visible (insufficient for a purely visual pose estimate), the IMU's prediction plus those few features' measurements can still constrain the pose. In contrast, an LC system might have failed to produce a visual pose and thus lost track until more features appear. This is why TC VIO has become increasingly preferred in modern systems. It is considerably more robust to motion blur, low-texture scenes, or rapid maneuvers where the camera alone struggles, because the IMU is fully integrated into the estimation of camera pose [2]. The improved observability [3] (which will be discussed next) of TC systems is the main reason why the TC outperforms the LC. By merging data at the lowest level, states that would be unobservable in one sensor can become observable.

The trade-off for TC is increased complexity. The state space is larger (you might be estimating sensor biases and feature coordinates together) and the mathematical model is more complex (directly relating raw measurements to state), so the cross-correlation between measurements needs to be considered. The computational cost can be significantly higher because the estimator (especially if batch optimization like an FGO) might need to consider many raw measurements and large state vectors simultaneously. For example, a TC SLAM system with a camera and IMU might end up optimizing thousands of feature points and poses in a window. This is something an LC system might avoid by only passing along a summarized pose from the vision module. Therefore, TC systems often require more careful implementation to achieve real-time performance (e.g., efficient solvers, sparsity exploitation, and parallel processing). The Kalman filter formulation of TC integration will have a larger Kalman gain matrix to compute (dimension equals state × measurements). If one sensor provides m measurements and another provides n, the

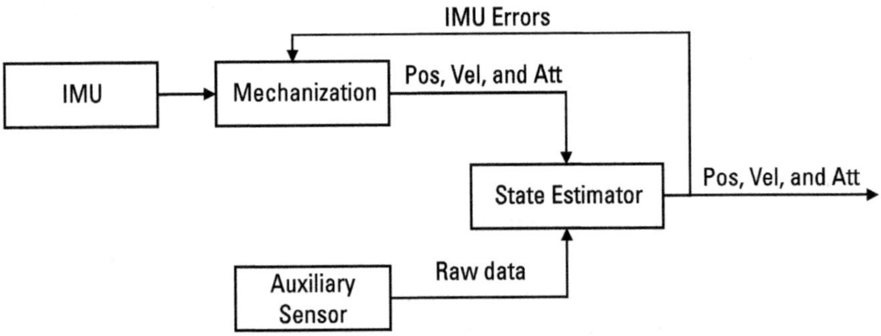

Figure 8.2 An example of a typical IMU/auxiliary sensor TC system.

combined system might have $m + n$ measurements in one update. This can be computationally heavy if m and n are large.

Despite the complexity, advancements in computing and algorithms have made TC fusion feasible for many applications. The benefit in accuracy and robustness is often well worth the extra effort. Finally, this section is concluded using the following example. Consider an autonomous drone in a tunnel: GPS signals are blocked, UWB ranging is sparse (and affected by multipath and nonline-of-sight (NLOS)), and lidar features may be repetitive (walls looking similar). A TC system could fuse IMU, barometer, lidar scan matches, and maybe vision (if active lighting is available), all in one graph optimization. Each sensor on its own might be unreliable (lidar scan matching could drift among similar walls, or visual features could be repetitive due to the wall markings), but, together in one estimation framework, they constrain each other. The IMU ensures continuity and short-term precision, the UWB (even 1 to 2 beacons giving partial information) provides a local reference, the barometer stabilizes altitude, and the lidar anchor local structure detects loop closures. Such a TC approach is resilient: losing one sensor temporarily does not break the entire solution; it just reduces the dimension of the measurements until that sensor returns. In LC, losing one sensor would remove an entire module's output from the fusion, potentially causing large errors if that module were providing vital corrections.

8.1.3 Comparative Summary

LC and TC sensor fusion methods each have distinct advantages suited for different indoor positioning and navigation scenarios. Table 8.1 summarizes their advantages and disadvantages. LC, characterized by simplicity, is advantageous when computational resources are limited, rapid prototyping is necessary, or off-the-shelf sensor solutions suffice. It performs effectively in ideal conditions but significantly degrades when sensors partially fail or data is sparse. In contrast, TC fusion integrates raw sensor data directly, achieving consistently higher accuracy, especially in challenging indoor environments such as visually sparse areas, by effectively leveraging partial measurements. However, this comes with increased computational complexity and a greater need for meticulous calibration and fault detection. LC methods excel in scalability and flexibility, enabling the easy integration of new sensors or rapid prototyping. TC systems, although more resilient in difficult conditions, demand robust fault management and careful tuning to handle sensor interdependencies. Thus, for high-performance applications

Table 8.1
Comparison Between LC and TC Sensor Fusions

Aspects	**LC Sensor Fusion**	**TC Sensor Fusion**
Data utilization	LC fuses data at the solution level.	TC fuses data at the measurement level, utilizing raw sensor outputs. This maximizes information usage.
Accuracy (performance in nominal versus degraded conditions)	LC performs optimally under nominal conditions when all sensors provide full data. In degraded environments, LC tends to lose accuracy or even fail when a sensor's data becomes unreliable or unavailable.	TC performs optimally under nominal conditions when all sensors provide full data. In degraded environments, TC achieves higher accuracy and reliability by fusing raw measurements compared with the LC integration. It can maintain performance with partial data.
Complexity (design, implementation, computational)	LC has a simpler design and implementation. Uses multiple smaller filters or modules in cascade/parallel, each handling a sensor independently. The modular approach has lower computational requirements. The development is more straightforward because sensors communicate through standard interfaces, making LC easier to implement on limited resources.	TC has a higher complexity with a centralized fusion filter. All sensor data is processed in one unified estimator, requiring a comprehensive system model. The design must handle a larger state vector (including sensor error states) and manage all sensor inputs simultaneously, which increases computational load. Its software is more intricate, needing careful calibration and synchronization of sensors. TC demands more processing power and expertise to develop.
Robustness (sensor failures and environmental challenges)	LC offers independent resilience of subsystems. Sensors operate semi-independently; each can fall back on their own if the other fails. This means that LC is robust to a complete failure of one sensor, but not to partial observation degradations.	TC is resilient with partial data, but sensitive to bad data. TC can continue operating with limited or degraded sensor inputs. Robustness in TC relies on quality sensor models and fault detection to handle sensor glitches (because all data feeds into a single filter, an outlier or fault in one sensor can contaminate the entire solution if not detected).

Table 8.1
(Cont.)

Aspects	LC Sensor Fusion	TC Sensor Fusion
Failure behavior	LC has a graceful degradation in the face of sensor failures. If one sensor degrades or fails entirely, LC experiences a gradual performance drop rather than an abrupt collapse. LC will rely on the remaining sensor(s) to continue operating.	TC has a critical dependency on all inputs. A failing sensor's data, if not handled, directly deteriorates the overall solution. In TC, sensor failures must be mitigated by the filter logic (e.g., disabling or down-weighting a faulty sensor's measurements) to avoid feeding wrong data into state estimation. This is why TC is a popular solution when companies wish to have technical barriers to their solution.
Scalability (ability to extend/ integrate additional sensors)	LC scales easily to additional sensors. New sensors can be integrated by adding modules or separate filters without redesigning the entire system architecture. The modular nature means the system can grow in a plug-and-play fashion (with an assumption in nominal condition).	TC is challenging to scale due to intertwined design. Adding or changing sensors in a tightly coupled system often requires significant rework, meaning the state vector and measurement models must be expanded and rederived to include the new sensor data. The high interdependence means increasing the sensors can introduce complexity and unexpected interactions that need careful handling.
Real-time performance	LC is well-suited for real-time operation, even on limited hardware. Each sensor's data is processed at its native rate and fused periodically, making it efficient. This approach has historically been favored when computational resources are limited.	TC can be achieved in real time with modern computing, but the computational load is higher. TC filters process more data in one cycle, which can introduce latency (depending on the problem) if the hardware is slow.

requiring precision, reliability, and robustness indoors (e.g., autonomous robots, augmented reality, or precise indoor navigation), TC sensor fusion is advantageous. Conversely, LC approaches remain practical in simpler indoor navigation systems where ease of implementation, computational efficiency, and flexibility are prioritized.

8.2 Observability in Sensor Fusion

Observability is a concept in sensor fusion that used to describe which states of the system can be determined (or observed) from the available sensor data. A system is observable if, given the sensor measurements over time, the internal state can be uniquely inferred. Conversely, an unobservable state is one that cannot be determined based on the sensors' outputs, leading to ambiguity or drift in that dimension. Understanding observability is essential for designing fusion algorithms.

8.2.1 Observability

In multisensor fusion, certain combinations of sensors and motions can result in degenerate scenarios where some aspect of the state is not observable. Each sensor has its own limitations, and, when fusing them, one must ensure they complement each other sufficiently to cover those limitations. Challenges often arise when the sensors provide redundant information in some dimensions but none in others.

Consider lidar-inertial odometry (LIO), a 3-D lidar provides rich spatial information, and an IMU provides motion dynamics. In theory, a well-calibrated lidar+IMU can estimate a platform's full 6-DoF pose. However, there are tricky cases. If the environment has low geometric variation (e.g., a tunnel or corridor with uniform walls), as shown in Figure 8.3(a), lidar scan matching might struggle with degeneracy, meaning that moving forward in a tunnel produces little unique 3-D feature change, so the lidar essentially cannot tell if it slid forward or backward. At the same time, the IMU can sense motion, but its accelerometer bias could cause drift. Similarly, pure rotation of the sensor (standing in place and turning) gives the lidar no translation data; translation in X or Y is unobservable from the scans because no parallax occurs, as shown in Figure 8.3(b). The IMU in that scenario senses rotation but might be very difficult to estimate position. As a result, certain LIO implementations notice increased uncertainty or drift in unobservable directions (like along a corridor's direction or around the vertical axis in place rotation). This is fundamentally an observability issue. The sensor data (scans + IMU readings) in that scenario do not inform all degrees of freedom.

VIO has a well-known observability issue. A monocular camera on its own cannot observe the absolute scale of the environment. When an IMU is added, the combination becomes observable in scale under general motion, because the accelerometer measures actual accelerations in m/s^2, which, when compared against the camera's relative pose change (in unknown units), yields the scale. However, if the motion is insufficient (e.g., camera only rotates but

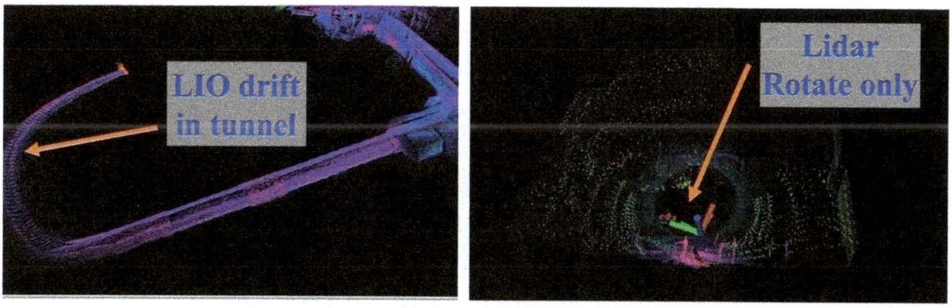

(a) (b)

Figure 8.3 (a) The LIO drift in long tunnel. The actual path follows a straight tunnel corridor. (b) The lidar has rotated motion only.

does not translate), then even monocular VIO cannot determine scale; for example, the accelerometer only senses gravity and the gyros sense rotation, and there is no translational acceleration to link image movement to physical units. This is a degenerate motion for VIO, causing scale to remain unobservable during that motion [4]. Another challenge in VIO is observing the IMU biases. Typically, accelerometer biases are observable only when the device's acceleration is varied (including holding it still, as then acceleration is zero and any measured acceleration must be bias). Gyro biases are observable when there is actual rotation (no rotation expected versus some rotation measured indicates bias). If the device's motion is too limited (e.g., it never experiences certain axis movements), bias in those axes might not be fully observable and can cause drift.

Integration challenges in LIO, VIO, and Wi-Fi, often mean "what can we actually learn about the state from these sensors?" When the answer is "not everything," the missing pieces manifest as drift or ambiguity. The next sections discuss drift resulting from unobservable states and how to handle unobservable modes.

8.2.2 Unobservable States, Drift, and Solutions

When certain state variables are unobservable, the estimates for those variables can drift arbitrarily. The fusion system may assign any value to an unobservable state and still get the same predicted measurements, that meaning sensor data does not pull that state back to truth. As a result, small errors in unobservable directions accumulate over time (drift) because nothing in the measurements can correct them.

To systematically identify unobservable states in a sensor fusion system, one can perform an observability analysis [3, 5, 6] of the system's mathematical model. This often involves computing the observability matrix (for linear systems) or using nonlinear observability tools for nonlinear systems. If the observability matrix is not full rank, it indicates that certain combinations of states are unobservable. For example, by analyzing the observability matrix for a VIO, one can pinpoint degenerate motions [7] that cause loss of observability in certain directions. In one study, Lee et al. [8] listed degenerate motions for multisensor navigation by examining the rank of the observability matrix under those motions. They could formally show, for instance, that, if a vehicle only undergoes constant velocity and constant heading (no change in those inputs), then certain biases or scale factors become unobservable. Similarly, in calibration problems, observability analysis can reveal if, say, moving with only one rotational axis active fails to observe some components of the extrinsic rotation [4] (this is related to the notion that you need excitation in all degrees of freedom).

In the observability analysis, we first derive the generally quantified rule for a nonlinear input/output model. Then we apply the rule to the Kalman filter and FGO for real state estimation applications. According to [9], the definition of locally observable (identifiable) is:

Definition 1. For a state estimation problem, a nonlinear model $\hat{\mathbf{y}} = h(\boldsymbol{\theta}, \mathbf{u}; x_0) : \Theta \rightarrow \mathcal{H}$, where $\hat{\mathbf{y}}$ is a prediction of $\mathbf{y} := [\boldsymbol{y}_1^T \; \dots \; \boldsymbol{y}_N^T]$ ($\boldsymbol{y}_k \in \mathbb{R}^p$), $\mathbf{u} := [\boldsymbol{u}_1^T \; \dots \; \boldsymbol{u}_N^T]$ ($\boldsymbol{u}_k \in \mathbb{R}^m$) is a given input vector and x_0 is an initial state vector, is called locally observable (identifiable) in $\boldsymbol{\theta}_m \in \Theta \subset \mathbb{R}^q$ if for all $\boldsymbol{\theta}_1$ and $\boldsymbol{\theta}_2$ in the neighborhood of θ_m hold that

$$\left\{h\left(\boldsymbol{\theta}_1, \mathbf{u}; x_0\right) = h\left(\boldsymbol{\theta}_2, \mathbf{u}; x_0\right)\right\} \Rightarrow \boldsymbol{\theta}_1 = \boldsymbol{\theta}_2 \tag{8.1}$$

Based on the nonlinear model, we usually consider parameter estimation by minimizing the LS function $V(\boldsymbol{\theta})$:

$$V(\boldsymbol{\theta}) := \frac{1}{2}\epsilon(\boldsymbol{\theta})^T P_v \epsilon(\boldsymbol{\theta}) \tag{8.2}$$

where $\epsilon(\boldsymbol{\theta}) = \mathbf{y} - \hat{\mathbf{y}} = \mathbf{y} - h(\boldsymbol{\theta}, \mathbf{u}; x_0)$ is the residual vector. It is trivial to derive the Jacobian matrix $\partial V(\boldsymbol{\theta})/\partial \boldsymbol{\theta}$ and the Hessian matrix $\partial^2 V(\boldsymbol{\theta})/\partial \boldsymbol{\theta}^2$ if $V(\boldsymbol{\theta})$ is derivable. Here the parameter estimation of $\boldsymbol{\theta}$ is equivalent to minimize the LS function $V(\boldsymbol{\theta})$. Therefore, the observability condition is written as $\partial V(\boldsymbol{\theta})/\partial \boldsymbol{\theta} = 0$ (first-order necessary condition) or $\partial^2 V(\boldsymbol{\theta})/\partial \boldsymbol{\theta}^2 > 0$ (second-order necessary condition) at $\hat{\boldsymbol{\theta}} = \arg\min_{\theta} V(\boldsymbol{\theta})$. Next, we mainly employ the first-order necessary condition, which can also be written as $\text{rank}(\partial V(\boldsymbol{\theta})/\partial \boldsymbol{\theta}) = q$.

Based on the rule, we can get the observability matrix for an EKF:

$$\mathcal{O}_{\text{EKF}} = \begin{bmatrix} \mathbf{H}_0 \\ \mathbf{H}_1\Phi(t_1, t_0) \\ \mathbf{H}_2\Phi(t_2, t_0) \\ \vdots \\ \mathbf{H}_T\Phi(t_T, t_0) \end{bmatrix} \tag{8.3}$$

where $\mathbf{H}_i$ is the observation matrix and $\Phi(t_i, t_0)$ is the state transition matrix of an EKF at time i. Unlike the EKF, the state transition model of FGO is included in the factors. Thus, the observability matrix of FGO is given as:

$$\mathcal{O}_{\text{FGO}} = \begin{bmatrix} \mathbf{J}_{t_0} \\ \mathbf{J}_{t_1} \\ \vdots \\ \mathbf{J}_{t_T} \end{bmatrix} \tag{8.4}$$

where $\mathbf{J}_{t_i}$ is the Jacobian matrix of residuals in FGO at time i. In conclusion, to check the observability of a state estimation model, we need to check the rank of the observability matrix $\mathcal{O}_{\text{EKF}}$ or $\mathcal{O}_{\text{FGO}}$. A full-rank observability matrix with respect to the state $\boldsymbol{\theta}$ implies that it is locally observable. If the observability matrix has rank-deficient columns for some state variables, the system cannot estimate these state variables, meaning that they are unobservable.

Once unobservable state dimensions are recognized, the next step is to mitigate their impact. The primary strategy is to add additional information to make those states observable. This could be additional sensors, as discussed: add a magnetometer to observe global heading (resolving yaw), add a GPS to observe global position and scale (resolving those drifts), and use wheel odometry to constrain motion in-vehicle navigation. Each extra sensor and/or information essentially provides equations that lock down the unobservable states. For monocular VIO, adding even occasional known distance measurements (such as detecting the floor for altitude or a known object size) can calibrate scale and prevent scale drift.

Another strategy is the use of constraints and assumptions. In some cases, certain unobservable states are not critical to the task and can be constrained by assumption. For example, in a quadrotor drone doing VIO indoors, global yaw (heading) might be unobservable and also might not be directly needed for local navigation. One could simply fix the initial yaw as the reference (basically assuming an arbitrary yaw = 0 at the start) and then the problem becomes observable in yaw relative to that reference (the system will still have yaw drift if there is bias, but one might accept slow drift or use periodic known heading fixes if possible). In indoor pedestrian tracking, one might assume the person's start orientation or occasionally use map alignment to reset heading drift (e.g., knowing building corridor directions to snap the orientation). This approach essentially chooses a convenient value for the unobservable state rather than letting it float. In EKF terms, one might set a high uncertainty for that state but also occasionally reset if possible.

However, when unobservable states do matter to the mission (such as scale in a mapping that needs metric units), then assumptions are not enough and real measurements or extra constraints are needed. If adding sensors is not feasible (like in pure vision SLAM in unknown environment with no priors), drift in unobservable directions is inevitable. In such cases, one might at least quantify the uncertainty in those directions growing over time, so the system is aware of its limitations. For example, a SLAM algorithm can report increasing covariance on global pose, indicating growing uncertainty.

A diagnostic flow for handling unobservable states is outlined in Figure 8.4:

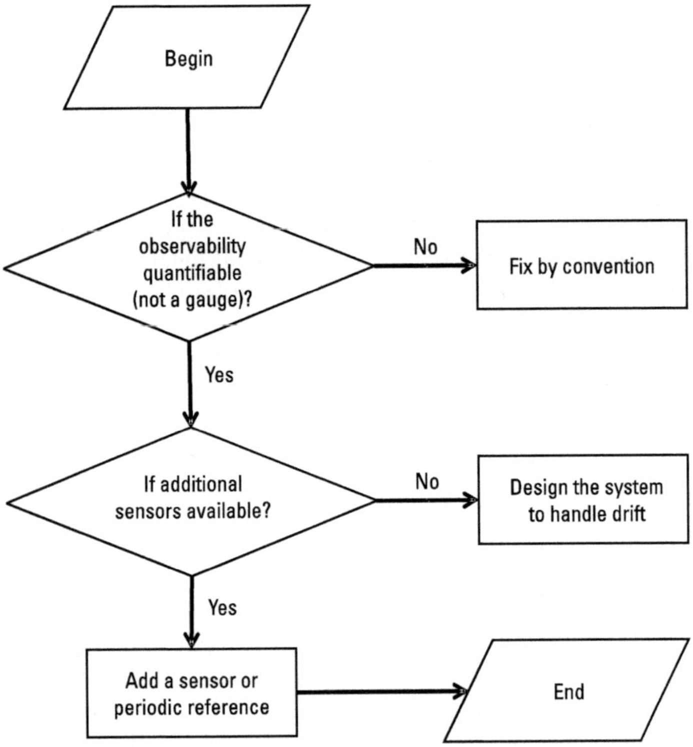

Figure 8.4 Flowchart to handle the unobservable state.

1. Determine if it is a critical physical quantity or a gauge (pure frame reference such as a missing yaw angle that we mentioned above).
2. If it is a gauge, just fix it by convention (no physical impact, just coordinate choice).
3. If it is a critical physical quantity, add a sensor or periodic reference to observe it (e.g., magnetometer for heading, altimeter for altitude, GNSS/Wi-Fi for position).
4. If an additional sensor is not possible, then use environmental or motion constraints (nonholonomic motion constraints for vehicles, planar motion assumption, zero-velocity updates) to supply pseudo-measurements. Please refer back to Section 6.6.2 for details.
5. If none of the above, accept that drift will occur and design the system to handle it gracefully (perhaps by resetting estimates occasionally or warning the user of increasing error).

Finally, one practical approach is to include redundant modalities specifically to address the unobservable aspects to make the system more robust.

8.3 Tuning of Sensor Fusion

Accurate sensor fusion relies not only on correct models and observability but also on properly tuning the assumed noise characteristics (covariances) of sensor measurements and process dynamics to make them nominal. The covariance values (process noise **Q** and measurement noise **R** in Kalman filtering terms) determine how the fusion algorithm weights different information sources. If these covariances are poorly chosen, the filter or optimizer can become overconfident in bad data or underutilize good data, leading to suboptimal or inconsistent estimates. Indoor environments pose special challenges to sensor fusion, often requiring careful tuning and additional compensation techniques. Key issues include multipath interference for RF sensors (Wi-Fi, UWB) and drift of inertial sensors due to lack of absolute references.

Tuning covariance is both a science and an art. It often requires experimentation and validation, but some techniques exist to estimate or adjust covariances systematically. Beyond initial covariance estimation, continuous adjustment and tuning of parameters is often needed as conditions change. The term "parameter" here included the noise covariances as discussed (**Q** and **R**), as well as other algorithm parameters such as outlier rejection thresholds and data association parameters. Ensuring that the fusion algorithm remains well-tuned in varied scenarios is challenging. Parameters that work well in one environment might not in another environment. Therefore, robust sensor fusion designs incorporate mechanisms for parameter adaptation and set parameters to values that offer a good balance across conditions.

It is desired to find a region in parameter space where performance is acceptable under various conditions. This means parameter robustness. In other words, sensor fusion algorithm should not be extremely sensitive to exact parameter values. Ideally, if covariances are off by some factor, the filter should still operate without diverging (maybe just with suboptimal performance).

Achieving this robustness is a goal of tuning. One approach is stress-testing the fusion with worst-case scenarios (e.g., drop one sensor, introduce large noise) and adjusting parameters to handle those. This often leads to conservative tuning; for example, set measurement noises higher than nominal so that if an unexpected error occurs, the filter is not thrown off. It is known that an EKF can diverge if covariances are badly tuned (especially if **R** is too low relative to actual sensor noise). Thus, many engineers will intentionally inflate **R** beyond the measured noise variance [10] to account for unmodeled errors and ensure stability. Figure 8.5 illustrates the impact: with a large initial state error, using a precise **R** in Figure 8.5(a) allows fast convergence but risks instability, while an overbounded **R** in Figure 8.5(b) slows convergence but improves robustness. The EKF is sensitive to this trade-off, whereas FGO remains stable in both cases.

8.3.1 Covariance Estimation Techniques

One approach to covariance tuning is to empirically estimate noise characteristics from real data. For example, for an IMU, one can collect data with

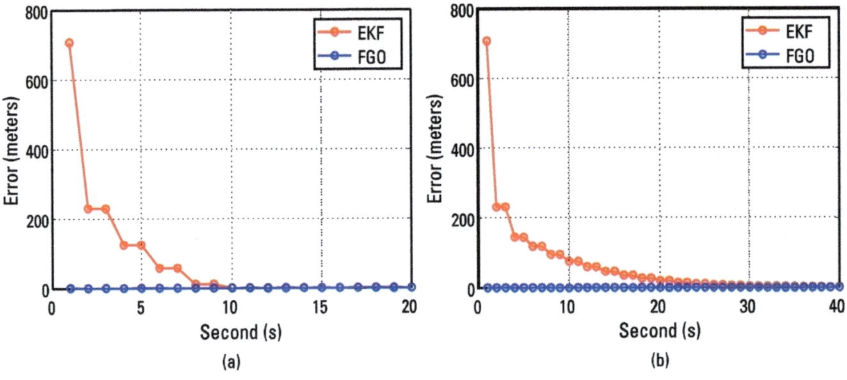

Figure 8.5 (a) Large initial guess and a precise **R**. (b) Large initial guess and an overbounded **R**.

the device static and calculate the variance of accelerometer and gyroscope readings to determine sensor noise covariance. Allan deviation plots are used in inertial sensor analysis to separate noise into bias instability, random walk, and so forth, which then informs the process noise model as we introduced in Chapter 6. For a Wi-Fi or UWB ranging sensor, one can measure the error distribution by placing the receiver at a known reference point and logging many measurements; the sample variance of the error gives the measurement noise covariance. This kind of characterization gives a starting point for **Q** and **R** values in a filter.

However, real-world conditions can deviate from the lab's precalibrated characterizations. Thus, adaptive techniques are often employed. A notable technique for state estimation filters is the adaptive EKF, which adjusts noise parameters on the fly. One such technique is innovation-based adaptive estimation: the filter monitors the statistical properties of the innovation (the difference between predicted measurement and actual measurement).

$$\tilde{y}_k = z_k - h\left(\hat{x}_{k|k-1}\right) \tag{8.5}$$

where $\tilde{y}_k$ is the innovation at time k, z_k is the actual measurement, and $h(\,)$ is the measurement function to the predicted state $\hat{x}_{k|k-1}$.

If the innovations are larger than expected (based on an assumed **R**), it indicates that the measurement noise might be higher than assumed, so **R** should be increased. Conversely, if innovations are consistently smaller, the filter might be able to decrease **R** to trust the sensor more. In practice, one can adjust **Q** and **R** in real time to keep the innovation covariance aligned with a theoretical expectation [11]. The innovation covariance is given by

$$\mathbf{S}_k = \mathbf{H}_k \mathbf{P}_{k|k-1} \mathbf{H}_k^T + \mathbf{R}_k \tag{8.6}$$

where $\mathbf{S}_k$ is the innovation covariance. $\mathbf{H}_k$is the Jacobian matrix of $h(\,)$ evaluated at the predicted state (linearizes the measurement function). $\mathbf{P}_{k|k-1}$ is the predicted state covariance. For instance, many Wi-Fi /INS filters will inflate the Wi-Fi measurement noise when the pedestrian is in a signal dark zone (RF signals are likely to be blocked) by detecting the increased innovation magnitude.

There is also an approach that incorporates using robust function, such as the Huber loss [12], which takes the innovation residual as input to it. Then the weighting given by the robust function will act as a scaling factor to the noise covariance, often formulated as

$$\mathbf{R}_k = \omega(\tilde{y}_k) \cdot \mathbf{R}_k \tag{8.7}$$

where $\omega(\tilde{y}_k)$ is the weight output by the robust function based on the current innovation. This approach helps to mitigate the impact of large outliers in the measurement residuals.

Another systematic approach is using the Expectation-Maximization (EM) algorithm to estimate noise covariances. EM can treat the true trajectory as hidden data and the sensor outputs as observed data and then iterate to find the covariance values that maximize the likelihood of the observations [13]. Researchers have applied EM in Kalman filter contexts [14] to auto-tune process and measurement noise, with some success in off-line or slow adaptation scenarios.

Some recently proposed methods use multiple models (multiple model adaptive estimation) with different noise assumptions and pick the best fit. Others leverage machine learning: for instance, using reinforcement learning (RL) to tune covariance [15] has been explored (the RL agent adjusts noise values to minimize state estimate error over time). In graph-based fusion, a similar approach is dynamic covariance scaling. If a residual is large, they scale up its covariance (down-weight that measurement) to prevent it from dominating the optimization (this is akin to a robust loss function in optimization).

For Kalman filter-based fusion, the primary parameters are indeed the covariance matrices $\mathbf{Q}$ and $\mathbf{R}$, as discussed earlier. Another important parameter is the Kalman filter initial covariance, $\mathbf{P}_0$, which represents the filter's initial uncertainty about the system state. This matrix plays a critical role during the early stages of estimation, particularly before the filter has had sufficient observations to converge. Some filters include a small stabilizing noise added to $\mathbf{Q}$ to ensure numerical stability (even if a model is deterministic). Adjusting that can affect how quickly the filter forgets past information. A common tuning trick is covariance inflation. If one suspects that the filter is too confident, one can artificially inflate $\mathbf{P}$ or $\mathbf{Q}$ slightly.

Finally, in multisensor setups, covariance estimation could also involve understanding correlations, if any. Usually, different sensors are treated as independent (no cross-covariance), but if they have common error sources (e.g., two-wheel encoders on left/right wheels have correlated errors when the surface is uneven), one might model that. However, many fusion frameworks assume a diagonal $\mathbf{R}$ matrix for simplicity (each sensor noise independent). If the correlation is significant, a more complex error model is needed (e.g., the TDOA model mentioned in Chapter 4).

8.3.2 Thresholds

Outlier rejection thresholds (such as gating thresholds for innovations that used to exclude measurements) are another important set of parameters. If too tight, good data might be discarded; if too loose, outliers slip in and disturb the estimate. Systems often adapt these based on the recent history of measurement residuals. For instance, if a particular sensor (say, a lidar odometry) has been consistent for a while, the system might tighten the gate threshold to be more sensitive; if it then experiences a sudden change (perhaps a difficult scene causing large residuals), an adaptive logic might widen the gate to allow data rather than reject everything (or conversely, realize lots of outliers are coming and widen to avoid triggering constant rejections). Assuming the nominal noise distribution and over a single fault at a time, a threshold based on chi-square hypothesis testing can be optimal. In this case, a measurement is accepted if the squared Mahalanobis distance,

$$d_k^2 = \tilde{y}_k^T \mathbf{S}_k^{-1} \tilde{y}_k \tag{8.8}$$

satisfies

$$d_k^2 < \chi^2_{m,\alpha} \tag{8.9}$$

where $\tilde{y}_k$ is the innovation at time k, $\mathbf{S}_k$ is the innovation covariance, m is the measurement dimension, and $\chi^2_{m,\alpha}$ is the upper α quantile of the chi-square distribution with m degrees of freedom (e.g., $\alpha = 0.95$ for 95% confidence). However, this chi-square threshold is challenging when there are multiple faults (bias and/or non-nominal distribution).

Beyond the threshold gating, robust estimation techniques provide a principled way to handle outliers by smoothly reducing their influence rather than outright rejecting them. A common class of such methods includes M-estimators [2, 16], which replace the squared error in the measurement update step with a robust loss function that saturates for large residuals. The Huber loss [12], for instance, behaves quadratically for small residuals (like standard LS) and linearly for large ones, reducing the impact of outliers while retaining sensitivity to inliers. Another widely used robust loss is the Cauchy loss [17], which aggressively suppresses large residuals, making it more robust in high-outlier scenarios. Its influence function decreases to zero as the residual grows, effectively down-weighting extreme outliers. More sophisticated approaches such as graduated nonconvexity (GNC) [18] estimators gradually transition from a convex to a nonconvex loss landscape, allowing robust convergence even in high-outlier regimes. These robust cost functions effectively tune the weight of each measurement in the update step, enabling the system to remain stable in the presence of unexpected disturbances without requiring manual threshold tuning.

8.4 Calibration Techniques

Accurate calibration of sensors is a fundamental prerequisite for effective sensor fusion. Calibration determines the parameters that map a sensor's raw output to real-world units or to a common reference frame. It also removes biases and systematic errors. Poor calibration can lead to large errors, because inconsistencies between sensors will propagate through the fusion algorithm. Historically, sensor calibration has been recognized as a foundational step. This includes that both an intrinsic calibration addresses each individual sensor's internal errors and an extrinsic calibration finds the spatial alignment between different sensors. Early multisensor systems often relied on careful manual calibration, meaning factory-set intrinsics and labor-intensive extrinsic using known reference objects or alignment instrumentation. Over time, techniques have evolved toward automated and online calibration, supported by deeper understanding of system observability and the development of algorithms that can estimate calibration parameters during a normal operation.

8.4.1 Intrinsic Calibration

Intrinsic calibration focuses on correcting errors and biases inherent to individual sensors, thereby ensuring each sensor's output is as accurate as possible before fusion. This involves estimating sensor-specific parameters that affect measurements in systematic ways. Intrinsic calibration is typically performed as a priori to extrinsic calibration and sensor fusion, because one must trust each sensor's readings individually before combining them. Many intrinsic parameters remain stable once determined (e.g., a camera's focal length or a lidar's laser firing angles are fixed after manufacturing), so one common approach is an offline calibration using controlled references.

8.4.1.1 Camera

For example, a camera's intrinsic parameters [19] include focal length, optical center, and lens distortion coefficients; calibrating these allows the camera to produce geometrically correct image measurements (e.g., undistorted pixel coordinates). A well-established method is to have the camera observe a known pattern such as a checkerboard from various angles and distances; from the correspondences between image points and known pattern geometry, algorithms (such as Zhang's method [20]) compute the camera's intrinsic matrix and distortion coefficients. Chapter 7 has briefly introduced this technique. These techniques have become standard and highly refined over years of research [21–23] and are now routinely available in computer vision libraries such as OpenCV [24].

8.4.1.2 Lidar

Lidars and depth sensors may be factory-calibrated for parameters such as laser beam angles or intensity response (for example, a Velodyne lidar is calibrated to detect a reference reflectivity level [25]). This calibration process typically compensates for manufacturing variations across laser diodes, ensuring uniformity in both angular resolution and intensity output. Some manufacturers also calibrate for temperature drift, synchronization lag, and even individual channel timing delays. While these factory settings are sufficient for many out-of-the-box applications. In real-world deployment environments, such as those involving wide temperature ranges, aggressive motion, or sensor remounting, may require recalibration or fine-tuning. Advanced systems may perform online calibration [26] to adjust for such shifts during operation, especially when fusing lidar data with other sensors such as IMUs or cameras in high-precision applications.

8.4.1.3 IMU

Similarly, an IMU has biases in accelerometers and gyroscopes, scale factor errors, and axis misalignments that can be calibrated so that its reported accelerations and rotations closely reflect true motion. For IMUs, manufacturers often provide nominal bias and scale values, but each unit may still require fine-tuning. A practical example of IMU intrinsic calibration is bias and noise tuning: the IMU can be placed stationary to measure the accelerometer biases (as the average stationary output minus gravity) and gyroscope biases (average turn-rate when still), and undergoing controlled rotations to detect any scale factor errors. Such procedures, sometimes aided by statistical tools such as Allan variance analysis, yield parameters to correct the IMU's raw outputs. This was introduced in Chapter 6.

8.4.2 Extrinsic Calibration

Beyond intrinsic factors, multisensor systems require spatial calibration to align sensors in space. Extrinsic calibration determines the rigid-body transformation (usually, 6-DoF) between sensors' coordinate frames. Accurate spatial calibration is essential for sensor fusion; the misalignment of a few centimeters can degrade fusion accuracy or even make the sensor data inconsistent.

This calibration finds the rotation and translation that map one sensor's coordinate system into that of another. For example, if a lidar and camera are mounted on a vehicle, extrinsic calibration will yield the 3-D rotation and translation {**R**, **t**} that transforms a point observed by the lidar into the camera's coordinate frame. These extrinsic parameters allow fusing of data, such as projecting lidar points onto camera images or combining their measurements for obstacle detection. There are two broad approaches for extrinsic calibration: target-based and motion-based methods, as shown in Figure 8.6.

8.4.2.1 Target-Based Calibration

In target-based calibration, one uses specially designed objects or patterns visible to multiple sensors to provide known reference points. Common targets include checkerboard patterns [19] (seen in camera images and also detectable as planes in 3-D lidar), geometric arrangements of retro-reflectors or LED markers, or calibration objects with multiple modalities (e.g., a planar board with an AprilTag [27] for camera and reflective surfaces for lidar). By collecting sensor data of the same target and running optimization, the relative pose of sensors can be solved. For example, a camera-lidar calibration might involve placing a checkerboard in view of both the camera and lidar; the camera sees the checkerboard corners while the lidar perceives the planar board. By aligning these, it yields the transform between camera and lidar frames. Target-based methods can achieve high accuracy but require a controlled setup.

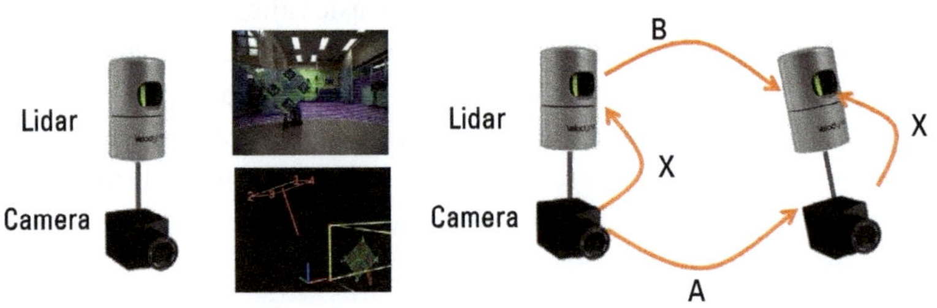

(a) (b)

Figure 8.6 (a) The target-based method. (b) The motion-based method where **A** and **B** are camera and lidar motions, and **X** is the extrinsic transformation solved from **AX** = **XB**.

8.4.2.2 Motion-Based Calibration

Motion-based calibration does not require a special object; instead, it leverages the sensors' own observations of the environment or their motion. A classic example is the hand-eye calibration [28] technique in a multisensor context. If one sensor can estimate its motion (e.g., an IMU provides relative orientation and position via integration, or a lidar SLAM gives pose changes), and another sensor also measures motion or has observations that depend on the same motion, the relative pose can be inferred by aligning the two motion estimates. These targetless methods are convenient as they use natural features (such as alignments of 3-D point clouds and camera features) or common motion. However, they can be sensitive to the environment. Both sensors must observe enough corresponding structure or motion for the calibration to converge [29]. Furthermore, certain motions can lead to ambiguities: for full 6-DoF extrinsic calibration, the platform should experience rotational and translational motion in all axes (to make sure the observability is strong enough); purely planar motion, for example, might leave the relative roll angle unobservable. As noted earlier, a lack of excitation in some degrees of freedom means those parts of the transform cannot be determined [4].

8.5 Temporal Calibration

In addition to spatial calibration, temporal calibration is equally critical in modern sensor fusion systems, particularly when fusing high-rate, asynchronous sensors such as cameras, IMUs, lidars, and UWB receivers. Even with perfectly calibrated intrinsics and extrinsics, discrepancies in timestamp alignment between sensors can introduce significant errors. Small timing offsets could lead to large spatial misalignments. Temporal calibration ensures that the data from different sensors is correctly synchronized, either by estimating fixed time delays or by modeling clock drift.

8.5.1 Synchronization

Proper time synchronization ensures that sensor measurements used together in the fusion algorithm indeed correspond to the same real-world moment. In a sensor fusion scenario with different sensor modalities, each device might have its own oscillator (clock) and sampling rate, leading to potential misalignment in time. Even a small timestamp error can introduce significant fusion errors, especially in high-speed or highly dynamic systems. For example, considering fusing an IMU reading taken slightly before a camera image, the

IMU might indicate a rotation that the camera has not yet observed, causing an inconsistency in estimation. Therefore, aligning data streams in time is crucial. In SLAM and other TC fusion systems, synchronization errors can result in ghosting or smearing of features and degraded state estimates [29], ultimately causing drift or divergence in the solution.

Figure 8.7 illustrates three representative time synchronization scenarios commonly encountered in VIO systems. In the ideal case (Case 1), the IMU and camera are well synchronized, with their measurements occurring in consistent alignment. This is typically achieved through precise hardware triggering or the use of a shared clock source. However, deviations from this ideal are common in practice. Case 2 shows a situation where the camera and IMU share the same clock but have different internal trigger times, resulting in a fixed offset between their measurement events. This misalignment, if uncorrected, can introduce significant temporal errors into the fusion process. Case 3 depicts a more challenging scenario in which the IMU and camera operate on independent time sources. This leads not only to a constant offset but also to time-varying drift, where the relative timing between sensor streams changes over time. To achieve accurate temporal calibration in such systems, two problems must be addressed: (1) a constant offset between sensor clocks, as shown in Case 3 in Figure 8.7 (one sensor's timestamps are always, say, 5 ms behind that of the other sensor), and (2) the drift (one sensor's clock runs slightly faster, so the offset changes over time). A straightforward approach to handle a known constant offset is to subtract it from timestamps, but finding that offset requires precise calibration. Often, a synchronization calibration is done by observing a common event. For example, if a camera sees an LED turning on at time t_{cam} and an IMU detects a vibration (from the LED's control circuit) at time t_{imu}, one can deduce an initial offset. More systematically, many systems adopt a common time base via hardware: all sensors are GPS timestamped or triggered by the same clock signal [30]. This hardware synchronization is ideal because it eliminates relative drift (all sensors effectively become one clock). It is commonly used in research platforms; for instance, an unmanned aerial vehicle (UAV) might distribute a 1-Hz pulse per second (PPS) from a GNSS receiver or use IEEE 1588 precision time protocol (PTP) [31] to all sensors, aligning their times to within microseconds.

Without hardware synchronization, a software one is needed. Filters and smoothing algorithms can estimate time offset as part of the state. For example, a VIO system might include a time offset parameter between a camera and IMU and adjust it by observing that IMU-reported motions consistently lead or lag the camera's observed motions, as shown in Figure 8.8. Another

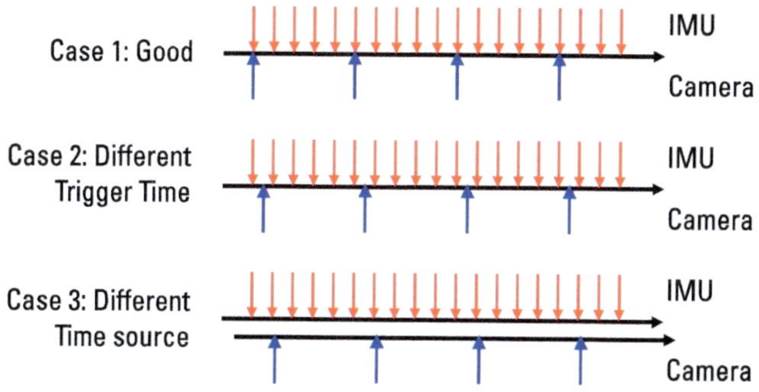

Figure 8.7 Illustration of different time synchronization cases in the VI system.

approach is using filtering on timestamps themselves. Techniques treat the sensor clocks as having unknown skew and offset and use timestamp exchange or common observations to estimate those parameters [32]. One study [33] introduced a passive sync algorithm that monitors the discrepancy in sensor timestamps and can significantly reduce sync error even when sensors have unknown latency and drift. In essence, it periodically recalculates how one clock is drifting relative to another and corrects future timestamps accordingly.

Sensor fusion algorithms can be made somewhat robust to small sync errors. For example, increasing the assumed measurement noise for asynchronous sensors can prevent the filter from overtrusting possibly mistimed data (essentially blurring the measurement in time). Some advanced algorithms even estimate an effective time offset online by examining innovation residuals. If the data consistently arrives "late" (causing a predictable residual pattern), the filter can infer a time delay and compensate. In critical systems, where an out-of-sync sensor reading could cause a hazard (e.g., an autonomous car mis-localizing itself), designers often include fail-safes, meaning that if synchronization uncertainty becomes too high, the system might temporarily reduce reliance on that sensor or issue an alert.

Figure 8.8 Illustration of time offset in a VIO system.

8.5.2 Measurement Timing

Due to differing sensor update rates, a fusion system often needs to interpolate measurements so that information is available on a common timeline or at desired intervals. For example, an IMU might provide readings at 200 Hz, while a Wi-Fi receiver outputs positions at 5 Hz. If the fusion algorithm operates at 100 Hz (perhaps tied to a control loop), it must integrate these inputs appropriately. It can use each new IMU reading as it arrives, but only occasionally get a Wi-Fi update. Measurement interpolation fills the temporal gaps by estimating what a sensor reading would be at an intermediate time. This is crucial when combining fast sensors with slower ones. Without interpolation, one might have to wait for the slow sensor or else process asynchronous data with complex handling of time delays.

8.5.2.1 Direct Measurement Interpolation

The simplest approach is linear interpolation in time. If a sensor provides readings at times t_0 and t_1, one can linearly interpolate to get an estimate at t in between. For instance, if a magnetometer gives heading at 1 Hz, but an attitude and heading reference system (AHRS) filter runs at 50 Hz, the magnetometer heading can be linearly interpolated for intermediate filter steps. Linear interpolation is computationally trivial and works reasonably if the sensor signal changes slowly or nearly linearly between samples. However, many sensor quantities are not linear over time (e.g., angles on a circle or an object accelerating). Linear interpolation of orientation can cause issues because rotations live on a manifold (a non-Euclidean space). In such cases, specialized interpolation [34] such as spherical linear interpolation (SLERP) for quaternions is used to smoothly interpolate orientations on the unit sphere. For example, given two orientation quaternions at times t_0 and t_1, SLERP provides a steady rotation at constant angular velocity between them, avoiding anomalies such as gimbal lock that could occur with naive linear interpolation of Euler angles (as we mentioned in Chapter 2). Similarly, one can linearly interpolate position coordinates if acceleration is mild; if acceleration is significant, linear interpolation will lag behind the actual curved trajectory.

8.5.2.2 Continuous-Time Method

To improve on linear assumptions, one can use polynomial splines (e.g., cubic splines) or other smooth interpolation methods. A spline fit through a sequence of sensor samples can provide continuously differentiable estimates and better capture the dynamics. These methods are known as continuous-time methods. For example, an odometer that outputs distance every second could be interpolated with a cubic spline to estimate distance at 0.1-second intervals, preserving continuity of velocity. Engineers often use splines for IMU orientation integration or to interpolate delayed measurements. A special case is orientation: techniques such as SQUAD (Spherical and Quadratic interpolation) [35] extend SLERP to multiple segments, ensuring a smooth orientation trajectory through several keyframes. Spline-based interpolation adds some computational cost (solving for polynomial coefficients) but offers greater accuracy when sensor data can be assumed to vary smoothly.

A more advanced method treats interpolation as a probabilistic regression problem. One can model the sensor's time series as a Gaussian Process (GP), which provides a continuous function that best fits the known samples and a measure of uncertainty. GP interpolation (also called Bayesian interpolation) [36] can be very powerful, especially if the sensor noise characteristics are known. For example, continuous-time SLAM approaches [37] sometimes use GP regression to represent a robot's trajectory as a continuous function, allowing sensors such as IMUs and cameras to be fused without discrete resampling. A GP can naturally incorporate uncertainty. If there is a large gap between sensor updates, the interpolation uncertainty grows, which the fusion algorithm can take into account. The downside is computational complexity. GP inference involves inverting covariance matrices, which can be expensive as the number of points grows. Researchers have developed efficient GP frameworks for sensor fusion, but they are heavier than simpler interpolation schemes. Figure 8.9 illustrates the difference between cubic spline interpolation and GP regression for continuous-time estimation from discrete sensor data samples.

8.5.2.3 Physics-Based Method

There is a classic trade-off when choosing an interpolation method. Direct linear interpolation is fast and often good enough for small gaps, but can introduce error if the sensor reading is highly nonlinear over the interval. High-order interpolation (splines and GP) can be more accurate but costs more CPU and potentially adds latency. In real-time systems, one usually cannot perform very heavy interpolation computations at a high rate. A practical strategy is to use physics-based models for interpolation. For example, in between Wi-Fi fixes, one can integrate IMU readings (assuming constant acceleration between IMU samples, effectively a linear model in acceleration), which are a form of model-based interpolation for the Wi-Fi position. This leverages the IMU to provide a physically realistic interpolation of position and velocity, rather than just a mathematical spline.

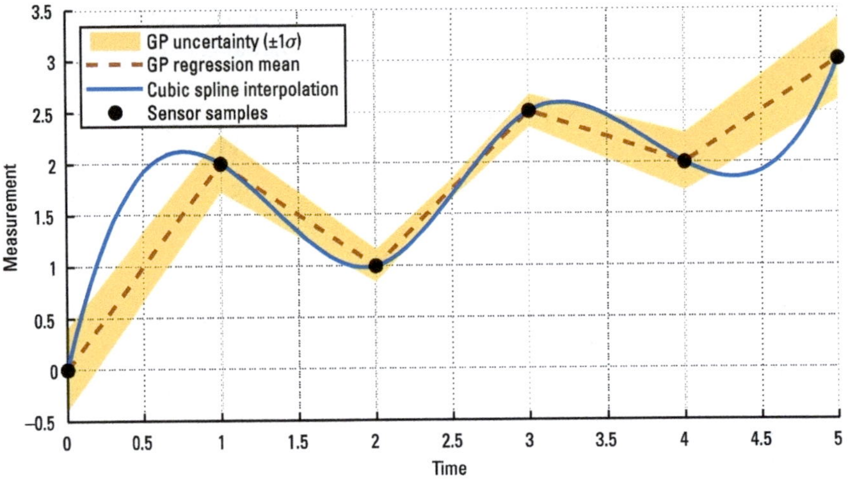

Figure 8.9 Illustration of continuous-time methods for sensor data interpolation. Discrete sensor samples (black dots) are interpolated using a cubic spline (solid blue curve), providing a smooth and continuously differentiable estimate. GP regression (orange dashed curve) provides both a continuous estimate and a measure of uncertainty (shaded region), which grows in areas with sparse samples. These approaches allow for high-resolution and uncertainty-aware trajectory estimation in sensor fusion systems.

8.6 Efficiency

In designing sensor fusion algorithms, especially for real-time positioning, efficiency is a key consideration. Efficiency can refer to computational complexity (how the algorithm scales with number of states or measurements), memory usage, and the ability to meet real-time deadlines with limited hardware. There is often a trade-off between the accuracy of an algorithm and its complexity. More accurate methods (such as batch optimization) tend to be more computationally expensive than simpler approximations (such as a basic Kalman filter). The choice between real-time versus offline processing also influences which algorithms are viable. This section discusses strategies to achieve real-time performance, as well as considerations for offline processing where latency is not an issue.

8.6.1 Real-Time Versus Offline

Real-time systems require algorithms that can guarantee an update within a fixed time budget for each sensor input. For example, if IMU comes at 200

Hz (5-ms period), the fusion algorithm must handle prediction at 200 Hz. If camera frames at 30 Hz provide updates, the system has ~33 ms to integrate the IMU between frames and incorporate the vision update. A VIO might do a lot of work for each frame (feature tracking, marginalization of the old keyframe, solving a small bundle adjustment (BA) problem). Systems such as VINS-Mono [32] achieve this by limiting the number of feature points and keyframes in the optimization (keeping it to a sliding window of maybe 5 to 10 keyframes) so the optimization remains small (state vector of maybe 50 variables, solved in tens of milliseconds). They also use multithreading to do feature tracking in parallel with state estimation. The complexity per frame is roughly proportional to the number of features tracked and number of iterations in the solver; with 100 features and few iterations, it fits in 33 milliseconds on mobile CPUs. If one tried to use thousands of features, it would not meet real-time. Thus, practical efficiency often requires trading off some accuracy (not using all features or marginalizing some poses) for speed.

Efficiency can be improved by using specialized hardware. Many modern sensor fusion tasks offload some computations to GPUs (which is specialized in visual processing and deep learning parts, if any). However, classical filter matrix operations can also use linear algebra libraries that exploit CPU vector instructions or multicore. Ensuring real-time performance often involves profiling the algorithm, identifying bottlenecks (such as a big matrix inversion or a resampling step), and optimizing those (maybe using block matrix math or reducing the frequency of heavy steps).

Offline processing (mapping or smoothing after data collection) can afford heavier computation. For example, one might run a full bundle adjustment on hours of data, which might take minutes or hours on a PC/server to get the most accurate map. For online use, one might run a reduced algorithm and then refine offline. Some mapping solutions indeed do a coarse real-time pose estimation and log data and then do a big optimization offline to refine the map (useful for applications such as surveying where final accuracy is needed but not in real time).

8.6.2 Memory Considerations

Efficiency also includes memory usage. A particle filter with many particles or an EKF-SLAM with large covariance matrix can be memory heavy. For instance, storing a $1,000 \times 1,000$ covariance matrix (for 1,000 landmarks) is 10^6 elements, which is fine (8 million bytes if double) on a PC but might be heavy on a microcontroller. Sparse methods alleviate memory by storing only nonzero entries. Factor graphs typically store factors in memory that scale with number of measurements (which can be large but manageable if using pointers to data and streaming methods). Two examples on quantifying memories are given:

- *Example 1:* Particle filter versus EKF for indoor localization—If the state is just (x, y, θ) for a person, an EKF is trivial to run (constant time basically per step). A particle filter with 1,000 particles might also be fine but is certainly heavier (1,000 times the motion and measurement evaluation). On a smartphone, 1,000 particles can be processed in a few milliseconds, which is still fine. If more complex sensors such lidar scans are used, evaluating a scan's likelihood for each particle is heavy. One might downsample the scan or reduce particle count to compensate. There is a known approach. A particle filter often is used when there are multimodal uncertainties, but, once localized, one might switch to an EKF for efficiency. Some systems implement a dual mode: use particle filter for global localization and then hand off to an EKF or a single-hypothesis smoother for precise tracking once confident. This leverages efficiency. A particle filter (with more cost) is only used when needed.
- *Example 2:* For a 2-D SLAM scenario with 500 3-DOF poses and 200 2-DOF landmarks, an EKF would manage a 1,900-dimensional state vector and a correspondingly dense 1,900 × 1,900 covariance matrix. The computational complexity of each EKF update is O(n2) (where n = 1,900), which critically limits scalability as the map grows. In stark contrast, FGO represents this problem using a sparse graph structure with 700 nodes (500 poses and 200 landmarks). This sparsity allows the full optimization problem to be solved efficiently using sparse factorization techniques, while incremental methods such as iSAM [38] can provide efficient online updates. Consequently, FGO methods offer substantially superior scalability compared to EKF as the environment size and trajectory length increase.

8.7 Conclusions

Sensor fusion has advanced significantly, enabling robust positioning and navigation in scenarios that would be impossible with any single sensor. By carefully calibrating sensors, synchronizing their data streams, and choosing appropriate fusion architectures (loose versus tight coupling), engineers have built systems that exploit the best aspects of each sensor to compensate for the weaknesses of others. Throughout this chapter, we discussed that effective fusion requires attention to practical details: calibration ensures that sensor outputs are comparable; time alignment prevents spurious errors; observability analysis tells us which quantities we can actually estimate; and the tuning of noise covariances and other parameters is key to filter consistency and performance in the field.

In conclusion, sensor fusion is a cornerstone of modern navigation and situational awareness systems. It embodies the saying "the whole is greater than the sum of its parts," as fused systems achieve robust performance that no individual sensor could accomplish. The progress in sensor fusion algorithms (from Kalman filters to graph optimizers) and computing power has continually expanded the horizons of what fused sensors can do, enabling everything from a smartphone guiding you through a mall to a robotics autonomously exploring a underground tunnel site. As sensors proliferate and diversify, effective fusion will remain a critical and evolving discipline, bridging the gap between raw sensor outputs and coherent understanding of position and movement in the world.

In terms of trends, computational improvements and theoretical advances are allowing more complex fusion to run in real time. We see autonomous robotics using factor graphs with hundreds of landmarks and states on the fly, and smartphones performing AI-based sensor fusion (such as Google's Visual Positioning Service, which fuses camera images with GNSS and map data). The line between mapping and localization has become a blur. Many systems simultaneously build a map and use it, effectively doing SLAM, which is a grand application of sensor fusion bringing together IMUs, lidar, vision, wheel encoders, Wi-Fi, and GNSS.

References

[1] Song, Y., and L. -T. Hsu, "Tightly Coupled Integrated Navigation System Via Factor Graph for UAV Indoor Localization," *Aerospace Science and Technology*, Vol. 108, 2021, p. 106370.

[2] Bai, X., W. Wen, and L.-T. Hsu, "Robust Visual-Inertial Integrated Navigation System Aided by Online Sensor Model Adaption for Autonomous Ground Vehicles in Urban Areas," *Remote Sensing*, Vol. 12, No. 10, 2020, p. 1686.

[3] Zheng, X., W. Wen, and L. -T. Hsu, "Tightly-Coupled Visual/Inertial/Map Integration with Observability Analysis for Reliable Localization of Intelligent Vehicles," *IEEE Transactions on Intelligent Vehicles*, 2024.

[4] Walters, C., et al., "A Robust Extrinsic Calibration Framework for Vehicles with Unscaled Sensors," *2019 IEEE/RSJ International Conference on Intelligent Robots and Systems (IROS)*, 2019, pp. 36–42.

[5] Hermann, R., and A. Krener, "Nonlinear Controllability and Observability," *IEEE Transactions on Automatic Control*, Vol. 22, No. 5, 1977, pp. 728–740.

[6] Yang, Y., and G. Huang, "Observability Analysis of Aided INS with Heterogeneous Features of Points, Lines, and Planes," *IEEE Transactions on Robotics*, Vol. 35, No. 6, 2019, pp. 1399–1418.

[7] Huang, G. P., A. I. Mourikis, and S. I. Roumeliotis, "Observability-Based Rules for Designing Consistent EKF SLAM Estimators," *The International Journal of Robotics Research*, Vol. 29, No. 5, 2010, pp. 502–528.

[8] Lee, W., C. Chen, and G. Huang, "Degenerate Motions of Multisensor Fusion-Based Navigation," *2024 IEEE International Conference on Robotics and Automation (ICRA)*, 2024, pp. 8113–8119.

[9] Van Doren, J. F., et al., "Identifiability: From Qualitative Analysis to Model Structure Approximation," *IFAC Proceedings Volumes*, Vol. 42, No. 10, 2009, pp. 664–669.

[10] Simon, D., *Optimal State Estimation: Kalman, H Infinity, and Nonlinear Approaches*, New York: John Wiley & Sons, 2006.

[11] Sun, W., P. Sun, and J. Wu, "An Adaptive Fusion Attitude and Heading Measurement Method of MEMS/GNSS Based on Covariance Matching," *Micromachines*, Vol. 13, No. 10, p. 1787, 2022, https://www.mdpi.com/2072-666X/13/10/1787.

[12] Huber, P. J., *Robust Statistical Procedures*, SIAM, 1996.

[13] Dempster, A. P., N. M. Laird, and D. B. Rubin, "Maximum Likelihood from Incomplete Data Via the EM Algorithm," *Journal of the Royal Statistical Society: Series B (Methodological)*, Vol. 39, No. 1, 1977, pp. 1–22.

[14] Gao, X., et al., "A Novel Residual-Based Bayesian Expectation–Maximization Adaptive Kalman Filter with Inaccurate and Time-Varying Noise Covariances," *Measurement*, Vol. 235, 2024, p. 114937.

[15] Gu, J., J. Li, and K. Tei, "A Reinforcement Learning Approach for Adaptive Covariance Tuning in the Kalman Filter," *2022 IEEE 5th Advanced Information Management, Communicates, Electronic and Automation Control Conference (IMCEC)*, Vol. 5, December 16–18, 2022, pp. 1569–1574.

[16] Lucas, A., "Robustness of the Student T Based M-Estimator," *Communications in Statistics-Theory and Methods*, Vol. 26, No. 5, 1997, pp. 1165–1182.

[17] Hampel, F., et al., *Robust Statistics: The Approach Based on Influence Functions*, Wiley, 1986.

[18] Yang, H., et al., "Graduated Non-Convexity for Robust Spatial Perception: From Non-Minimal Solvers to Global Outlier Rejection," *IEEE Robotics and Automation Letters*, Vol. 5, No. 2, 2020, pp. 1127–1134.

[19] Oth, L., et al., "Rolling Shutter Camera Calibration," *Proceedings of the IEEE Conference on Computer Vision and Pattern Recognition*, 2013, pp. 1360–1367.

[20] Zhang, Z., "Camera Calibration with One-Dimensional Objects," *IEEE Transactions on Pattern Analysis and Machine Intelligence*, Vol. 26, No. 7, 2004, pp. 892–899.

[21] De la Escalera, A., and J. M. Armingol, "Automatic Chessboard Detection for Intrinsic and Extrinsic Camera Parameter Calibration," *Sensors*, Vol. 10, No. 3, 2010, pp. 2027–2044.

[22] Gennery, D. B., "Generalized Camera Calibration Including Fish-Eye Lenses," *International Journal of Computer Vision*, Vol. 68, No. 3, 2006, pp. 239–266.

[23] Muglikar, M., et al., "How to Calibrate Your Event Camera," *Proceedings of the IEEE/CVF Conference on Computer Vision and Pattern Recognition*, 2021, pp. 1403–1409.

[24] Bradski, G., "The OpenCV Library," *Dr. Dobb's Journal: Software Tools for the Professional Programmer*, Vol. 25, No. 11, 2000, pp. 120–123.

[25] R. Bergelt, O. Khan, and W. Hardt, "Improving the Intrinsic Calibration of a Velodyne Lidar Sensor," *2017 IEEE SENSORS*, 2017, pp. 1–3.

[26] Lv, J., et al., "Observability-Aware Intrinsic and Extrinsic Calibration of Lidar-IMU Systems," *IEEE Transactions on Robotics*, Vol. 38, No. 6, 2022, pp. 3734–3753.

[27] Huang, J. K., and J. W. Grizzle, "Improvements to Target-Based 3D Lidar to Camera Calibration," *IEEE Access*, Vol. 8, 2020, pp. 134101–134110.

[28] Horaud, R., and F. Dornaika, "Hand-Eye Calibration," *The International Journal of Robotics Research*, Vol. 14, No. 3, 1995, pp. 195–210.

[29] Yeong, D. J., et al., "Sensor and Sensor Fusion Technology in Autonomous Vehicles: A Review," *Sensors*, Vol. 21, No. 6, 2021, p. 2140.

[30] Geiger, A., et al., "Vision Meets Robotics: The Kitti Dataset," *The International Journal of Robotics Research*, Vol. 32, No. 11, 2013, pp. 1231–1237.

[31] Idrees, Z., et al., "IEEE 1588 for Clock Synchronization in Industrial IoT and Related Applications: A Review on Contributing Technologies, Protocols and Enhancement Methodologies," *IEEE Access*, Vol. 8, 2020, pp. 155660–155678.

[32] Qin, T., P. Li, and S. Shen, "Vins-Mono: A Robust and Versatile Monocular Visual-Inertial State Estimator," *IEEE Transactions on Robotics*, Vol. 34, No. 4, 2018, pp. 1004–1020.

[33] Olson, E., "A Passive Solution to the Sensor Synchronization Problem," *2010 IEEE/RSJ International Conference on Intelligent Robots and Systems*, October 18–22, 2010, pp. 1059–1064.

[34] Buss, S. R., and J. P. Fillmore, "Spherical Averages and Applications to Spherical Splines and Interpolation," *ACM Transactions on Graphics (TOG)*, Vol. 20, No. 2, 2001, pp. 95–126.

[35] Robeson, S. M., "Spherical Methods for Spatial Interpolation: Review and Evaluation," *Cartography and Geographic Information Systems*, Vol. 24, No. 1, 1997, pp. 3–20.

[36] Peršić, J., et al., "Spatiotemporal Multisensor Calibration Via Gaussian Processes Moving Target Tracking," *IEEE Transactions on Robotics*, Vol. 37, No. 5, 2021, pp. 1401–1415.

[37] Anderson, S., and T. D. Barfoot, "Towards Relative Continuous-Time SLAM," *2013 IEEE International Conference on Robotics and Automation*, 2013, pp. 1033–1040.

[38] Kaess, M., A. Ranganathan, and F. Dellaert, "iSAM: Incremental Smoothing and Mapping," *IEEE Transactions on Robotics*, Vol. 24, No. 6, pp. 2008, 1365–1378.

9

Advanced Indoor Positioning and Indoor Navigation Techniques

Indoor positioning and navigation technologies have rapidly evolved beyond the basic solutions that we mentioned in the previous chapter. The future indoor positioning systems incorporates crowdsourced data, collaborative device networks, AI, and even emerging smart environments. This chapter explores advanced techniques that push the frontier of indoor localization accuracy and robustness.

The crowdsourcing mapping technique will first be introduced from basic concepts to application examples. Next, the developments of collaborative positioning will be introduced, including its architecture, technologies used, estimation method, applications, and limitations. Then the employment of AI for PDR and motion analysis will be introduced, from its key features to implementation examples. Lastly, the concept of radio sensing and reconfigurable intelligent surface (RIS) will be introduced with an outlook on their potential in indoor positioning and indoor navigation.

9.1 Crowdsourcing Mapping

Crowdsourcing leverages data contributed by many users (typically via their smartphones or IoT devices) to build indoor maps and positioning databases.

Compared to labor-intensive traditional surveying (the traditional fingerprinting method introduced in Chapter 4), crowdsourcing offers low-cost, flexible, and scalable indoor mapping [1]. Figure 9.1 contrasts a lone handset update with a fully crowdsourced update.

Modern smartphones carry multiple sensors (Wi-Fi, Bluetooth, IMU, cameras), making them ideal mobile mappers. Over the last decade (2012–2022), numerous solutions have emerged to aggregate user-collected data into indoor maps [2]. Techniques vary from grid-based radio heatmaps to landmark-based maps and semantic maps with labeled rooms or points of interest [2]. By intelligently fusing crowdsourced measurements, systems can construct maps for large venues without dedicated survey teams.

9.1.1 Database: Smartphone Data, Beacon Positions, Classification, and Graph Optimization

A crowdsourced indoor mapping database typically organizes data hierarchically by site (area, building, floor, or room) and by sensor modality. Smartphones contribute multisensor logs including radio frequency readings (Wi-Fi service set identifiers (SSIDs), signal strengths), inertial tracks (accelerometer/gyroscope steps and turns), magnetometer readings (for magnetic-field maps), and images of the environment. In addition, beacon infrastructure (Wi-Fi AP, Bluetooth beacons, UWB anchors) can be logged, including their identifications and positions if known or their signal signatures if unknown. All this

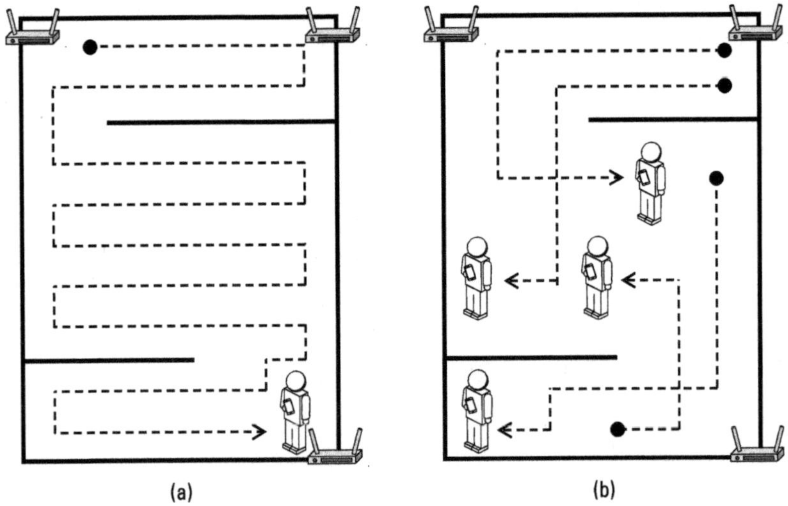

Figure 9.1 Database update by single agent (a) or crowdsourcing agents (b).

data must be classified and tagged with location context. For example, each crowdsourced datapoint is labeled by building, floor, and perhaps room or corridor if that can be inferred (sometimes the mapping system uses phone GPS at building entrances or known elevator locations as anchor points for initial labeling). Data platforms such as Google Indoor Maps and Apple Indoor Mapping Data Format (IMDF) use such hierarchical schemas to integrate crowd contributions with building metadata [3–5].

To merge many users' contributions into a coherent map, graph optimization techniques are widely employed. Each user's sensor trace can be modeled as a path graph (with waypoints connected by step/heading and observations (Wi-Fi AP detection, a photograph of a sign, a Bluetooth beacon ping) as constraints. Mapping algorithms then align and optimize a global graph of all users' data, as shown in Figure 9.2. For example, Li et al. constructed a graph with nodes representing waypoints (pedestrian trajectory points) as well as landmarks (Wi-Fi APs), and edges representing observed relationships [2]. By running a SLAM optimization on this crowdsourced graph, the collective "venue map" is refined and yields much higher accuracy than any single-user trace. Academics have gone further by combining crowdsourced Wi-Fi with other sensors to create dynamic radio maps. One example is GraphIPS, which forgoes any preinstalled calibration and instead computes a radio map on the fly from users' walking data. As each smartphone user walks, GraphIPS uses the phone's inertial sensors to estimate relative movement (e.g., PDR) and collects Wi-Fi round-trip times or signal strength. Multiple users' data are fused in a graph-based algorithm with multidimensional scaling to solve for both the users' step positions and the Wi-Fi AP locations simultaneously. The result is an automatically generated Wi-Fi coverage map that can localize devices with accuracy comparable to laborious survey-based methods [6].

With the abundance of crowd data, AI techniques assist in interpreting and describing the indoor space. Machine learning models can classify

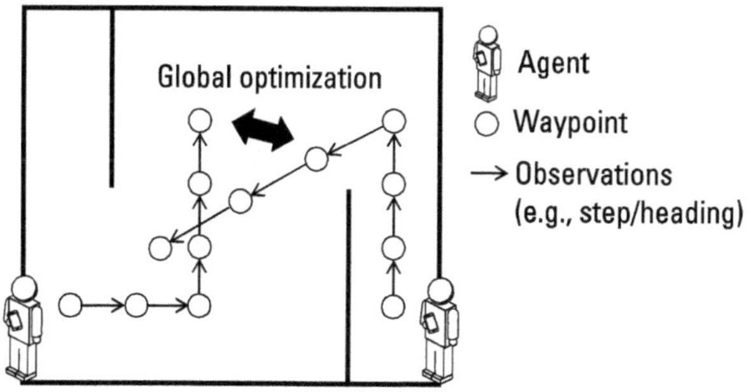

Figure 9.2 Illustration of a global graph optimization for multiple agents.

areas into semantic categories (corridor versus room versus open lobby) based on sensor patterns. For instance, a deep neural network might learn that a long sequence of similar magnetic readings and step counts likely indicates a hallway, whereas a cluster of unique Wi-Fi signals might indicate a particular room. Graph neural networks (GNNs) have been applied to indoor positioning graphs to improve localization accuracy by learning the structure of signal relationships [7, 8]. In essence, an AI "description" layer can augment the raw geometric map with semantic labels (such as "Room 101" or "North Entrance") and points of interest recognized from photos or sensor cues. This rich description makes the resulting map more useful for navigation (e.g., allowing queries like "find the nearest elevator"). As AI continues to improve, future crowdsourced mapping platforms may rely on neural networks to perform more of the mapping task autonomously, meaning crowd sensed data with minimal manual intervention. These complementary collection modes are visualized in Figure 9.3.

9.1.2 Applications: Wi-Fi Maps, and Visual Positioning Systems

Crowdsourcing has enabled two particularly influential indoor mapping applications: Wi-Fi radio map databases [6] and visual positioning systems (VPS) [9].

- *Crowdsourced Wi-Fi mapping:* An early example is Google's Floor Plan Marker app that launched in 2012 [10]. It invites users (especially venue owners or enthusiasts) to walk their building and collect Wi-Fi and cellular data [11]. As a user walks following on-screen instructions, the app passively scans for nearby Wi-Fi hotspots and measures cell tower signals and then uploads these to Google's servers [11]. Over time, as multiple users traverse an area, the system accumulates a dense signal map. Google reported that this crowdsourced approach

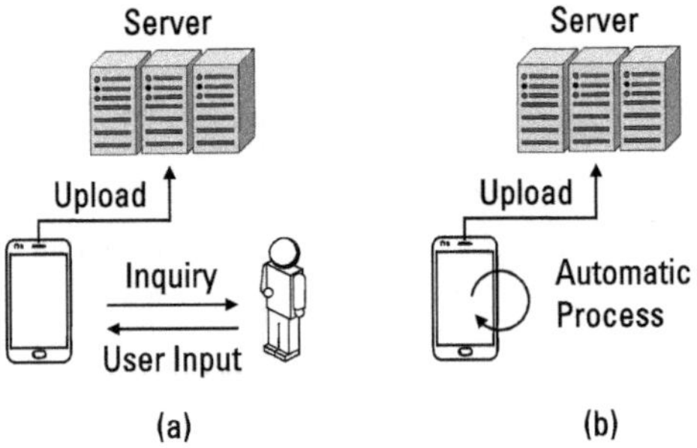

Figure 9.3 Illustration of a crowdsourcing database update in active manner (a) and passive manner (b).

helped to "fine-tune" indoor positioning in mapped venues by calibrating the geolocation of Wi-Fi access points [12]. Similarly, Apple uses its hundreds of millions of devices to continually scan for Wi-Fi networks; although primarily for outdoor location services, this data can improve indoor positioning when correlated with user-confirmed locations. In addition, companies such as Cisco have also introduced cloud systems that learn Wi-Fi RF patterns from device telemetry in large enterprise buildings, continually updating indoor location estimates without site visits.

- *VPS:* Crowdsourcing is also powering image-based indoor localization. A VPS uses a database of images or visual features of the environment, as the example that collecting using Mappedin App [13] shown in Figure 9.4; a user's smartphone camera feed is matched against this database to determine where the user is and in what direction he or she is facing. Building and maintaining the VPS image database are

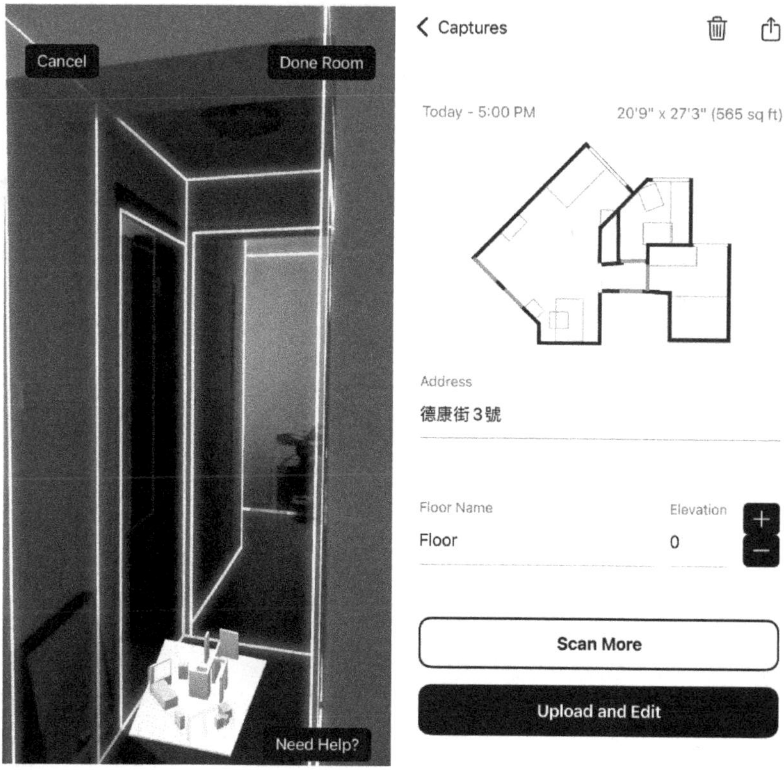

Figure 9.4 Indoor map generation by personal devices.

a massive crowdsourcing effort. Google's VPS, for example, leverages over 15 years of globally collected Street View imagery and supplements this with indoor photos [11]. Google's AR engineers have developed deep neural networks that identify stable visual features (e.g., edges, corners, or signage) in tens of billions of images and aggregate them into a 3-D point cloud map with trillions of points covering most countries. Initially, Google extended VPS to interiors of airports, malls, and transit stations through partnerships and user contributions. Retail partners provided scans of their aisles, and, for smaller venues, Google can crowdsource imagery from Local Guides or even AR game players. Every time users use the Live View AR navigation in Google Maps, for instance, their camera may capture new angles of a space that are fed back (with privacy controls) into the VPS dataset to refine it. An example is given in Figure 9.5.

Figure 9.5 Demonstration of Google's VPS [14].

In addition, Niantic (developer of Pokémon GO) likewise built a crowdsourced AR map. Players scan real-world locations with their phone cameras, uploading short videos that Niantic's map calls the Lightship Visual Positioning System. Niantic's VPS now spans many parks and city landmarks, enabling AR games to know exactly where a player is looking. Major tech companies (Apple, Meta) are also investing in AR cloud mapping, which is largely fueled by crowd data, to enable future AR glasses to instantly recognize the user's surroundings. In short, VPS turns the world's collective camera images into a massive indoor/outdoor positioning system.

Crowdsourcing does come with challenges: ensuring data quality (filtering out erroneous user data), preserving privacy (anonymizing contributors and sensitive imagery), and managing big data (Google's VPS processes trillions of points). Nonetheless, it has proven indispensable for mapping the complex, ever-changing indoor world. As crowdsourcing techniques mature, indoor maps will become richer (with semantic information and up-to-date changes) and available in any building where enough smartphone-toting people have gone before.

9.2 Collaborative Positioning

Traditional positioning and navigation techniques have been developed with a focus on single-user (agent) scenarios, where observations from a specific sensor are used exclusively to estimate the location of the corresponding agent. To improve positioning performance, it is common to increase either the number of observations (e.g., more access points) or the variety of observations (e.g., incorporating different types of measurements). This approach aims to enhance the capability to compensate for measurement errors. Beyond extending the observation dimension for a single agent, the concept of collaborative positioning has been proposed in recent years. This approach expands the observation dimension by incorporating inter-agent measurements from connected agents in close proximity, as illustrated in Figure 9.6.

In the twenty-first century, communication technology has undergone rapid development, significantly expanding bandwidth capacity. This expansion enables the transmission of large data (or measurement) packages either to a centralized server or between connected agents. In recent years, 5G mobile networks have been developed, aiming to provide data rates of up to 20 gigabits per second for civilian use. This advancement makes the transmission of large amounts of data with low latency feasible, thereby facilitating collaborative positioning in practical applications.

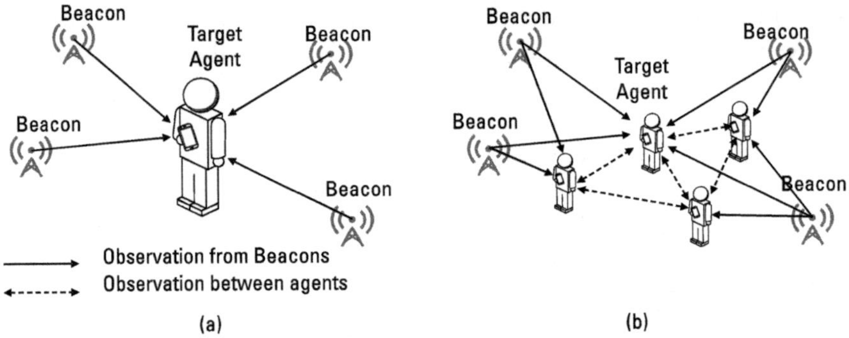

Figure 9.6 Single-agent positioning (a) and collaborative positioning (b).

In general, collaborative positioning utilizes observations shared by neighboring agents through network connections to estimate the position of a specific agent or to simultaneously optimize the positions of all connected agents. The early method of collaborative positioning was proposed in the field of robotics, where there was a demand for localizing a team of robots with limited observations [15]. With diverse observations from different agents, positioning estimation can achieve greater accuracy and robustness compared to the conventional standalone agent positioning method [16]. The concept of collaborative positioning has since been extended to various technologies, such as GNSS positioning in outdoor environments [17], and signals of opportunity (SOPs) for GNSS-denied environments [18], among others. This technique is particularly beneficial for indoor positioning and navigation, which are often employed in environments with severe observation limitations, including issues related to measurement availability, geometry, and accuracy [19]. Collaborative positioning techniques for indoor positioning and navigation can be categorized by system architecture, employed technologies, applied techniques, and estimation methods. A comprehensive classification of these techniques can be found in [20].

9.2.1 System Architecture: Centralized and Decentralized Collaboration

In terms of network composition, collaborative positioning can be categorized into two types of system architectures: centralized collaboration and decentralized collaboration, as illustrated in Figure 9.7.

In the centralized collaboration architecture, all agents in the network connect to a master unit, typically a central server. Each individual agent

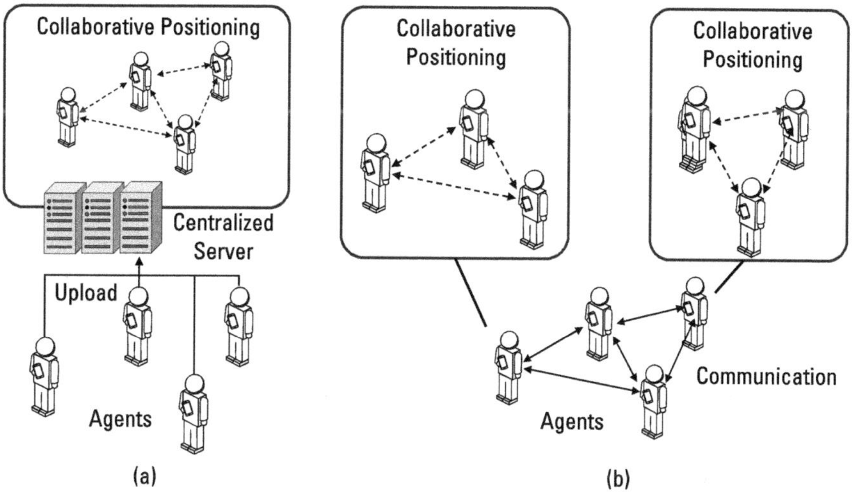

Figure 9.7 Centralized collaboration (a) and decentralized collaboration (b).

transmits its observations to this central server. The server is responsible for receiving and processing these observation data, which includes tasks such as synchronization, estimation, and analysis. The collaborative positioning algorithm is applied by the central server once the estimation network is determined and the measurements are ready. The central server focuses not on a specific agent but responds to all connected agents. Therefore, centralized collaborative positioning usually optimizes the positions of all involved agents simultaneously, as follows:

$$\mathbf{X}_{\text{agents}} = f_{\text{CCP}}\left(\mathbf{Z}_{\text{agents}}\right) \tag{9.1}$$

where $\mathbf{Z}_{\text{agents}} = [\mathbf{z}_1 \quad \cdots \quad \mathbf{z}_n]$ is the vector consisting of the observation vectors from each connected agent n, within the network, meaning $n \in$ network. This vector is applied with the centralized collaborative positioning algorithm, f_{CCP}, to estimate the $\mathbf{X}_{\text{agents}} = [\mathbf{x}_1 \quad \cdots \quad \mathbf{x}_n]$ consist of the locations of all the agents. Once their positions are estimated, the solution for each agent is fed back through their individual connections. As all decisions are made by the central server, its capacity significantly affects the performance of collaborative positioning. A major benefit of this system is the transfer of the computational load to a master unit, which typically has greater flexibility in terms of cost, space, and power. Additionally, a centralized service is generally more trustworthy in terms of user privacy, as it is less likely to disclose the location of individual users to third parties. However, the centralized architecture also has a significant drawback: long-range communication links may degrade connectivity and increase communication latency. One notable existing application that employs centralized collaborative positioning is the use of cobots (collaborative robots) in constrained environments, such as warehouses. For example, Amazon has developed and deployed the mobile robot, Titan, which collaborates with multiple robots to reduce repetitive motions for cargo transportation in its fulfillment centers [21]. In these applications, the robots communicate with each other to avoid collisions and optimize their tasks in terms of both time and cost efficiency.

In the decentralized collaboration architecture, the target agent establishes communication links with neighboring agents to form a local network [22]. These neighboring agents respond to the target agent and share their observations related to the target. After receiving these external observations, the collaborative positioning algorithm is applied on the target agent, along with its own observations, in order to estimate the target agent's position, as follows:

$$\mathbf{x}_{\text{target}} = f_{\text{DCP}}\left(\mathbf{Z}_{\text{agents}}\right) \tag{9.2}$$

where $\mathbf{x}_{\text{target}}$ is the position of the target agent, determined by applying the decentralized collaborative positioning algorithm f_{DCP}. Such a decentralized architecture offers the benefit of communication efficiency, where the position solution can be directly utilized by the target agent without the need for additional communication. Moreover, communications between agents are typically within close proximity, offering higher flexibility in communication methods and facilitating timely solutions. However, this architecture has its drawbacks. The estimation process, which involves a large computational load, is handled by a specific agent with limited resources. This limitation can reduce the capacity for algorithm complexity as well as the size of the network. Additionally, data privacy is a concern in practice, as the location information of other agents is disclosed to the target agent. The most common application of the decentralized method is in autonomous systems operating over wide areas, such as fleets.

The system architecture significantly influences the setup of collaborative positioning, both in terms of hardware and software. These aspects should be decided based on the specific needs of the application, the operating environment, and the available resources.

9.2.2 Technologies and Techniques Using Collaborative Positioning

In terms of technologies, specifically the observation types from different sensors, collaborative positioning can be classified into two categories: homogeneous networks and heterogeneous networks [24], as illustrated in Figure 9.8.

In a homogeneous network, all agents are assumed to use identical setups, encompassing both the hardware and the technology employed. The entire collaboration system focuses on a specific kind of measurement (or feature), with the observation extension primarily concerning the number of observations. The trustworthiness of measurements is easier to ascertain in a homogeneous network, as each agent can be regarded as identical.

However, in practical collaborations, it is often difficult to ensure identical performance or characteristics for each agent. The heterogeneity of the collaboration network introduces additional challenges in estimation. However, this diversity in technologies, where each agent provides various types of observations, enhances the diversity that contributes to more accurate estimations. The combination of different sensor measurements from various agents can be complementary. For instance, integrating infrared and visible light measurements [25], UWB with pedestrian dead reckoning (PDR) [26], or acoustic measurements with PDR [27], is particularly beneficial in indoor environments where single-sensor performance is limited. It is important to note that collaboration in a heterogeneous network may require additional knowledge about the reliability of observations and a proper integration framework for effective estimation.

In the context of indoor positioning and navigation techniques, the architecture of collaborative positioning offers a greater variety of applications. Collaboration can occur between agents using the same type of technique, for example, integrating RSS ranging with TOA ranging techniques [28].

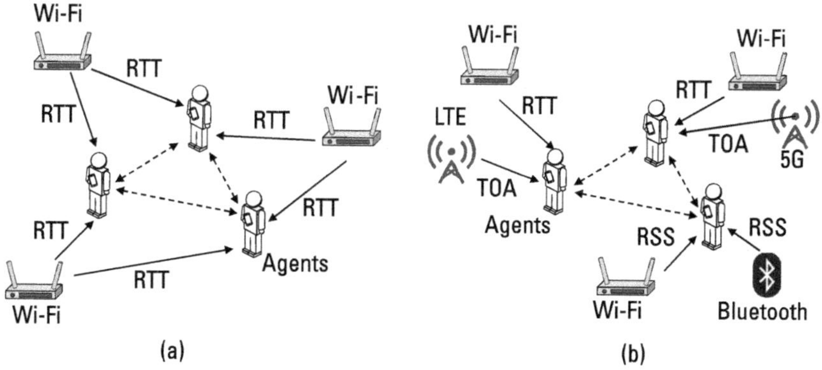

Figure 9.8 Homogeneous network (a) and heterogeneous network (b).

Alternatively, collaboration can involve agents employing complementary techniques, such as combining AOA with ranging measurements [29].

Furthermore, collaboration between different agents improves the spatial diversity of observations, enabling new features for advanced applications. For instance, signal propagation geometry is location-dependent and varies among connected agents. This means that the time of arrivals for signals will differ based on their AOA among different agents. Such a property can be utilized to form a collaborative network akin to an antenna array, providing additional functions such as line-of-sight (LOS) signal detection [30]. Additionally, the diversity in signal reception geometry among agents can offer features to distinguish measurements from abnormal sources, thereby aiding in spoofing detection [31, 32].

In general, collaborative positioning extends observations by diversifying technologies and techniques across multiple agents. This approach not only offers benefits beyond traditional single-agent positioning methods but also introduces new functions for advanced applications.

9.2.3 Estimation Method for Collaborative Positioning

After selecting the appropriate technologies and techniques, the position of the agent needs to be determined by applying various estimation methods. These methods are based on the system model that correlates observations with the agent's position as the unknown variable. Among the various estimation methods, the most popular ones applicable to collaborative positioning include multidimensional scaling [33], LS [28], Kalman filter [27], and FGO [34].

Multidimensional scaling (MDS) has been widely used in data analysis, particularly for data dimension reduction, bearing similarities to principal component analysis. MDS is capable of visually representing data in a 2-D or 3-D graph by mapping similarities or dissimilarities into Euclidean distances within a defined space. In the context of collaborative positioning, MDS aims to determine the coordinates of each agent (e.g., in a 2-D Euclidean space) to match the observed Euclidean distances between pairs of agents as closely as possible. Because the method only constrains inter-agent distances, the set of agent coordinates can be translated, rotated, or mirrored while preserving their relative geometry. Consequently, coordinates derived from MDS require further calibration using the absolute positions of several anchor agents. Generally, collaborative positioning using MDS involves three steps: (1) constructing a similarity matrix incorporating the relative distance observations between pairs of agents, (2) decomposing this matrix to obtain new space coordinates based on eigenvectors and corresponding eigenvalues, and (3) aligning the resulting agent coordinates with the desired coordinate system through translation, rotation, or mirroring, based on the absolute positions of selected anchor agents. A comprehensive introduction and mathematical derivation of MDS are available in [35].

The LS estimation method is a popular and efficient tool for solving overdetermined problems, fitting well with the collaborative positioning task where the number of inter-agent measurements usually exceeds the number of agent coordinates to be estimated. Let $\mathbf{x}_{\text{agents}}$ be the vector that incorporates the agent positions to be estimated and $\mathbf{z}_n$ represent the available observations between different pairs of agents. By examining the geometrical relationship (e.g., angular relationships [28]) between agents and observations, a transformation $\mathbf{F}$ can be formulated to map coordinates from the agent position space to the measurement space. The objective is to find $\mathbf{x}_{\text{agents}}$ that, under the defined relationship $\mathbf{F}$, most closely matches the measurements $\mathbf{z}_n$. This can be conceptualized as minimizing the difference between $\mathbf{F}\mathbf{x}_{\text{agents}}$ and $\mathbf{z}_n$. This objective can be transformed into minimizing the sum of squared errors and solved by forming an LS estimation task.

Most collaborative positioning tasks involve continuous estimations over a certain time span, which introduces additional time-related correlations. Consequently, the Kalman filter framework can be employed in collaborative positioning by extending the state and observation to encompass multiple agents. The system matrix must also be updated to reflect the geometrical relationship between different agents in conjunction with their observations. The trustworthiness of different agents (or their corresponding measurements) can be addressed through a carefully designed covariance matrix. The positions of the agents can then be determined by applying standard Kalman filter estimation procedures. Similar extensions can be made with FGO, which has gained popularity as an estimation method in recent years. In this approach, different agents are represented as nodes, while observations act as constraints (edges) between these nodes (agents). The positions of agents are then determined by optimizing the cost function, seeking nodes that align with the edge constraints. A significant advantage of FGO in collaborative positioning is its flexibility in adding and removing nodes, making it well-suited for dynamic networks in practical applications.

Numerous variants of the aforementioned estimation methods have been developed and employed for collaborative positioning. The selection of a suitable method should be based on the specific requirements of the application. A systematic approach to evaluate the performance of different estimation methods is to analyze the Cramer-Rao lower bound. This analysis quantitatively provides an indication of the best achievable performance for various methods.

9.2.4 Applications

Collaborative positioning has actually moved into mainstream products, exploiting the sheer density of connected devices plus short-range precision radios. The applications include:

- *Finding lost items and devices (crowd networks):* Apple's Find My and Samsung/Tile analogs turn hundreds of millions of smartphones into passive "finder nodes." When a node hears an opt-in Bluetooth beacon from a lost phone, laptop, or tag, it relays the beacon's identification plus its own GNSS fix to the cloud, end-to-end encrypted so neither Apple or Samsung nor the passer-by learns who owns the item [36]. Because even powered-off iPhones and offline Galaxy devices emit rotating BLE identifications, the mesh can localize trackers that have no GPS or data plan.
- *Precision finding with UWB:* Short-range UWB chips (e.g., Apple U1, Samsung UWB) exchange nanosecond pulses to measure time-of-flight and achieve ≈10-cm accuracy indoors [37]. In "Precision Finding," an iPhone multilaterates its AirTag (or a Galaxy phone its SmartTag+) and overlays an arrow and distance in AR. The same ranging lets two phones prioritize the closest peer for Nearby Share or trigger context-aware actions such as hands-free car unlocking.
- *Team localization:* Collaborative ranging is equally useful for people. Firefighters or factory staff can form an ad hoc UWB/BLE mesh: edge units that still see GNSS seed a relative pose graph so that every teammate retains situational awareness even deep indoors. Consumer apps are experimenting with similar meshes for friend-tracking at concerts and stadiums.
- *Large-scale device networks (smart cities and venues):* Samsung's SmartThings Find surpassed 300 million opt-in Galaxy devices within 2 years, combining BLE for coarse crowd detection and UWB for near-field guidance [38]. Both Apple and Samsung rotate beacon identifications and encrypt location updates to curb stalking abuse, illustrating how privacy engineering is integral to collaborative networks. By fusing dense BLE measurements with subdecimeter UWB ranges, crowd

networks already deliver capabilities unattainable by any single sensor. As Bluetooth 5.4 AOA, FiRa-certified UWB, and 6G integrated sensing mature, virtually every phone, vehicle, and IoT node will contribute to (and benefit from) a global, privacy-preserving indoor positioning.

9.2.5 Limitations in Practice

Despite being developed and studied for over 20 years, collaborative positioning has seen very limited implementation in practice due to several unresolved practical issues. One major concern is the computation load required for collaborative positioning [39]. In a centralized collaboration architecture, the central server often has the capacity to handle the significant computational load during estimation, especially in a large network. However, in a decentralized collaboration network, the agent is typically limited by space and cost constraints, resulting in a reduced capacity to handle the computational demands of a large network. The computation load increases in a squared manner with the number of agents, as collaborative positioning usually involves pairwise observations between agents. Consequently, many collaborative positioning systems have an estimation complexity of $O(n^2)$. Moreover, these systems often employ observational diversity in geometry to enhance estimation performance, which may only converge at a certain network size. Therefore, a trade-off must be considered between power consumption and performance enhancement. Advanced algorithm design can also contribute to reducing the complexity of estimation [40].

Another significant issue is the performance of the estimation algorithm, particularly concerning estimation convexity [41]. As observations in collaborative positioning are extended across the agent space, the convexity of the estimation task can be diminished compared to single-agent tasks. Furthermore, many collaborative positioning systems do not use anchors, necessitating the estimation of all agents' positions, akin to a SLAM task with a large number of unknowns [41]. The observability of the estimation task is another critical factor that should be considered before implementing collaborative positioning, as it can lead to significant degradation in estimation accuracy [42].

Last but not least, the feasibility of applying collaborative positioning in specific environments also warrants consideration. For instance, radiolocation-based collaborative positioning methods can experience significant degradation due to signal obstructions and reflections in environments dense with obstacles [43]. These types of errors can be detrimental to methods developed for normal environments. Therefore, additional algorithmic considerations are necessary, such as implementing agent selection beforehand [33], or incorporate the multipath effect during the positioning estimation [44].

9.3 Data-Driven PDR and Human Pose Estimation

We will explore how AI transforms inertial PDR and sparse-sensor human pose estimation. This chapter introduces the core idea behind data-driven personalization, drift compensation, and transfer learning for robust PDR and then details deep-learning frameworks that reconstruct full-body kinematics from minimal wearables, merging biomechanics with continual learning to enable pervasive motion-tracking systems.

9.3.1 AI for Motion Data Analysis

Modern motion tracking leverages AI and deep learning to automatically extract meaningful patterns from raw inertial sensor data. Traditional signal processing for IMUs required expert-designed features (e.g., step detection thresholds or handcrafted filters). In contrast, deep learning can learn rich representations directly from raw signals, eliminating the need for manual feature engineering. For example, a deep network can take accelerometer and gyroscope time series and internally learn features corresponding to footsteps, turns, or other motion signatures. This data-driven approach has significantly boosted performance and robustness in tasks such as human activity recognition and inertial navigation. The AI-based motion data analysis framework is illustrated in Figure 9.9. Key AI architectures that are used in motion data analysis include feed-forward deep networks, CNNs, RNNs (notably LSTMs), transformers, and emerging self-supervised learning strategies.

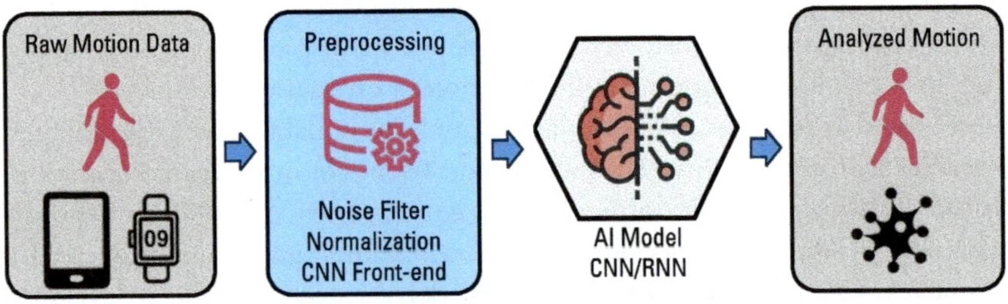

Figure 9.9 The framework of AI-based motion data analysis.

CNNs learn local patterns directly from IMU streams. One-dimensional filters slide across raw accelerometer/gyroscope data to pick up step cycles or abrupt motion changes, and stacked layers turn these low-level traces into high-level motion events [45]. When signals are mapped to spectrograms, 2-D CNNs exploit frequency-time structure just as effectively. RNNs, especially LSTM units, excel at modeling temporal context. Their hidden state integrates information over many samples, enabling accurate velocity, displacement, and activity estimates [46]. A common design first extracts per-window CNN features and then feeds them to an LSTM for sequence-level inference. Transformers replace recurrence with self-attention, processing an entire window in parallel while weighing each time step's relevance [47]. Hybrid blocks combine a local CNN front-end with transformer layers, capturing both fine-grained footfall cues and long-range gait rhythm [48]. Self-supervised learning (SSL) further reduces the need for labeled data. By masking or permuting segments and training the network to reconstruct them, large unlabeled sensor logs yield rich representations that transfer to downstream tasks with minimal annotation [48].

Together, CNNs for feature extraction, RNN/Transformer modules for temporal reasoning, and SSL for representation learning form a streamlined pipeline that converts raw motion signals into actionable insights such as activity labels, step counts, or trajectory estimates, all with little manual engineering [49].

9.3.2 AI for PDR

Traditional PDR algorithms face challenges such as sensor bias, varying user gait, and cumulative error, as mentioned in Chapter 6. AI-driven PDR systems address these challenges through personalization, adaptive modeling, and learned error correction, as illustrated in Figure 9.10. A key advantage of AI is the ability to personalize the motion model to each user. Every individual has unique movement characteristics (e.g., stride length, walking speed, device carrying style). Instead of using one-size-fits-all parameters, a data-driven PDR can learn a custom model for each profile. For example, a network can infer a user-specific step length model from a few minutes of that person's walking data, adapting to whether the user is a long-legged adult or a small child. Similarly, models can adjust for context. For example, if an IMU is worn on the belt versus in a pocket, the AI can recognize the pattern shift and recalibrate its outputs accordingly. This adaptability extends PDR to diverse users and device placements without requiring extensive manual calibration.

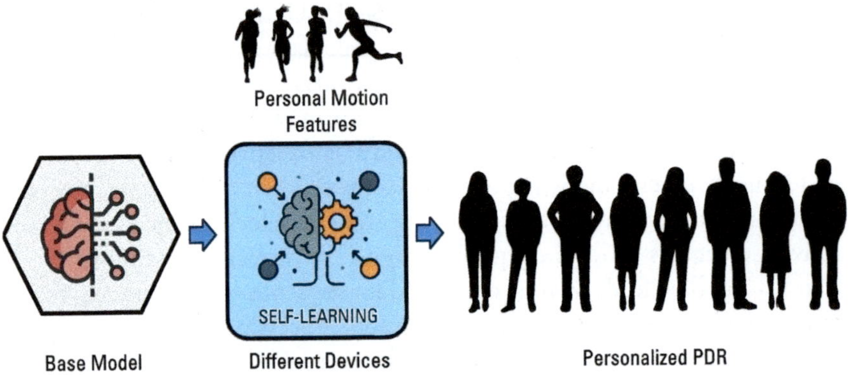

Figure 9.10 The personalized PDR.

AI could also tackle drift and calibration in PDR with modules that learn error behavior instead of relying on fixed heuristics. For foot-mounted INS, the key is to fire a ZUPT only when the shoe is truly stationary. Wagstaff et al. replaced the classical SHOE test with an LSTM classifier that directly recognizes zero-velocity events from raw IMU data. By fusing its outputs in an EKF, they achieved better accuracy during walking, running, and stair-climbing [50]. End-to-end inertial odometry nets (IONet, RoNIN, iMoT) segment the IMU stream into short windows and learn the 2-D/3-D displacement of each window with CNN, RNN, or Transformer. By concatenating the window outputs, the navigation state is effectively reinitialized every few hundred milliseconds, preventing unbounded integration drift on smartphones carried in arbitrary positions [51–53]. Learning-based stride-length estimators personalize step length and heading to each walker and handset. The EL-SLE network combines CNN, LSTM and an online-learning head; after a few dozen strides it adapts to the user/device across heterogeneous phones and gaits [54]. In effect, the AI layer becomes a continuously learning error-compensator that keeps PDR accurate across users, walking speeds, and sensor placements.

A major future goal is achieving these improvements without extensive labeled data for every new user or environment. Transfer learning and domain adaptation techniques are instrumental here. A model trained on a broad population can be fine-tuned with a small amount of data from a new user to personalize the PDR for them. For example, adapting a pretrained step length predictor to a new walker's style with just a few calibration walks. Cross-domain adaptation is used to handle changes like new sensor positions or different motion types (walking vs. running) by adjusting the model rather than retraining from scratch. Generative approaches (e.g., using GANs) have even been explored to synthesize motion data for unseen users, helping networks learn invariances and reduce the burden of collecting real user data. The importance of transfer learning is highlighted by the fact that collecting extensive inertial data for every individual or context is impractical. Thus, these models must generalize and adapt with minimal new data. Another promising direction is continual learning, where the PDR model incrementally updates itself as it receives new data during daily use. Over time, the system "learns" the user's habits or any sensor drifts (for example, if a shoe sensor's orientation slowly changes, the model can adjust). This lifelong learning capability ensures sustained accuracy without manual re-calibration sessions. By incorporating these AI capabilities, we could expect the next-generation PDR systems are more robust and accurate.

9.3.3 Reconstructing Human Pose from Wearables

Another frontier in motion tracking is full-body human pose estimation using only a few wearable sensors. Traditional motion capture requires many sensors or cameras to capture each limb's movement as a VICON system. Now, researchers work on developing AI-enabled sparse-sensor motion capture, estimating a person's whole-body pose with just a smartwatch on the wrist and smart glasses or earbuds on the head. This is achieved through intelligent sensor fusion and learned kinematic models. Even with limited IMUs, advanced algorithms can infer unmeasured joint angles by leveraging patterns learned from large motion datasets. For example, by observing the motion of one arm and the head, a model can predict the likely pose of the other arm and legs during walking, since human gait has rhythmic, coordinated patterns. The AI essentially fills in the gaps in sensor coverage, reconstructing a full skeleton pose in real time.

A typical system for sparse IMU-based pose estimation, as illustrated in Figure 9.11, uses two to six wearable nodes placed on strategic body parts, such as the wrists, ankles, trunk, and head. Deep learning models, often sequence-to-sequence networks or autoencoders, process the streams from these few sensors to estimate the angles of all major joints. A recent study demonstrated predicting upper-body or full-body kinematics from only 3–6 IMUs, with average angular errors around 10° to 15° per joint [55]. The approach combined a transformer-based sequence model with a training dataset of full motion-capture data, using self-supervision to learn the correlations between the motion of instrumented segments and the rest of the body. The result is a plausible pose sequence, even during unscripted daily activities, using an order of magnitude fewer sensors than a conventional motion capture suit. Crucially, these models could integrate physical knowledge. For example, knowing that if the wrist swings forward, the opposite leg likely swings back (to conserve momentum), the network can infer leg motion from arm motion patterns.

Sensor fusion and kinematic constraints are key to making this work. AI models incorporate not just learned patterns but also basic physics and human anatomy constraints to ensure realistic poses. One strategy is combining data-driven predictions with a physics-based optimizer. For example, MobilePoser is a system that takes any subset of common consumer IMUs (from a smartphone, smartwatch, and earbuds) and produces full-body pose estimates [56]. It uses a multistage deep neural network to estimate limb orientations and then a physics-based motion optimizer to refine these estimates and ensure consistency in global translation. By blending neural predictions with biomechanical constraints (such as joint limits or ground contact assumptions), such systems achieve accurate and physically plausible motion reconstruction. In practice, this means that even if you only wear a couple of everyday devices, AI can track your overall posture. The network might first guess your likely skeleton configuration from the partial data, and then an optimization step adjusts the pose so that, for instance, feet remain on the floor and limbs lengths stay constant.

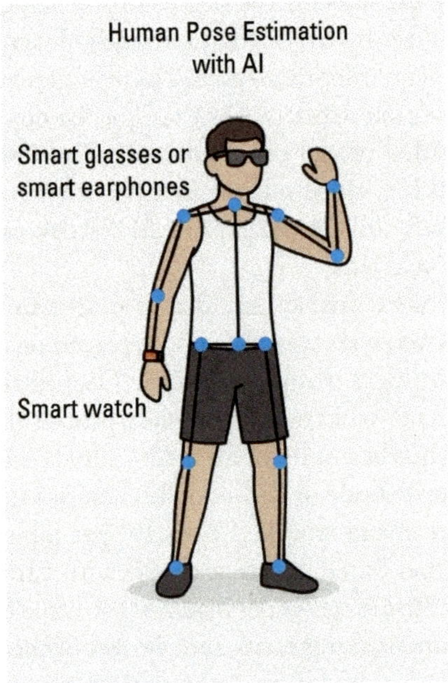

Figure 9.11 Human pose estimation with AI using sparse IMU data.

The applications of sparse wearable pose estimation are wide-ranging. In indoor navigation and AR/VR, a few wearables can provide a user's full pose for avatar animation or intuitive control in virtual environments. For instance, indoor AR navigation could overlay directions on smart glasses while a wrist IMU helps to determine if the user is crawling, walking, or gesturing, enabling context-aware guidance. In gaming and fitness, minimal sensors let users see their movements mirrored in-game or get real-time coaching on exercise form. Rehabilitation and healthcare stand to benefit as well. For example, a patient who had a stroke could wear a smartwatch and ankle monitor, and AI would reconstruct his or her gait to assess improvements in symmetry and balance, without needing a full gait lab. The feasibility of these scenarios has been demonstrated by systems achieving near-real-time pose tracking with just consumer devices. For example, a combination of two smartwatches (one per wrist) plus a head-mounted sensor has been used for lightweight motion capture in daily settings [57]. As deep learning models continue to improve, we may see single-sensor pose estimation become feasible.

9.4 Radio Sensing: Communication and Sensing

Beyond the RF use in communication and positioning, radio signals themselves can act as sensors of the environment. This concept of radio sensing leverages wireless signals (Wi-Fi, cellular, radar) to detect and interpret what is happening in a space, such as the presence of people, their movements (even gestures), and changes in the physical environment. With advances in high-frequency RF and antenna arrays, the line between communication and sensing is blurring. In 6G visions, the slogan is "communication as a sensing system": infrastructure simultaneously provides data service and scans the environment like a radar. In indoor positioning, radio sensing can complement or even replace dedicated sensors for mapping and tracking. This section will briefly introduce the concept of radio sensing and discuss a few applications.

9.4.1 Concept of Radio Sensing and Integrated Sensing

Radio waves are affected by the environment they propagate through. By analyzing characteristics of radio signals (such as signal strength, phase, time-of-flight, frequency shifts), we can infer properties of the environment. Radio sensing entails using standard communication signals to gather information. Crucially, wireless signals could transmit through obstacles that light cannot. For example, Wi-Fi signals go through some thin walls. Researchers have exploited this to "see" through walls using Wi-Fi by capturing the reflections from a person behind the wall [58]. This is basically a bistatic radar using Wi-Fi frequencies.

Technically, the integrated sensing and communication might work as the following example. A Wi-Fi AP sends normal data frames, but the receiver also captures the Channel State Information (CSI), which is essentially a channel "fingerprint" (introduced in Chapter 5). Changes in CSI over time indicate movement of objects. For example, if a person walks in the room, his or her body reflections will alter the multipath pattern seen in CSI. By applying signal processing to CSI data, the system can detect events (like someone walking in) and even classify specific motions or identities (if combined with machine learning algorithms). Similarly, phased array antennas in 5G/6G base stations can sense the AOA of echoes, forming a sort of RF image of the space.

Briefly speaking, the concept is to treat radio waves as a source of sensory data about the physical world, sometimes called integrated sensing and communication (ISAC) in research [59]. A 6G base station might not only give your phone a data link but might also continuously scan like lidar/radar enabling SLAM for many potential applications.

9.4.2 Applications of Radio Sensing

At the environmental scale, dense Wi-Fi or UWB links act like the beams of a tomographic scanner. MIT's Wi-Vi and its successor RF-Pose push this further [60], using antenna arrays and phase-cancellation tricks to image and track people through walls. Their capabilities are prized for search-and-rescue or smoke-filled fire grounds where optical SLAM collapses.

Researchers repurposed Wi-Fi CSI to estimate crowd density, flag breaches of social-distancing thresholds, and verify that quarantined individuals remained within their assigned rooms. This application was pushed by the COVID-19 pandemic [61]. It achieves remote compliance checks without cameras or intrusive wearables. The same signal analysis now informs smart-building dashboards that throttle HVAC or issue congestion alerts during outbreaks [62].

Health monitoring may prove even more transformative. Submillimeter chest movements modulate reflected Wi-Fi or FMCW radar waves, letting systems like Vital-Radio log respiration and heart rate unobtrusively. Sleep-tracking smart displays extend the idea, classifying coughs, snores, and restlessness for nightly wellness reports. In elder care, continuous CSI analysis catches gait slowdowns or the telltale "thud and stillness" of a fall, summoning help when the occupant cannot.

In the context of indoor positioning, a challenge is how to fuse radio sensing outputs with other indoor positioning. For instance, if a Wi-Fi based system "senses" a person's motion, how does it integrate that with that person's smartphone PDR or BLE position? Possibly the AI network could treat sensed objects as additional "virtual devices" and run a joint tracker? This is an underexplored territory.

9.5 Reconfigurable Intelligent Surface

A reconfigurable intelligent surface (RIS) is a thin panel of thousands of programmable "meta-atoms" that impose controllable phase (and sometimes amplitude) shifts on impinging radio waves [63]. By coordinating those tiny elements, the surface can beam-form reflections, bend signals around obstacles, or create an artificial multipath. In effect, the radio channel itself becomes a software-defined asset, making RIS a flagship technology for 6G communications [64] and centimeter-level indoor positioning [65].

As illustrated in Figure 9.12, a controller programs the RIS to redirect the transmitter's signal along a precisely known angle toward the handset, creating a strong, "labeled" reflection in addition to (or instead of) the weak direct path. That extra, geometry-controlled path supplies the receiver with a distinct time-of-flight or AOA measurement [66]. These measurements tighten multilateration and yielding centimeter-level indoor positioning accuracy even when the true LOS is blocked. For indoor positioning this capability is

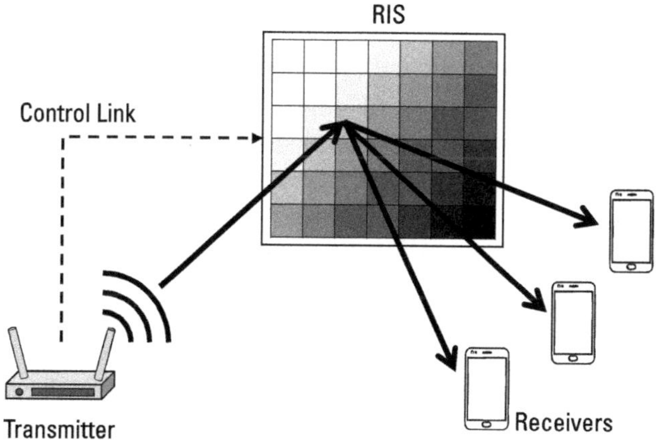

Figure 9.12 Demonstrate of a RIS can dynamically manipulate reflections (via a controller) to enhance coverage and localization accuracy indoors.

transformative. A factory, warehouse, or shopping mall hung with a handful of panels can steer existing 5G, Wi-Fi RTT, or UWB signals into corners that would otherwise be radio dead zones, ensuring that devices always have a strong ranging path. By cycling through known reflection patterns, a single surface can behave like several virtual anchors. Thus, the moment a particular pattern produces the strongest return. The receiver learns its angle or distance to the panel, even when the true LOS is blocked. Early laboratory prototypes (a 1.3-m^2, 4,096-element surface is a notable example) have already slashed 2-D errors by an order of magnitude under such NLOS conditions.

However, there are many challenges ahead [66]. Tens of thousands of elements must be wired or wirelessly addressed and their phase states refreshed within milliseconds, yet today's control buses and optimization algorithms are only beginning to cope with that scale. Standards offer little guidance on how a handset distinguishes an RIS-reflected path from a direct one, or how multiple panels should coordinate when several users demand incompatible beams. Cost, calibration, and physics add further friction. Despite these hurdles, the momentum to push RIS forward is strong. EU programs such as RISE-6G, telecom vendors from Nokia to Ericsson, and metamaterial start-ups alike are moving RIS from theory to field trials. If the technology clears the engineering and economic barriers, smart panels could become as commonplace as Wi-Fi routers, silently converting troublesome multipath into a cooperative ally and opening a new, programmable chapter in indoor navigation.

9.6 Conclusions

Indoor positioning and navigation techniques have advanced dramatically, evolving from single-sensor methods into sophisticated, integrated systems. In this chapter, we explored cutting-edge approaches that leverage crowdsourced data, device collaboration, artificial intelligence, and novel radio hardware to tackle the challenges of indoor localization. A recurring theme is aggregation of information, for example, multiple devices cooperating (crowd mapping, collaborative networks such as Apple's Find My), multiple algorithms' fusion (AI enhancing classical PDR, graph optimization merging many traces), and multiple sensors integration (cameras with IMUs, Wi-Fi with vision). The result is a new generation of indoor positioning systems that are far more accurate, resilient, and scalable than what was possible a decade ago. As we can see now, major technology standards bodies are clearly invested in this future. IEEE 802.11 and Bluetooth have added fine-ranging and sensing capabilities to enable meter-level and even centimeter-level positioning. 5G

NR has built-in positioning reference signals and is demonstrating submeter accuracy in industrial settings, with 6G targeting even integrating sensing surfaces. These developments mean that the infrastructure of tomorrow's wireless networks will double as indoor localization infrastructure, blurring the line between communication and navigation systems.

This chapter also highlighted how AI and data-driven approaches advanced techniques, including optimizing crowdsourced maps, to AI-based INS that learns human motion patterns. AI is crucial in handling the complexity of indoor environments (with their irregular layouts, dynamic changes, and diverse signal behaviors). At the same time, two important aspects are getting more attention: (1) the need to address privacy (e.g., ensuring crowdsourced mapping data or RF sensing does not violate individuals' privacy), and (2) the security (e.g., preventing spoofing of collaborative networks or RIS configurations).

In conclusion, the indoor positioning field is moving toward "ambient intelligence," meaning that the environment (through sensors, surfaces, and networks) actively assists in localization (this is actually what enables human-centric navigation). At the same time, personal devices become multisensor hubs that intelligently fuse whatever information is available. This improvement can finally lead to the fusion of indoor and outdoor positioning into seamless user experiences. As you walk from outside (GNSS) to inside (Wi-Fi/UWB), your phone or wearable device will smoothly transition between systems. The major companies, standards, and research communities are aligned in this vision. In the foreseeable future, we will likely see the gap between indoor and outdoor positioning not just narrow but effectively disappear as technology achieves a unified global localization service, indoors, outdoors, and hopefully everywhere in between.

References

[1] Zhou, B., et al., "Crowdsourcing-Based Indoor Mapping Using Smartphones: A Survey," *ISPRS Journal of Photogrammetry and Remote Sensing*, Vol. 177, 2021, pp. 131–146.

[2] Li, C., et al., "Crowdsourcing-Based Indoor Semantic Map Construction and Localization Using Graph Optimization," *Sensors*, Vol. 22, No. 16, 2022, p. 6263, https://www.mdpi.com/1424-8220/22/16/6263.

[3] "Indoor Mapping Data Format," Apple Inc., 2021, https://register.apple.com/resources/imdf/.

[4] "Maps SDK for Android," Google for Developers, 2025, https://developers.google.com/maps/documentation/android-sdk/overview.

[5] Chang, A., "Inside Google's Fascinating Stash of 10,000 Indoor Maps," *WIRED*, 2012, https://www.wired.com/2012/07/17-organizations-that-made-their-floorplans-public-on-google-maps/.

[6] Zhao, Y., et al., "GraphIPS: Calibration-Free and Map-Free Indoor Positioning Using Smartphone Crowdsourced Data," *IEEE Internet of Things Journal*, Vol. 8, No. 1, 2021, pp. 393–406.

[7] Vishwakarma, R., R. B. Joshi, and S. Mishra, "IndoorGNN: A Graph Neural Network Based Approach for Indoor Localization Using WiFi RSSI," in *Big Data and Artificial Intelligence*, V. Goyal, et al., (eds.), New York: Springer Nature Switzerland, 2023, pp. 150–165.

[8] Etiabi, Y., et al., "MetaGraphLoc: A Graph-Based Meta-Learning Scheme for Indoor Localization Via Sensor Fusion," arXiv preprint, 2024, https://arxiv.org/abs/2411.17781.

[9] Prisacariu, V., and P. Fite-Georgel, "Lightship VPS: Building Our 3D Map From Crowdsourced Scans," Niantic, https://nianticlabs.com/news/vps-part-2?hl=en.

[10] "Google Announces Google Maps Floor Plan Marker App," *GPS World*, 2012, https://www.gpsworld.com/lbsindoor-positioningnewsgoogle-announces-google-maps-floor-plan-marker-app-12863/.

[11] "ARCore Developer Documentation," Google Developers, 2024, https://developers.google.com/ar/develop.

[12] Dupont, C., "Help Improve the 'Blue Dot' for People Visiting Your Venue," Google Blog, 2012, https://maps.googleblog.com/2012/04/help-improve-blue-dot-for-people.html.

[13] "Indoor Positioning," Mappedin, https://www.mappedin.com/indoor-positioning/.

[14] Shankar, B., "Google Reveals New Visual Positioning System for Maps," 2018, https://mobilesyrup.com/2018/05/08/google-visual-positioning-system-maps/.

[15] Mourikis, A. I., and S. I. Roumeliotis, "Performance Analysis of Multirobot Cooperative Localization," *IEEE Transactions on Robotics*, Vol. 22, No. 4, 2006, pp. 666–681.

[16] Patwari, N., et al., "Locating the Nodes: Cooperative Localization in Wireless Sensor Networks," *IEEE Signal Processing Magazine*, Vol. 22, No. 4, 2005, pp. 54–69.

[17] de Ponte Müller, F., "Survey on Ranging Sensors and Cooperative Techniques for Relative Positioning of Vehicles," *Sensors*, Vol. 17, No. 2, 2017, p. 271.

[18] Morales, J. J., J. J. Khalife, and Z. M. Kassas, "Information Fusion Strategies for Collaborative Inertial Radio SLAM," *IEEE Transactions on Intelligent Transportation Systems*, Vol. 23, No. 8, 2022, pp. 12935–12952.

[19] Conti, A., et al., "Network Experimentation for Cooperative Localization," *IEEE Journal on Selected Areas in Communications*, Vol. 30, No. 2, 2012, pp. 467–475.

[20] Pascacio, P., et al., "Collaborative Indoor Positioning Systems: A Systematic Review," *Sensors*, Vol. 21, No. 3, 2021, p. 1002, https://www.mdpi.com/1424-8220/21/3/1002.

[21] Jarrett, C., "Introducing Titan, Amazon's New Mobile Robot That Can Lift Up to 2,500 Pounds," https://www.aboutamazon.com/news/operations/amazon-unveils-titan-fulfillment-center-robot.

[22] Saeedi, S., et al., "Neural Network-Based Multiple Robot Simultaneous Localization and Mapping," *IEEE Transactions on Neural Networks*, Vol. 22, No. 12, 2011, pp. 2376–2387.

[23] "Vehicle-to-Vehicle Communication," United States Department of Transportation, https://www.nhtsa.gov/technology-innovation/vehicle-vehicle-communication.

[24] Fang, S. H., and T. N. Lin, "Cooperative Multi-Radio Localization in Heterogeneous Wireless Networks," *IEEE Transactions on Wireless Communications*, Vol. 9, No. 5, 2010, pp. 1547–1551.

[25] Keskin, M. F., O. Erdem, and S. Gezici, "Cooperative Localization in Hybrid Infrared/Visible Light Networks: Theoretical Limits and Distributed Algorithms," *IEEE Transactions on Signal and Information Processing over Networks*, Vol. 5, No. 1, 2019, pp. 181–197.

[26] Zhu, J., and S. S. Kia, "Decentralized Cooperative Localization with LoS and NLoS UWB Inter-Agent Ranging," *IEEE Sensors Journal*, Vol. 22, No. 6, 2022, pp. 5447–5456.

[27] Yang, Z., X. Feng, and Q. Zhang, "Adometer: Push the Limit of Pedestrian Indoor Localization Through Cooperation," *IEEE Transactions on Mobile Computing*, Vol. 13, No. 11, 2014, pp. 2473–2483.

[28] Xiong, H., et al., "A Novel Hybrid RSS and TOA Positioning Algorithm for Multi-Objective Cooperative Wireless Sensor Networks," *IEEE Sensors Journal*, Vol. 18, No. 22, 2018, pp. 9343–9351.

[29] Eren, T., "Cooperative Localization in Wireless Ad Hoc And Sensor Networks Using Hybrid Distance and Bearing (Angle of Arrival) Measurements," *EURASIP Journal on Wireless Communications and Networking*, Vol. 2011, No. 1, 2011, p. 72.

[30] Wang, X., et al., "CA2T: Cooperative Antenna Arrays Technique for Pinpoint Indoor Localization," *Procedia Computer Science*, Vol. 34, 2014, pp. 392–399.

[31] Mismar, T., J. Kim, and M. Alam, "Indoor Antispoofing Cooperative Localization in Cellular Networks," *IEEE Transactions on Aerospace and Electronic Systems*, Vol. 51, No. 4, 2015, pp. 2823–2833.

[32] Bhamidipati, S., and G. X. Gao, "GPS Multi-Receiver Joint Direct Time Estimation and Spoofer Localization," *IEEE Transactions on Aerospace and Electronic Systems*, 2018, p. 1-1.

[33] Chen, L., K. Yang, and X. Wang, "Robust Cooperative Wi-Fi Fingerprint-Based Indoor Localization," *IEEE Internet of Things Journal*, Vol. 3, No. 6, 2016, pp. 1406–1417.

[34] Zhang, G., et al., "3D Mapping Database Aided GNSS Based Collaborative Positioning Using Factor Graph Optimization," *IEEE Transactions on Intelligent Transportation Systems*, Vol. 22, No. 10, 2021, pp. 6175–6187.

[35] Saeed, N., et al., "A State-of-the-Art Survey on Multidimensional Scaling-Based Localization Techniques," *IEEE Communications Surveys & Tutorials*, Vol. 21, No. 4, 2019, pp. 3565–3583.

[36] Apple, "Apple's Find My Network Now Offers New Third-Party Finding Experiences," Apple Inc., 2021.

[37] Samsung, "What Is Ultra-Wideband and How Does It Work?" *Samsung Insights*, 2021, https://insights.samsung.com/2021/08/25/what-is-ultra-wideband-and-how-does-it-work-3.

[38] "Samsung SmartThings Find Rapidly Expands with over 300 Million Nodes Helping to Locate Devices," Samsung Electronics Co., 2023, https://partners.smartthings.com/press-release/samsung-smartthings-find-rapidly-expands-with-over-300-million-nodes-helping.

[39] Sathyan, T., and M. Hedley, "Fast and Accurate Cooperative Tracking in Wireless Networks," *IEEE Transactions on Mobile Computing*, Vol. 12, No. 9, 2013, pp. 1801–1813.

[40] Tomic, S., M. Beko, and R. Dinis, "RSS-Based Localization in Wireless Sensor Networks Using Convex Relaxation: Noncooperative and Cooperative Schemes," *IEEE Transactions on Vehicular Technology*, Vol. 64, No. 5, 2015, pp. 2037–2050.

[41] Naseri, H., and V. Koivunen, "Cooperative Simultaneous Localization and Mapping by Exploiting Multipath Propagation," *IEEE Transactions on Signal Processing*, Vol. 65, No. 1, 2017, pp. 200–211.

[42] Kassas, Z. M., and T. E. Humphreys, "Observability Analysis of Collaborative Opportunistic Navigation with Pseudorange Measurements," *IEEE Transactions on Intelligent Transportation Systems*, Vol. 15, No. 1, 2014, pp. 260–273.

[43] Vaghefi, R. M., and R. M. Buehrer, "Cooperative Source Node Tracking in Non-Line-of-Sight Environments," *IEEE Transactions on Mobile Computing*, Vol. 16, No. 5, 2017, pp. 1287–1299.

[44] Kulmer, J., et al., "Anchorless Cooperative Tracking Using Multipath Channel Information," *IEEE Transactions on Wireless Communications*, Vol. 17, No. 4, 2018, pp. 2262–2275.

[45] Zeng, M., et al., "Convolutional Neural Networks for Human Activity Recognition Using Mobile Sensors," *6th International Conference on Mobile Computing, Applications and Services*, November 6–7, 2014, pp. 197–205.

[46] Ordóñez, F. J., and D. Roggen, "Deep Convolutional and LSTM Recurrent Neural Networks for Multimodal Wearable Activity Recognition," *Sensors*, Vol. 16, No. 1, 2016, p. 115, https://www.mdpi.com/1424-8220/16/1/115.

[47] Vaswani, A., et al., "Attention Is All You Need," *Proceedings of the 31st International Conference on Neural Information Processing Systems*, Long Beach, CA, 2017.

[48] Rao, B., et al., "CTIN: Robust Contextual Transformer Network for Inertial Navigation," *Proceedings of the AAAI Conference on Artificial Intelligence*, Vol. 36, No. 5, 2022, pp. 5413–5421.

[49] Chen, C., and X. Pan, "Deep Learning for Inertial Positioning: A Survey," *IEEE Transactions on Intelligent Transportation Systems*, Vol. 25, No. 9, 2024, pp. 10506–10523.

[50] Wagstaff, B., V. Peretroukhin, and J. Kelly, "Robust Data-Driven Zero-Velocity Detection for Foot-Mounted Inertial Navigation," *IEEE Sensors Journal*, Vol. 20, No. 2, 2020, pp. 957–967.

[51] Chen, C., et al., "IONet: Learning to Cure the Curse of Drift in Inertial Odometry," *Proceedings of the Thirty-Second AAAI Conference on Artificial Intelligence and Thirtieth Innovative Applications of Artificial Intelligence Conference and Eighth AAAI Symposium on Educational Advances in Artificial Intelligence*, New Orleans, LA, 2018.

[52] Herath, S., H. Yan, and Y. Furukawa, "RoNIN: Robust Neural Inertial Navigation in the Wild: Benchmark, Evaluations, & New Methods," *2020 IEEE International Conference on Robotics and Automation (ICRA)*, 2020, pp. 3146–3152.

[53] Nguyen, S. M., D. V. Le, and P. Havinga, "iMoT: Inertial Motion Transformer for Inertial Navigation," *Proceedings of the AAAI Conference on Artificial Intelligence*, Vol. 39, No. 6, 2025, pp. 6209–6217.

[54] Shu, M., G. Chen, and Z. Zhang, "EL-SLE: Efficient Learning Based Stride-Length Estimation Using a Smartphone," *Sensors*, Vol. 22, No. 18, 2022, p. 6864, https://www.mdpi.com/1424-8220/22/18/6864.

[55] Geissinger, J. H., and A. T. Asbeck, "Motion Inference Using Sparse Inertial Sensors, Self-Supervised Learning, and a New Dataset of Unscripted Human Motion," *Sensors*, Vol. 20, No. 21, 2020, p. 6330, https://www.mdpi.com/1424-8220/20/21/6330.

[56] Xu, V., et al., "MobilePoser: Real-Time Full-Body Pose Estimation and 3D Human Translation from IMUs in Mobile Consumer Devices," *Proceedings of the 37th Annual ACM Symposium on User Interface Software and Technology*, Pittsburgh, PA, 2024.

[57] Lee, J., and H. Joo, "Mocap Everyone Everywhere: Lightweight Motion Capture with Smartwatches and a Head-Mounted Camera," *2024 IEEE/CVF Conference on Computer Vision and Pattern Recognition (CVPR)*, June 16–22, 2024, pp. 1091–1100.

[58] Conner-Simons, A., and R. Gordon, "Artificial Intelligence Senses People Through Walls," *MIT News*, 2018, https://news.mit.edu/2018/artificial-intelligence-senses-people-through-walls.

[59] Cui, Y., et al., "Integrating Sensing and Communications for Ubiquitous IoT: Applications, Trends, and Challenges," *IEEE Network*, Vol. 35, No. 5, 2021, pp. 158–167.

[60] Adib, F., and D. Katabi, "See Through Walls with WiFi!" *Proceedings of the ACM SIGCOMM 2013 Conference on SIGCOMM*, Hong Kong, China, 2013.

[61] Sharma, A., et al., "Passive WiFi CSI Sensing Based Machine Learning Framework

for COVID-Safe Occupancy Monitoring," *2021 IEEE International Conference on Communications Workshops (ICC Workshops)*, June 14–23, 2021, pp. 1–6.

[62] Samareh Abolhassani, A., et al., "Improving Residential Building Energy Simulations Through Occupancy Data Derived from Commercial Off-the-Shelf Wi-Fi Sensing Technology," *Energy and Buildings*, Vol. 272, 2022, p. 112354.

[63] Liu, Y., et al., "Reconfigurable Intelligent Surfaces: Principles and Opportunities," *IEEE Communications Surveys & Tutorials*, Vol. 23, No. 3, 2021, pp. 1546–1577.

[64] Basar, E., et al., "Wireless Communications Through Reconfigurable Intelligent Surfaces," *IEEE Access*, Vol. 7, 2019, pp. 116753–116773.

[65] Elzanaty, A., et al., "Reconfigurable Intelligent Surfaces for Localization: Position and Orientation Error Bounds," *IEEE Transactions on Signal Processing*, Vol. 69, 2021, pp. 5386–5402.

[66] Ma, T., et al., "Reconfigurable Intelligent Surface-Assisted Localization: Technologies, Challenges, and the Road Ahead," *IEEE Open Journal of the Communications Society*, Vol. 4, 2023, pp. 1430–1451.

About the Authors

Li-Ta Hsu received BS and PhD degrees from National Cheng Kung University (NCKU), Taiwan, in 2007 and 2013, respectively. He is currently an associate professor in the Department of Aeronautical and Aviation Engineering (AAE) at the Hong Kong Polytechnic University (PolyU), where he has held his faculty role since 2016. He was a visiting researcher and a JSPS postdoctoral researcher at UCL and University of Tokyo in 2012 and 2014, respectively. In 2023, he was a visiting research scientist at Google. Dr. Hsu has been teaching short courses in indoor positioning and indoor navigation on ION GNSS+ since 2020. He also hosted the IPIN Conference in 2024 in Hong Kong, where he served as the General Chair. He is an associate editor of *NAVIGATION* and *IEEE Transactions on Aerospace and Electronic Systems.* His research interests are estimation and optimization theories, urban GNSS positioning, indoor positioning, and multisensor integrated navigation systems for smart devices and unmanned autonomous systems.

Guohao Zhang received a bachelor's degree in mechanical engineering and automation from the University of Science and Technology, Beijing, People's Republic of China, in 2015. He received a master's degree in mechanical engineering and a PhD degree in aeronautical and aviation engineering from The Hong Kong Polytechnic University, Hong Kong, in 2017 and 2022, respectively, and started his teaching career as a research assistant professor in the Department of Aeronautical and Aviation Engineering at the same university in 2022. He was a visiting researcher at Nanyang Technological University in

2024. He is a member of the steering committee of the International Conference on Indoor Positioning and Indoor Navigation. He is an active member of the Institute of Navigation. His research interests include GNSS urban positioning, collaborative positioning, machine learning-aided GNSS, signal propagation modeling, and remote sensing.

Weisong Wen is an assistant professor at the Department of Aeronautical and Aviation Engineering, Hong Kong Polytechnic University (PolyU). He earned a PhD from PolyU in 2020 and was a visiting scholar at the University of California Berkeley. His research spans the full stack of autonomous systems, particularly drones, covering artificial intelligence-enabled perception, multisensor fusion (GNSS/LiDAR/IMU/Vision), navigation, and control. With over 100 publications in prestigious venues, his innovations in 3-D LiDAR-aided GNSS positioning have gained significant industry recognition. Dr. Wen is ranked among Stanford's top 2% most-cited scientists and is an active member of the IEEE RAS, IEEE ITSS, and ION.

Index

Artech House
GNSS Technology and Applications Library

Elliott Kaplan and Christopher Hegarty, Series Editors

A-GPS: Assisted GPS, GNSS, and SBAS, Frank van Diggelen

All Source Positioning, Navigation, and Timing, Rongsheng (Ken) Li

Applied Satellite Navigation Using GPS, GALILEO, and Augmentation Systems, Ramjee Prasad and Marina Ruggieri

Cryptography and Satellite Navigation, Joseph J. Rushanan and James T. Gillis

Digital Terrain Modeling: Acquisition, Manipulation, and Applications, Naser El-Sheimy, Caterina Valeo, and Ayman Habib

Geographical Information Systems Demystified, Stephen R. Galati

GNSS Applications and Methods, Scott Gleason and Demoz Gebre-Egziabher

GNSS Interference Threats and Countermeasures, Fabio Dovis, editor

GNSS Markets and Applications, Len Jacobson

GNSS Receivers for Weak Signals, Nesreen I. Ziedan

GNSS for Vehicle Control, David M. Bevly and Stewart Cobb

GPS/GNSS Antennas, B. Rama Rao, W. Kunysz, R. Fante, and K. McDonald

Implementing e-Navigation, John Erik Hagen

Inertial Navigation Systems Analysis, Kenneth Britting

Introduction to GPS: The Global Positioning System, Second Edition, Ahmed El-Rabbany

Location-Based Services in Cellular Networks: From GSM to 5G NR, Adrián Caldaldo García, Stefan Maier, and Abhay Phillips

MEMS-Based Integrated Navigation, Priyanka Aggarwal, Zainab Syed, Aboelmagd Noureldin, and Naser El-Sheimy

Navigation Signal Processing for GNSS Software Receivers, Thomas Pany

Navigation and Tracking in Space: Analysis and Algorithms, Sanat K. Biswas and Andrew G. Dempster

The Present and Future of Indoor Navigation, Laura Ruotsalainen, Martti Kirkko-Jaakkola, and Jukka Talvitie

Principles of GNSS, Inertial, and Multisensor Integrated Navigation Systems, Second Edition, Paul D. Groves

Principles of Indoor Positioning and Indoor Navigation, Li-Ta Hsu, Guohao Zhang, and Weisong Wen

Radionavigation Systems, Borje Forssell

RF Positioning: Fundamentals, Applications, and Tools, Rafael Saraiva Campos and Lisandro Lovisolo

Server-Side GPS and Assisted-GPS in Java™, Neil Harper

Spread Spectrum Systems for GNSS and Wireless Communications, Jack K. Holmes

Understanding GPS/GNSS: Principles and Applications, Third Edition, Elliott Kaplan and Christopher Hegarty, editors

Ubiquitous Positioning, Robin Mannings

Wireless Positioning Technologies and Applications, Second Edition, Alan Bensky

For further information on these and other Artech House titles, including previously considered out-of-print books now available through our In-Print-Forever® (IPF®) program, contact:

Artech House Publishers
685 Canton Street
Norwood, MA 02062
Phone: 781-769-9750
Fax: 781-769-6334
e-mail: artech@artechhouse.com

Artech House Books
16 Sussex Street
London SW1V 4RW UK
Phone: +44 (0)20 7596 8750
Fax: +44 (0)20 7630 0166
e-mail:
artech-uk@artechhouse.com

Find us on the World Wide Web at: www.artechhouse.com